A Field Guide to the
SNAKES
OF
AUSTRALIA

Tie Eipper & Scott Eipper

A Field Guide to the
SNAKES
OF
AUSTRALIA

Tie Eipper & Scott Eipper

JOHN BEAUFOY PUBLISHING

Dedication

For Scott,
I'm not good with words (well soppy ones, I can describe a snake or two well enough). I was a simple gal when I met you, never owned a tent with more than one room. I wouldn't be who I am today without your love, support, (lectures!), terrible sense of humour and annoying promptness. You're my rock, my biggest supporter, my never-ending chocolate supply, and the one who catches me when I fall down escalators and cliffs. I love you. Thank you for everything (especially for putting up with my quirks – there's a dinner set or two at least still to come!) FYI – this book dedication totally makes up for all the wedding anniversaries I forgot and all the ones I will forget.
Xo

For Andre Kember,
Families are like branches on a tree. We grow in different directions, yet our roots remain the same.

First published in the United Kingdom and Australia in 2024 by John Beaufoy Publishing Ltd
11 Blenheim Court, 316 Woodstock Road, Oxford OX2 7NS, England
www.johnbeaufoy.com

Copyright © 2024 John Beaufoy Publishing Limited
Copyright in text © 2024 Tie Eipper and Scott Christopher Eipper
Copyright in photographs © as specified on individual photographs
Copyright in maps © 2024 Tie Eipper and Scott Christopher Eipper

Photo Credits
Front cover: Rosen's snake, *Suta fasciata,* Scott Eipper
Spine: Green tree python, *Morelia viridis,* Shane Black
Back cover, top to bottom: Water python, *Liasis fuscus,* Scott Eipper; Common tree snake, *Dendrelaphis punctulatus,* Scott Eipper; Common tiger snake, *Notechis scutatus scutatus,* Scott Eipper

All rights reserved. No part of this publication may be reproduced, stored in a retrieval system or transmitted in any form or by any means, electronic, mechanical, photocopying, recording or otherwise, without the prior written permission of the publishers.

Great care has been taken to maintain the accuracy of the information contained in this work. However, neither the publishers nor the authors can be held responsible for any consequences arising from the use of the information contained therein.

ISBN 9781913679637

Edited by Krystyna Mayer
Project management by Rosemary Wilkinson
Designed by Gulmohur Press, New Delhi
Printed and bound in Malaysia by Times Offset (M) Sdn. Bhd.

Contents

Introduction	6
Snake Myths and Fallacies	7
Snake-bite Prevention	8
Snake Bites & First Aid	8
Habitats	10
Using This Book	15
Glossary	21
Snake or Lizard?	23
Species Descriptions & Keys	**24**
Snakes and Snake-like Families	24
File Snakes	25
Pythons	28
Colubrids	58
Water Snakes	72
Elapids	74
Sea Snakes	230
Homalopsids	281
Blind Snakes	292
Encountering a Snake in a House or Garden	346
Encountering a Stranded Sea Snake	346
The World's Most Venomous Snakes	348
Checklist of the Snakes of Australia	350
References & Further Resources	354
Acknowledgements	361
Index	362
Notes	366

Introduction

Australia is currently home to between 221 and 227 species and 13 subspecies of snake, 160 of which are endemic. It is not clear whether six species currently occur in Australia. These snakes have subsequently been included in both the keys and species descriptions to enable users of this book to identify the species if encountered.

One python has been included that has historically been part of the Australian snake fauna. There is conjecture as to whether it occurs on an Australian island because no specimens were collected and it has not been seen recently despite searches. Five additional species of sea snake have been recorded adjacent to Australian waters. It is likely that they may at times enter Australian territorial waters and consequently they are included here. Furthermore, several Papuan species are found remarkably close to Australian territories, so it is expected that they could occur in Australia too.

Many species are being investigated, and invariably more will be described and resurrected from synonymy. During the writing of this book, a new species of snake was discovered in central Australia, the Desert whip snake *Demansia cyanochasma*.

Occasionally other exotic species are found locally, due to accidental and illegal, deliberate imports.

Snakes form the suborder Serpentes, and along with lizards (Lacerta) form the order Squamata. Squamates are characterized by their skull structure and by males having two hemipenes. According to the fossil record, snakes first appeared around 112 million years ago in Laurasia. There are about 32 snake families. Australia is currently home to seven of these: blind snakes, pythons, file snakes, elapids – including sea snakes and sea kraits – homalopsids, colubrids and natricids.

This book is divided into sections with keys for family, genus and species-level identification. For each family and genus, there is a short overview of the taxa incorporated. All genus and species descriptions are arranged in alphabetical order. To make it easiest for the reader when identifying sea snakes, the marine elapids have been separated into their own section. This, however, does not reflect the fact that marine elapids are not closely related to their terrestrial counterparts; they are, in some cases, more closely related to terrestrial species than to other marine species.

Illegally kept exotic escape, Boa imperitor, Penrith, NSW Scott Eipper

Acanthophis sp., an undescribed species from Camooweal, QLD, Will Scott

Snake Myths and Fallacies

- Snakes are not repelled by vibration – vibrating snake repellers are completely ineffective.
- Venomous snakes and pythons cannot interbreed and form 'super snakes'
- *Only harmless snakes climb* – many venomous snakes are found off the ground, particularly while hunting prey.
- Snakes do not bite their tails and form hoops to go down hills.
- Snakes do not chase people, if they have perceived a threat they will defend themselves, but given the opportunity, they will almost always flee.
- Snakes cannot outrun a person – the fastest snakes in the world can move at about 12km per hour in short bursts compared to an average person being able to run at 20km per hour.
- Snakes are not attracted to milk.
- Flowering plants do not repel snakes, nor do essential oils, statues or traditional pets.
- No known chemicals/home remedies used in pest control are effective snake deterrents.
- *Sea snakes cannot bite* – just like their terrestrial relatives, all sea snakes are capable of biting people.
- *It is possible to snake-proof your yard.* It is not. The best thing you can do is have a yard that is debris free, low-cut lawns with pets inside and pet food not left lying around.

Snake next to vibration device, Scott Eipper

Hoplocephalus bitorquatus, Lake Broadwater, QLD, Scott Eipper

Snake-bite Prevention

- Never attempt to restrain, capture or handle a wild snake.
- Never attempt to kill a snake – this is not only usually illegal, but it also places you at serious risk (20 per cent of all Australian fatalities from snake bites have occurred because the victim tried to handle or kill the snake).
- Never place your limbs where you cannot see them.
- Always walk around at night with a torch.
- Always wear shoes and long trousers in locations where snakes may be present.
- Take care when turning over potential shelter sites such as corrugated iron sheets, rocks and logs, as well as other debris.
- Avoid attracting snakes to your residence or workplace by keeping the grass short, gardens tidy, rubbish to a minimum and by keeping pets and their food inside to minimise attracting rodents.

Snake Bites & First Aid

Note that this information is a guide for Australian snake bites only and is correct at the time of printing. If someone is unlucky enough to be bitten by a snake, the early application of the correct first-aid practices can greatly increase the chances of them making a full recovery.

- Seek urgent medical help in Australia, by calling triple zero '**000**'.
- **DO NOT** wash the bite site.
- Apply a pressure immobilization bandage and splint if the bite is on a limb, or a pressure pad if it is to the head, neck or torso, and minimize movement.
- **DO NOT** catch, chase or kill the snake – this is extra movement (if done by the patient increasing blood flow) and could result in further bites.
- **DO NOT** eat or drink alcohol, tea, stimulants, food or medication without expert advice.
- **DO NOT** apply hot or cold packs, electrical shocks, suction devices or tourniquets/ligatures.
- APPLY a P.I.B as per the instructions opposite.

In the first instance DRSABCD must be followed, but it is usually not required immediately after a snake bite. The DRSABCD action plan:

D – look for **D**anger.
R – check for a **R**esponse.
S – **S**end for help.
A – clear the **A**irway.
B – sustain **B**reathing.
C – start **C**PR (if required).
D – apply a **D**efibrillator if indicated.
(*DRSABCD is vital if a person has collapsed and is unresponsive*).

SNAKE BITES & FIRST AID

Pressure Immobilization Bandaging (PIB) First Aid

Move away from the area where the bite occurred (if required), lay the patient down and keep them calm. Any movement of the limb quickly results in venom absorption and must be prevented; therefore first aid must be an immediate priority after a snake bite.

Where possible, do not allow the patient to walk. In the case of a snake bite to a lower limb, splinting of both legs should be carried out to completely immobilize the lower half of the body.

Two components must be satisfied – pressure over the bite site and bitten limb and general immobilization. This involves the application of:

Starting at the end of the limb, apply a broad (minimum 75mm wide) elastic bandage to the entire bitten limb at a very firm pressure of at least 40mmHg for an arm and 55mmHg for a leg. The Australian Venom Research Unit (AVRU) recommends SETOPRESS TM High Compression Bandages as these bandages lose minimal tension with prolonged application.

Splints effectively immobilize the entire limb, in combination with laying the patient down and ensuring they are completely still to minimize any movement. Do not use a sling.

In rare cases, a person may be bitten on the body, face or neck. In these instances, direct pressure should be applied over the bite site with a pressure pad made from a cloth (a hand towel, T-shirt, etc.) and held firmly in place until medical attention can be sought.

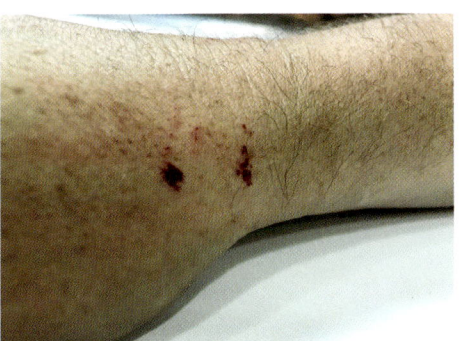

Step 0 Bite site

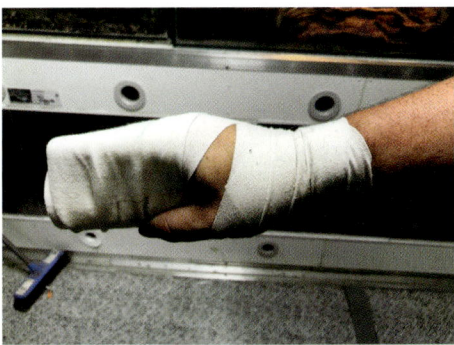

Step 1 Wrap bandage over bite site

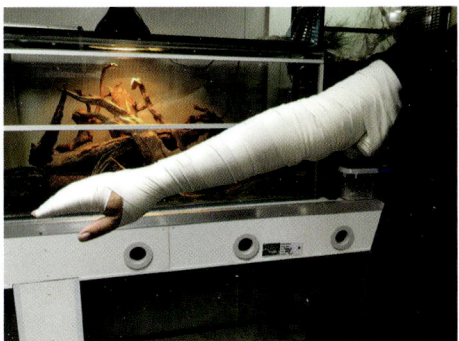

Step 2 Extend bandage up entire limb

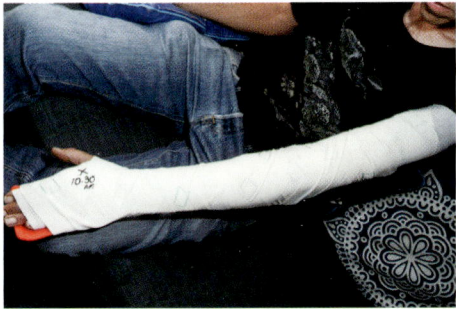

Step 3 Splint applied to keep immobile, time of bite and site marked on bandage

Respiratory Insufficiency Management

The bite from any venomous snake has the potential to cause envenomations that may result in an allergic reaction. In its most severe instance, this can lead to difficulty in breathing, which can be a result of anaphylaxis.

Symptoms include acute, rapid-onset illness, with typical clinical symptoms appearing as:
- Tingling around the mouth
- Swelling of the lips, tongue and face
- Tightness in the throat
- Difficult or noisy breathing
- Rash and hives
- Difficulty talking
- Coughing
- Dizziness
- Vomiting or abdominal pain
- Being pale and floppy (infants and young children)

If **anaphylaxis** is suspected:
- Seek urgent medical help in Australia by calling triple zero **'000'**.
- Lay the person flat on the back (or seated if breathing is difficult).
- Obtain and follow the instructions on an epinephrine (adrenaline) pen – usually injected in the outer thigh.
- **Do Not** allow the patient to stand or walk around.
- Monitor breathing and commence CPR (30 chest compressions followed by two rescue breaths) as per DRSABCD, if required and continue until the patient's breathing is normal and stable.
- Monitor the patient for at least four hours.

Habitats

Australian snakes are found in a wide variety of habitats. Some species are generalists, while others are restricted to small microhabitats within larger ecological communities. A sample of the variety of important habitats for Australian snakes is provided on the following pages.

It is outside the scope of this book to provide overviews of what forms a particular habitat type. This is due to the influence of climate, elevation, geology, hydrology, ocean currents, plant diversity, rainfall and paleogeography. More recently (in the past 65,000 years) humans have had significant influence over habitats, with changes to fire regimes and bringing in new fauna to Australia resulting in changes to the flora and fauna and therefore ecosystems. The most significant changes occurred with the arrival of European settlers and the urbanization of Australia, which has brought a myriad of changes, from introduced plants and animals, to land clearing, mining and farming.

HABITATS 11

Beach, Scott Eipper

Alpine grassland, Scott Eipper

Blacksoil grassland – dry season, Scott Eipper

Blacksoil grassland – wet season, Scott Eipper

Brigalow woodland, Scott Eipper

Coastal heath, Scott Eipper

Cool temperate woodland, Scott Eipper

Coral reef, Blanche d'Anastasi

Dry open woodland, Scott Eipper

HABITATS

Estuary, Scott Eipper

Estuary bottom, Josh Jensen

Gibber desert, Adam Elliott

Granite exfoliations in heath, Scott Eipper

Littoral rainforest, Scott Eipper

Mallee woodland, Scott Eipper

Mangroves, Scott Eipper

Modified habitat – degraded farmland, Tie Eipper

Modified habitat – habitat destruction, Scott Eipper

HABITATS

Modified habitat – urban environment, Scott Eipper

Modified habitat – rubbish piles, Cody Eipper

Modified habitat – sleeper pile, Bailey Eipper

Mulga woodland, Scott Eipper

Rainforest, Scott Eipper

Rocky grassland, Scott Eipper

Rocky shrubland, Scott Eipper

Rocky woodland, Scott Eipper

Saltbush and mulga woodland, Scott Eipper

14 HABITATS

Wallum, Scott Eipper

Sand ridge desert, Scott Eipper

Sand ridge desert, Scott Eipper

Sandstone escarpment with heath, Scott Eipper

Subtropical woodland, Scott Eipper

Swamp, Scott Eipper

Tropical wetland, Scott Eipper

Tropical woodland, Scott Eipper

Vine forest, Dean Purcell

Using This Book

The taxonomy adheres to Eipper & Eipper, 2023, exclusive of any newly described taxa.

Distinguishing many of the species depends on the use of scale counts, distribution, scale position and sometimes internal organ position. This can be a difficult task, so dichotomous keys are included, as are images of every species in the corresponding species descriptions to help the reader with identification. In the keys, characteristics are used to lead the user to the next choice or arrive at the identification.

The values presented in the keys are typical for the species, unless specified. Occasionally the term '(part)' may be used in genus and family keys. This occurs where a suite of characteristics identifies an individual with a genus or family but due to the diversity of that genus or family it may also be found in other couplets. There are always exceptions to rules, so caution must be taken when arriving at a conclusion.

The species descriptions should be looked at closely to assist in confirmation. In recognition of the fact that some species are more problematic than others, making it difficult to enable a conclusive, correct identification, species-level identification difficulty categories are also included with each genus or family summary. The ease in arriving at the correct identification within a family or genus has been quantified in a ranking of 1–5, one being simple and five being very difficult.

Two species easily confused – Acanthophis pyrrhus, Tie Eipper

Two species easily confused – Acanthophis wellsei, Scott Eipper

Sometimes snakes do not follow the typical characteristics in the keys. If you come across a snake (or any reptile) that does not do so, or does not visually agree with the typical description, the authors would be interested in knowing more about these unusual individuals. Relatively recently, some species have been primarily split on

Aberrant individual or unusual appearance, albinism, Acanthophis rugosus, Tie Eipper

Aberrant individual or unusual appearance, pied, Pseudonaja textilis, Tie Eipper

Aberrant individual or unusual appearance, albinism, Pseudechis porphyriacus, Scott Eipper

Aberrant individual or unusual appearance, algal growth, Aipysurus duboisii, Claire Goiran

Aberrant individual or unusual appearance, albinism, Morelia spilota spilota, Scott Eipper

Aberrant individual or unusual appearance, scale fusion of nasal and loreal scales, Tropidonophis mairii, Tie Eipper

phylogenetic placement. In some cases, conclusive identification by way of external characteristics is very difficult, if not impossible in the field.

The name or names of the first person or persons to describe a species are provided after the scientific names for each species. In cases where such names are in parentheses, this indicates that that there has been a species name change since the original discovery.

Some species descriptions include a section called 'similar species'. This is used to aid readers with species that are readily confused with each other. Species identification is primarily determined using the keys but if a similar, easily confused species, is found in a different genus, a quick way to determine the correct identification is provided.

Both species and subspecies have their own descriptions. Subspecies are clearly defined by the additional subspecific name in a description's title. Pronunciation of the

Threat display, Pseudechis colletti, Tie Eipper

Threat display, Vermicella vermiformis, Scott Eipper

scientific names has also been provided.

Additionally, ecological information and distribution comments are included, along with maps that can assist in determining species identification.

Disposition is included for each species featured — it is a generalization based on the experiences of the authors with both wild and captive individuals, and on closely

Threat display, Oxyuranus microlepidotus, Tie Eipper

related species. In some cases the descriptions have been drawn from the experience of trusted colleagues who have worked with species that the authors have not. Note, however, that every animal has variation in temperament and may behave in an unusual manner. Any suspected venomous or harmful species must be treated with the appropriate caution and, generally speaking, snakes must never be handled unless you hold the appropriate permits to do so.

Pronunciation

ay as in day
a as in sat
ah as in hah
ee as in see
e as in let
er as in stern

ie as in lie
i as in sit
air as in lair
oh as in low
o as in lot
or as in sore

aw as in jaw
ue as in due
u as in shut
oo as in hood
ow as in wow
oy as in toy

Etymology

The composition of scientific names often has meanings that relate to species traits. These traits often have Latin or Greek origins. This in turn can influence how the names themselves are constructed, such as gender rules. In some cases, names are seemingly arbitrary arrangements of letters forming a meaningless name. In these cases, this is indicated accordingly in the generic descriptions. Meanings and derivations have come from the descriptions themselves or from the examination of appropriate references.

Etymology — the subspecies name 'niger' means black, Notechis scutatus niger, Scott Eipper

IUCN Listings

The IUCN Red List of Threatened Species was created in 1964 and has been revised many times since then. It ranks the world's animals and other organisms according to the following criteria, which are given at the end of each species description:

Extinct (EX) Exceeding reasonable doubt that a species exists any longer.
Extinct in the Wild (EW) Survives only in captivity, presumed extinct in the wild after extensive surveys.
Critically Endangered (CR) In a grave state of becoming extinct in the wild.
Endangered (EN) Very high likelihood of becoming extinct in the wild and complies with any of the IUCN's criteria A to E for Endangered.
Vulnerable (VU) Meets one of the five Red List criteria, making it a high risk of unnatural (human-generated) extinction, unless humans intervene.
Near Threatened (NT) Near to becoming endangered in the near future.
Least Concern (LC) Unlikely to become endangered or extinct in the immediate future.
Data Deficient (DD) Not enough data to evaluate.
Not Evaluated (NE) Has not been evaluated and a listing has not been made.

Potential Danger Rankings

The level of risk posed by a snake to a healthy person is quantified in this book according to the hazard posed by an adult individual of the species. Juvenile snakes have a lower venom yield than adult individuals, but despite the myths, they can control their venom output. The toxicity of the venom in some species exhibits ontogenetic change, but that does not mean that a species is more or less toxic as an adult. In the case of non-venomous species, large constrictors can overpower a person and therefore the hazard posed by these species is reflected as applicable. All snakebites should be treated by a medical professional.

DANGEROUSLY VENOMOUS Venom can kill a healthy adult or child.
VENOMOUS Venom is unlikely to kill a healthy adult but can yield severe systemic effects.
DANGEROUS Could kill a person by a mechanical injury or by oxygen deprivation to vital organs.
HARMFUL Venom does not usually cause death or systemic symptoms. This ranking also applies to a bite that could warrant stitching, such as that from a large python.
HARMLESS Bite may puncture the skin. While bites can be harmless, a bite site can become infected and cause a secondary concern.

Measurement & Key Abbreviations

Sizes quoted in species descriptions for body measurements (where available) are average maximum sizes, but exceptions can occur. Breeding information (for example clutch/litter sizes) is taken from current literature and should be treated as an indicative value, as ongoing research can change the values provided.
TL Total length, from tip of head to tip of tail
MB Midbody scale rows range
SUB Subcaudal scale range
VENT Ventral scale range
DSR Dorsal scale rows

Visual Identification Aids

Dorsal 1 – pointed

Dorsal 2 – trilobed

Dorsal 3 – rounded

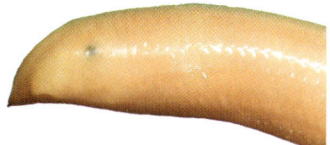
Profile 1 – pointed

Profile 2 – bluntly angular

Profile 3 – angular

Profile 4 – rounded

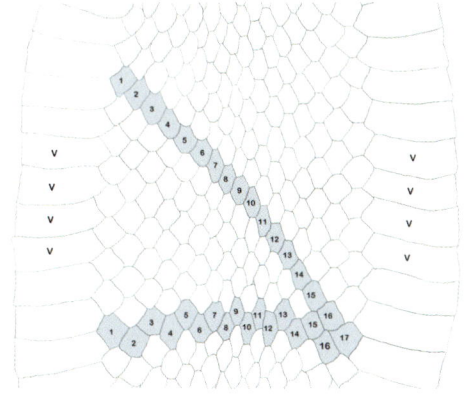
Scalation, midbody, demonstrating counting methods

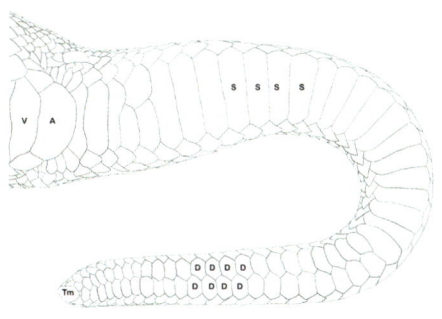
Scalation, tail

Body Scalation
V – Ventral
S – Single subcaudal
D – Divided subcaudal
A – Anal scale
Tm – Terminal

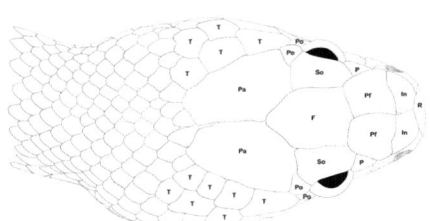
Scalation, head – dorsal elapid

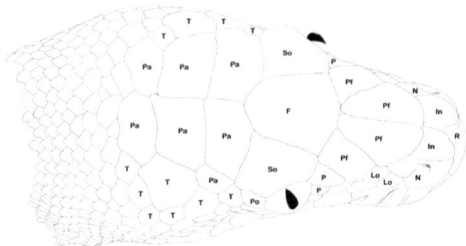
Scalation, head – dorsal python

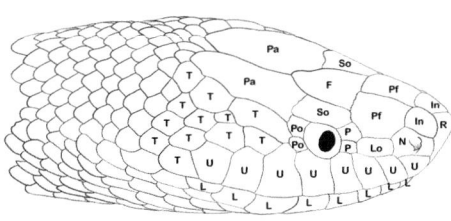
Scalation, head – profile colubrid

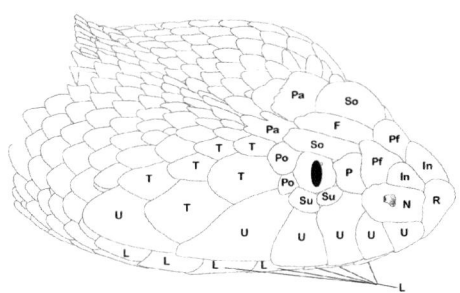
Scalation, head – profile elapid

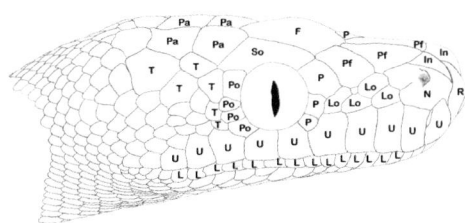
Scalation, head – profile python

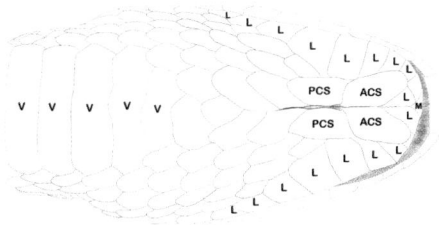
Scalation, head – ventral

Head Scalation

ACS – Anterior Chin Shield (Pregenial)
F – Frontal
In – Internasal
L – Lower Labial (Infralabial)
Lo – Loreal
M – Mental
N – Nasal
P – Preocular
Pa – Parietal

PCS – Posterior Chin Shield (Postgenial)
Pf – Prefrontal
Po – Postocular
R – Rostral
So – Supraocular
Su – Subocular
T – Temporal
U – Upper Labial (Supralabial)
V – Ventral

Glossary

anal scale The scale or scales that are anterior to the cloaca (see diagram, p. 19).
anaphylaxis Severe, fast-acting allergic reaction that can be fatal.
anterior Front of body; towards front.
anticoagulant Substance that prevents clotting of blood.
antivenom Medication used to treat venomous bites and stings.
apical spine The spine on terminal scale on most blind snakes; also known as a terminal spine.
aquatic Living in or near water.
arboreal Refers to individuals that live above the ground, for example in vegetation or trees.
basking Act of a snake exposing itself to increased temperature to raise its core body temperature.
buccal cavity Oral cavity.
cathemeral Active at any time of day or night.
caudal At or near rear half of body.
chenopod shrublands Semi-arid plains vegetated with saltbush, samphire and similar.
clutch size Number of eggs laid by female snake in a single reproductive event.
coagulant Component of venom that causes the blood to clot.
conical Shaped like a cone.
constriction Act of coiling tightly around an animal, causing asphyxiation.
couplet A pair of contrasting characters used in a key for identification.
crepuscular Active at dawn and dusk.
cryptic Refers to disguised appearance, either through colour and pattern or habits.
diurnal Active during the day.
dorsal Of, on or relating to upper half or top of a structure or body.
elliptic Shape of an ellipse in reference to pupil shape.
etymology History of a name shown by tracing its origin or development.
family Taxonomic rank above genus and below order.
fangs In snakes, sharp, grooved or hollow teeth modified to pierce living tissue and inject venom.
fossorial Living or active beneath the soil surface.
gene Basic unit of genetic control. Each gene has a specific function and is found on a specific section of a specific chromosome.
genus (genera) Taxonomic group above species and below family.
gravid Pregnant, with abdominal cavity containing formed eggs or young.
haemolytic/haemolysin Components of venom that cause red blood cells to be destroyed.
heliothermic Refers to individual that must bask to raise its body temperature.
herpetofauna Collective term referring to a group of amphibians and reptiles.
holotype The single specimen on which a species is based.
hybrid Genetic combination due to mating of two different species or subspecies.
juvenile Young individual.
juxtaposed In snakes, condition where scales abut, without overlapping.
lanceolate Shape of spearhead or lance – in reference to scale shape.
LD50 Required dose of toxin to kill 50 per cent of test subjects.
lectotype Specimen nominated from a syntype series to become the name-bearing type.

litter Number of young born by female snake in a single reproductive event.
maxilla Jaw (upper).
midbody scale rows Pertains to diagonal line of scales counted from a ventral scale over the body to the ventral scale on the other side (see diagram, p. 19).
monotypic Refers to a genus with a single recognized species.
morphological Pertaining to the form or structure of an animal, especially its external appearance.
musk Fluid produced by glands with a strong odour, usually used during defence.
myotoxic Refers to components of venom that attack the muscle tissue.
nape Back of neck.
nasal cleft Crease in nasal scale that centres on nostril (see images, p.19).
nasal scale Scale that contains nostril (see diagrams, p. 20).
necrosis Premature death of cells in living tissue.
neonate Newborn animal up to the age of six weeks.
neurotoxin Components of venom that attack nervous system.
nocturnal Active during the night.
nuchal Area where head and neck join.
order Taxonomic rank above family and below class.
oviparous Refers to animal that reproduces by laying eggs.
ovoviviparous Refers to animal that reproduces by forming unshelled egg sacs to house developing young, which are held inside female until ready to hatch, then expelled either still within egg sac or after leaving it.
pelagic Occurring in the open ocean.
phylogeny Relationships of groups of animals based on their evolutionary history. Used in taxonomy as a tool to determine lineages.
poison Substance that is harmful once ingested, inhaled or absorbed through skin.
posterior Rear of body; towards rear.
ptosis Uncontrollable drooping of eyelids.
rhomboid Diamond shaped.
rostral Scale on end of snout (see diagrams, p. 20).
species Basic unit of taxonomic classification.
species complex Group of animals that are composed of both described and undescribed taxa, which are currently lumped under a single species name.
stippling Patterns or markings created by grouping numerous small dots.
subcaudal scales The scales that are posterior to cloaca (see diagram, p. 19).
subspecies Taxonomic category that is a variation in a primary (nominate) species brought about by geographical or genetic isolation, usually characterized by a variation in morphological or genetic features.
supralabial scales The scales that are only on the side of the head running along the upper lip (see diagrams, p. 20).
sympatry Refers to individuals that share the same geographical area.
synonymy Chronological record of scientific names applied to a taxonomic unit.
syntypes Specimens of a type series that collectively apply to be the name-bearing type for a species.
taxonomy Study of plants and animals leading to their description, classification and naming.
terrestrial Living on or near the ground surface.

tines Forks of a bifurcated tongue.
torpor Dormant period of physical inactivity, usually characterized by reduced body temperature and slowed metabolic rate.
trilobed Having three lobes or projections – used to describe some blind snake snouts when viewed from above (see image, p. 19).
troglodyte Cave dwelling or living in subterranean environments.
type locality Place where a type specimen was discovered.
type species A species that is the name-bearing type for a genus.
venom Toxin injected by an animal to subdue prey or for defence against predators.
ventral Undersurface or belly of an animal.
ventral scales Enlarged scales running along underside of a snake (see diagrams, p. 19).
vertebral Along line of spine.
vertebrates Animals that have a backbone.
vestigial Refers to remnant of an appendage or other structure that has lost its original purpose through evolution.
viviparous Reproducing by giving birth to live young.
waif Individual that is transported by natural events such as storms, currents or floods into waters or regions outside the typical range for the species.

Snake or Lizard?

These are the basic differences between snakes and lizards (both members of the same order, Squamata):

- Lizards normally have eyelids, while snakes have a scale called a spectacle or brille overlaying the eye. Exempt from this are geckos, legless lizards and some skinks.

Legless Lizard – *Lialis burtonis*, Scott Eipper

Legless Lizard – *Lialis burtonis*, Scott Eipper

Legless Lizard – *Paradelma orientalis*, Scott Eipper

Legless lizard – *Aprasia parapuchella*, Scott Eipper

- Lizards normally have ear openings; snakes do not have ears and are inherently deaf to airborne sounds.
- Almost all Australian lizards have a broad, fleshy tongue. Exempt from this are varanoid lizards, such as monitors/goannas.
- Snakes are legless. Pythons and blind snakes have a vestigial pelvic girdle. Pythons have cloacal spurs. Most lizards have limbs, but in some fossorial species and legless lizards, these have become vestigial or non-existent.

Legless lizard – *Delma torquata*, Scott Eipper

Legless lizard – *Ophioscincus ophioscincus*, Scott Eipper

Species Descriptions & Keys

Key to Australian Snakes & Snake-like Families

1. Presence of limbs, eyelids, a broad fleshy tongue or ear openings .... Lizards
   No eyelids, ear openings or limbs 2

2. Eyes well developed, body not worm-like in appearance 3
   Eyes poorly developed, scaled body is worm-like in appearance .... Blind Snakes (p. 292)

3. Tail round in cross-section 4
   Tail flattened and paddle-like .... Sea Snakes (p. 230)

4. Scales typical in shape, overlapping, not file-like 5
   Scales conical in shape, not overlapping and file-like .... File Snakes (p. 25)

5. One or more loreal scale present; or if not present, divided anal scale with 23 or more midbody scale rows 6
   No loreal scale present .... Elapids (p. 74)

6. No rostral or labial pits, midbody scale rows less than 30 rows 7
   Rostral or labial pits usually present (can be hidden behind rostral scale in Aspidites), midbody scale rows greater than 30 rows .... Pythons (p. 28)

7. Scales smooth or keeled, if keeled, anal scale is divided 8
   Scales keeled; anal scale is single .... Water Snakes (*Tropidonophis mairii*) (p. 72)

8. Subcaudal scales 51 or more, scales smooth, with a loreal scale; .... Colubrid Snakes (p. 58)
   Subcaudal scales 50 or less, scales keeled or smooth; if smooth, loreal may or may not be present; .... Homalopsids (p. 281)

Family Acrochordidae (File Snakes)

File snakes are an old lineage of snakes comprising three species, two of which are found in Australia. They are exclusively aquatic, using their amazing rough scalation to hold on to fish, which they constrict before swallowing. When stranded on land, they are almost helpless. They are completely harmless.

File Snakes, genus *Acrochordus*

The marine-adapted Little file snake is placed in the genus *Chersydrus* by some authors. Outside Australia, these snakes are often referred to as 'wart snakes' due to their distinctive scalation. **Species-level identification difficulty – 2.**

ETYMOLOGY *Acrochordus*: pointed scale.
TYPE SPECIES *Acrochordus javanicus*.

Key to the Australian *Acrochordus*

1 5–7 scales between nasal and eye .... *A. granulatus* (p. 27)
    11–14 scales between nasal and eye; found in fresh water .... *A. arafurae* (p. 26)

Acrochordus granulatus, with algal growth, Puerto Galera, Oriental Mindoro, Philippines, Mark Rosenstein

Acrochordus arafurae, O'Shannassy River, QLD, Scott Eipper

ARAFURA FILE SNAKE *Acrochordus arafurae* McDowell, 1979
(Elephant Trunk Snake)

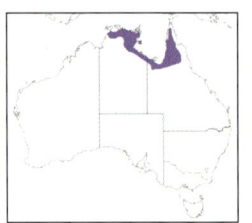

PRONUNCIATION *Ak-row-cord-us ah-rah-few-ree.*
ETYMOLOGY Arafura's pointed scale.
TYPE LOCALITY Lake Daviumbo, Western Province, PNG.
APPEARANCE Large, robust-bodied snake with head distinct from body. Dorsum grey to brown or blackish, marked with darker pigments that form cross-bands and spots. Tongue dark; mouth lining predominantly pink. Markings more defined on juveniles and subadults than on adults. Underside grey. Skin loose on land and at rest. Underside forms a keel while swimming. Adult females larger than males, reaching 220cm TL. **Scalation** MB 120–180 rows. Midbody scales conical in shape. **Similar species** *A. granulatus* (opposite).
RANGE Encountered in northern QLD, west of the GDR to WA border. Also in southern PNG. Occasionally encountered in estuarine environments, where it is vagrant.
COMMENTS Nocturnal. Aquatic. Inhabits lowland fresh water such as rivers, streams, pools and billabongs. Ambush predator, anchoring itself among submerged tree roots. **Diet** Fish, eels, including carrion. **Reproduction** Viviparous. 11–27 per litter. Neonates approximately 36cm TL and born in February–April. Parthenogenesis recorded. **Disposition** Inoffensive, but will bite if harassed.

BITE/VENOM HARMLESS
IUCN LISTING Least Concern.

Adel Grove, QLD, Scott Eipper

Gregory River QLD, Scott Eipper

O'Shannassy, River QLD, Scott Eipper

Gregory River QLD, Scott Eipper

LITTLE FILE SNAKE *Acrochordus granulatus* (Schneider, 1799)
(Marine File Snake, Little Wart Snake)

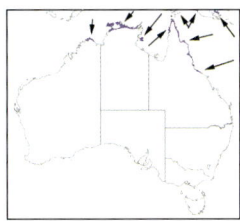

PRONUNCIATION *Ak-row-cord-us gran-u-la-tuss.*
ETYMOLOGY Granular pointed scale.
TYPE LOCALITY India.
APPEARANCE Medium-sized, robust snake with head distinct from body. Dorsal grey to brown or blackish, marked with lighter pigment forming pale cross-bands. Markings lose their definition with age. Pupils round, iris blue. Tongue dark. Ventral colouration grey, and tail strongly compressed. Underside forms a keel while swimming. Adult females larger than males, reaching 120cm TL. **Scalation** MB 90–160 rows. Midbody scales conical in shape. **Similar species** *A. arafurae* (opposite).
RANGE Encountered in coastal areas from Townsville, QLD, north and west to Kalumburu, WA. Also encountered in waters surrounding PNG and SE Asia.
COMMENTS Nocturnal. Aquatic. Inhabits brackish and marine ecosystems. Occasionally enters freshwater zones of river systems. Shelters in crab burrows or buries itself in sand by day. **Diet** Fish, chiefly gobies, blennies and mudskippers, and periodically small crabs, snails and carrion. **Reproduction** Viviparous. 1–12 young per litter. Neonates approximately 33cm TL and born in March–June. **Disposition** Inoffensive, but may bite when harassed.
BITE/VENOM HARMLESS
IUCN LISTING Least Concern.

Weipa, QLD, Hal Cogger

Juvenile, Cairns, QLD, Reid Newell

Juvenile, Cairns, QLD, Reid Newell

Cairns, QLD, Matt Summerville

Family Pythonidae (Pythons)

Pythons are thought to have evolved in either Australia or Indonesia. This is due to the diversity in the region. They are some of the world's best-known snakes, including some truly massive species such as the Burmese python *Python bivittatus*, Australian scrub python *Simalia kinghorni*, Reticulated python *Malayopython reticulatus* and African rock pythons *Python sebae*, all of which, despite their lack of venom, have used their amazing strength to cause the deaths of humans. This, however, is incredibly rare. Most pythons never attempt to harm humans unless provoked, and live alongside them in urban centres such as Sydney, Brisbane, Cairns, Darwin and Perth, where they provide pest control by consuming rodents and birds. Australian members of this family are diverse in their ecology and found over much of the country.

All Australian species have heat-sensitive pits (two species have pits hidden behind the labial and rostral scales), which are used to locate prey. The pits are sensitive to approximately 0.003 of a degree Celsius change, which allows a python to sense the presence of warm-blooded prey. Pythons have a pelvic girdle. This is a remnant from when their earlier relatives had legs. The girdle is evident in the form of cloacal spurs. These vestigial limbs are used by male pythons during courtship. Many Australian species, such as carpet and green pythons, are among the world's common species kept as pets.

Eleven species are endemic to Australia.

Key to the genera of Australian Pythons

1 Premaxilla with teeth, labial pits present 2
   Premaxilla without teeth, no labial pits .... *Aspidites* (p. 36)

2 5 or less infralabial pits 3
   6 or more infralabial pits 5

3 A single loreal 4
   Two or more loreals .... *Antaresia* (opposite)

4 Two pairs of prefrontal scales .... *Liasis* (p. 39)
   A single pair of prefrontal scales .... *Leiopython fredparkeri* (p. 44)

5 Scales on the top of the head large and plate-like 6
   Scales on top of the head are mostly small .... *Morelia* (p. 45)

6 Found in NT; more than 140 subcaudal scales; .... *Nyctophilopython oenpelliensis* (p. 54)
   Found in QLD; fewer than 140 subcaudal scales .... *Simalia* (p. 55)

Aspidites ramsayi, Uluru, NT, Scott Eipper

Children's Pythons, genus *Antaresia* Wells & Wellington, 1983

Antaresia currently comprises four species and two subspecies. These small pythons are often kept in captivity, as pets. The genus contains the world's smallest python, *A. perthensis*. Historically the genus has been placed in *Bothrochilus* and *Liasis*. The species have been split primarily by genetic separation. Three species are endemic to Australia. **Species-level identification difficulty** − 4.

ETYMOLOGY *Antaresia:* named after Antares, a yellow giant star in the constellation Scorpio.
TYPE SPECIES *Nardoa gilbertii*.

Key to *Antaresia*

1. Midbody rows more than 34 rows 2
   Midbody rows fewer than 34 rows .... *A. perthensis* (p. 35)

2. No pale lower lateral stripe along the anterior third of the body 3
   If patterned, a pale lower lateral stripe along the anterior
   third of the body 5

3. Patterns dark, usually strongly contrasting (except on island
   populations) consisting of large, ragged edge blotches 4
   Found north of Lockhart River with pattern not strongly
   contrasting, consisting of small, ragged edge blotches and flecking .... *A. papuensis* (p. 34)

4. Found south of Cairns (See note in species account) .... *A. maculosa maculosa* (p. 32)
   Found north of Innisfail (See note in species account) .... *A. m. peninsularis* (p. 33)

5. Fifth and sixth supralabial beneath the eye .... *A. childreni childreni* (p. 30)
   Sixth and seventh supralabial beneath the eye .... *A. c. stimsoni* (p. 31)

Antaresia childreni childreni, Mt Carbine, QLD, Shane Black

CHILDREN'S PYTHON *Antaresia childreni childreni* (Gray, 1842)

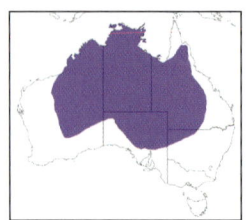

PRONUNCIATION *An-tah-ree-sah chil-dren-e.*
ETYMOLOGY Pertains to J. G Children, English naturalist.
TYPE LOCALITY Unknown.
APPEARANCE Medium-sized, slender snake with head distinct from body. Reddish or yellowish-brown above, frequently with darker brown blotched markings adorning dorsum. Some individuals are without patterns, and some island populations are melanistic. Pupil elliptic; eye colouration usually matches head. Tongue pink; inside of mouth pinkish. Markings more defined in juveniles than in adults. Ventral colouration white to cream. Adult females larger than males, reaching 100cm TL. **Scalation** MB 37–49 rows, 255–300 VENT, SUB 30–45 the first few SUBs anteriorly single, with remaining divided, and anal scale single. Scales matt in appearance and smooth. **Similar subspecies** *A. c. stimsoni* (opposite).

RANGE Encountered from northeastern QLD, to the Pilbara region of WA, across central Australia into arid SA, including northern Flinders Ranges, NSW, NT, and across most of arid QLD.

COMMENTS Mainly nocturnal. Terrestrial, inhabiting open woodland, spinifex-dominated grassland, black soil plains and rocky gorges. Shelters in rock crevices and recesses, tree hollows, and under bark. **Diet** Lizards, small mammals, frogs, bats. **Reproduction** Oviparous. 5–24 per clutch. Neonates approximately 30cm TL and hatch in October–February. Has been known to hybridize with Spotted pythons. **Disposition** Inoffensive but will bite if threatened.

BITE/VENOM HARMLESS

IUCN LISTING Least Concern.

Mt Guide, QLD, Scott Eipper

Normanton, QLD, Scott Eipper

Juvenile, Stanley Chasm, NT, Scott Eipper

Tennant Creek, NT, Scott Eipper

STIMSON'S PYTHON *Antaresia childreni stimsoni* (L. A. Smith, 1985)

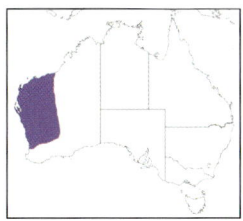

PRONUNCIATION *An-tah-ree-sah chil-dren-e stim-son-e.*
ETYMOLOGY Pertains to A. F. Stimson of the British Museum.
TYPE LOCALITY 15km SE of Nullagine, WA.
APPEARANCE Medium-sized, slender snake with head distinct from body. Reddish or occasionally yellowish-brown above, with darker spotting covering dorsum. Pale lateral line runs from neck, extending along first third of body. Tongue pink; inside of mouth pinkish. Pupil elliptic; eye colouration matches head. Markings more defined in juveniles than in adults. Ventral colouration white to cream. Adult females larger than males, reaching 120cm TL. **Scalation** MB 35–50 rows, 260–302 VENT, SUB 30–45 mainly divided (sometimes a few single anteriorly) and anal scale single. Scales shiny in appearance and smooth.
Similar species *A. c. childreni* (opposite), *A. perthensis* (p. 35).

RANGE Encountered in WA, from Perth to the Pilbara, and east across to Kalgoorlie in south and Onslow in north.

COMMENTS Nocturnal. Terrestrial, living in grassland, heaths with rock outcrops, open woodland and rocky gorges. Shelters under rocks, inside ant mounds and in lizard burrows. **Diet** Lizards, frogs, mammals. **Reproduction** Oviparous. 6–16 per clutch. Neonates approximately 30cm TL hatching in October–March. **Disposition** Inoffensive but will bite if threatened.

BITE/VENOM HARMLESS
IUCN LISTING Least Concern.

Bruce Rock, WA, Scott Eipper

Bruce Rock, WA, Scott Eipper

Beebingarra Creek, WA, Brian Bush

Southern Spotted Python
Antaresia maculosa maculosa (W. C. H. Peters, 1873)

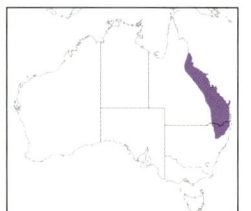

PRONUNCIATION *An-tah-ree-sah mac-u-low-sah mac-u-low-sah.*
ETYMOLOGY Spotted.
TYPE LOCALITY Port Mackay, QLD.
APPEARANCE Medium-sized, moderately built snake with head distinct from body. Dark or yellowish-brown above, with darker brown to black blotched markings covering dorsum. Some island populations have lighter markings. Tongue dark; inside of mouth pinkish. Pupil elliptic; eye colouration matches head. Ventral colouration white to cream. Adult females larger than males, reaching 150cm TL. **Scalation** MB 35–40 rows, 245–290 VENT, SUB 30–45 mainly divided (sometimes a few single anteriorly) and anal scale single. Scales shiny in appearance and smooth. **Similar subspecies** *A. m. peninsularis* (opposite). Animals found in overlap zone between Cairns and Innisfail are unable to be reliably split without genetic sequencing.
RANGE Encountered in eastern QLD to NE NSW. Between Cairns, QLD and Tamworth, NSW.
COMMENTS Primarily nocturnal. Terrestrial, living in brigalow, rainforest, rocky outcrops and woodland. Often uses cave entrances as ambush locations to capture bats. Shelters in rock crevices and beneath rocks and logs. **Diet** Lizards, frogs, bats, rodents. **Reproduction** Oviparous. 4–21 per clutch. Neonates approximately 32cm TL hatching in October–April. **Disposition** Inoffensive but will bite if threatened.
BITE/VENOM HARMLESS
IUCN LISTING Least Concern.

Brisbane, QLD, Scott Eipper

Mt Etna, QLD, Scott Eipper

Brisbane, QLD, Scott Eipper

CAPE YORK SPOTTED PYTHON *Antaresia maculosa peninsularis*
Esquerré, Donnellan, Pavón-Vázquez, Fenker & Keogh, 2021

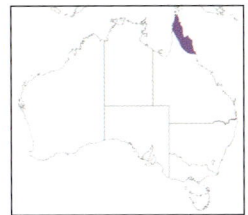

PRONUNCIATION *An-tah-ree-sah mac-u-low-sah pe-nin-sue-lah-riss.*
ETYMOLOGY Spotted peninsula; pertains to distribution centred on Cape York Peninsula.
TYPE LOCALITY Cooktown, QLD.
APPEARANCE Medium-sized, moderately built snake with head distinct from body. Colouration yellowish to light brown above, with darker brown to black blotched markings covering dorsum. Some island populations have lighter markings. Tongue dark; inside of mouth pinkish. Pupil elliptic; eye colouration matches head. Ventral colouration white to cream. Adult females larger than males, reaching 143cm TL. **Scalation** MB 35–40 rows, 252–287 VENT, SUB 40–48 mainly divided (sometimes a few single anteriorly) and anal scale single. Scales shiny in appearance and smooth. **Similar subspecies** *A. papuensis* (p. 34), *A. m. maculosa* (opposite). Animals found in overlap zone between Cairns and Innisfail cannot be reliably split without genetic sequencing.

RANGE Encountered in eastern QLD, between Lockhart River and Gordonvale. Populations north of Lockhart River on to the Torres Strait Islands require further investigation as it is unclear if these are *A. papuensis* or *A. m. peninsularis*.

COMMENTS Predominantly nocturnal. Terrestrial, living in brigalow, rainforest, rock outcrops and woodland. Shelters in rock crevices, and beneath rocks and logs. **Diet** Lizards, frogs, rodents. **Reproduction** Oviparous. 4–19 per clutch. Neonates approximately 33cm TL hatching in December–March. **Disposition** Inoffensive but will bite if threatened.

BITE/VENOM HARMLESS
IUCN LISTING Least Concern.

Juvenile, Cooktown, QLD, Scott Eipper

Mission River, QLD, Scott Eipper

Lake Morris, QLD, Scott Eipper

Mission River, QLD, Scott Eipper

Papuan Spotted Python *Antaresia papuensis*
Esquerré, Donnellan, Pavón-Vázquez, Fenker & Keogh, 2021

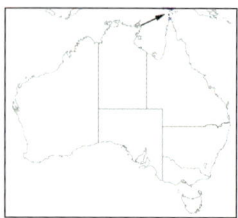

PRONUNCIATION *An-tah-ree-sah pap-u-en-sis*.
ETYMOLOGY In reference to being found in Papua New Guinea.
TYPE LOCALITY Badu Island, QLD.
APPEARANCE Medium-sized, moderately built snake with head distinct from body. Dorsal colouration yellowish to light brown, with fine to small darker brown to black blotched markings. Tongue dark; inside of mouth pinkish. Pupil elliptic; eye colouration matches head. Ventral colouration white to cream. Adult females larger than males, reaching 118cm TL. **Scalation** MB 39–41 rows, 253–284 VENT, SUB 40–48 mainly divided (sometimes a few single anteriorly) and anal scale single. Scales shiny in appearance and smooth. **Similar species** *A. maculosa peninsularis* (p. 33).
RANGE Encountered in far NE QLD on Moa, Badu, Hammond, Sabai and Thursday Islands and southern PNG. Likely to occur on northern Cape York Peninsula.
COMMENTS Predominantly nocturnal. Terrestrial, living in tropical open woodland. Has been found sheltering beneath man-made debris. **Diet** Lizards, frogs, rodents. **Reproduction** Oviparous. Nine per clutch. Neonate data unknown. **Disposition** Inoffensive but will bite if threatened.
BITE/VENOM HARMLESS
IUCN LISTING Least Concern.

Horn Island, QLD, Alexander Davies

Horn Island, QLD, Wes Read

Horn Island, QLD, Alexander Davies

Pygmy Python *Antaresia perthensis* (Stull, 1932)
(Anthill Python)

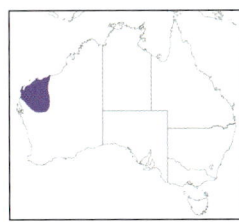

PRONUNCIATION *An-tah-ree-sah per-thenn-siss.*
ETYMOLOGY After the erroneous type locality.
TYPE LOCALITY Perth, WA.
APPEARANCE Small, slender python with head distinct from body. Dorsal colouration reddish or occasionally yellowish-brown, with darker flecking. Some individuals almost patternless. Tongue pink; inside of mouth pinkish. Pupil elliptic; eye colouration matches head. Ventral colouration white to cream. Adult females larger than males, reaching 80cm TL. **Scalation** MB 31–35 rows, 205–255 VENT, SUB 30–45 mainly divided (sometimes a few single anteriorly), and anal scale single. Scales matt in appearance and smooth. **Similar species** *A. c. stimsoni* (p. 31).
RANGE Encountered in WA, from Goldsworthy to Mt Magnet, including the Pilbara, Gascoyne and Murchison regions.
COMMENTS Nocturnal. Terrestrial. Inhabits spinifex-dominated grassland with rocky outcrops and rocky gorges. Shelters in termite mounds. **Diet** Geckos, skinks, frogs. **Reproduction** Oviparous. 2–12 per clutch. Neonates approximately 20cm TL and hatch in October–March. **Disposition** Inoffensive but will bite if threatened.
BITE/VENOM HARMLESS
IUCN LISTING Least Concern.

Capricorn, WA, Brian Bush

Karijini NP, WA, Jake Meney

Pannawonica, WA, Angus McNab

Paynes Find, WA, Scott Eipper

Black-headed & Woma Pythons, *genus Aspidites* Peters 1876

The *Aspidites* genus currently comprises two species. They are endemic to Australia and are the only two species of python without visible heat-sensing pits along their jaws. The heat-sensing pits are covered by the rostral scale. However, they are visible inside the mouth, behind the rostral scale. These snakes often lift their heads off the ground and angle the snout upwards; theoretically, this is done to use the rostral pits more effectively. Bites, while completely harmless, can register on venom-detection kits, from the residue of a snake's oral secretions. A bite from a large adult can be painful. Both species are endemic to Australia. **Species-level identification difficulty** – 1.

ETYMOLOGY Shield bearer, in reference to the head's symmetrical scales.

TYPE SPECIES *Aspidiotes melanocephalus*.

Key to *Aspidites*
1 Head and neck glossy black.. *A. melanocephalus* (opposite).
  Head and neck orange-brown to yellow... *A. ramsayi* (p. 38).

Aspidites ramsayi, Uluru, NT, Scott Eipper

Aspidites melanocephalus, Charters Towers, QLD, Scott Eipper

BLACK-HEADED PYTHON *Aspidites melanocephalus* (Krefft 1864)
(Black-headed Rock Python, Tar Pot)

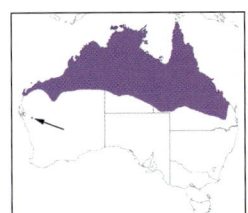

PRONUNCIATION *As-pid-die-tees mel-an-o-keff-ah-lus.*
ETYMOLOGY Black-headed shield bearer.
TYPE LOCALITY Bowen, QLD.
APPEARANCE Large, robust python with indistinct head. Dorsal colouration reddish to yellowish-brown or white, with dark brown to black cross-bands. Banding fades with age. Head and neck black. Labial pits covered by scales. Pupil elliptic; eye colouration matches head. Tongue and mouth lining dark. Ventral colouration cream to yellow, with orange, with brown and black markings. Adult females larger than males, reaching TL 300cm. **Scalation** MB 50–65 rows, 315–359 VENT, SUB 60–75 all single, occasionally a few divided posteriorly, and anal scale single. Scales matt in appearance and smooth.
RANGE Encountered from Mundubbera, QLD, across northern Australia to Woolen Station, WA.
COMMENTS Terrestrial, nocturnal, but occasionally basks during the day. Usually found on rocky or clay loam soils in open woodland, rainforest margins to spinifex-dominated grassland, black soil plains and rocky gorges. Shelters in burrows, rock crevices, deep soil cracks and caves. **Diet** Lizards, snakes, birds, mammals. **Reproduction** Oviparous. 5–20 per clutch. Neonates approximately 63cm TL and hatch in January–April. **Disposition** Inoffensive, but will defensively bluff strike at a perceived threat.
BITE/VENOM HARMLESS
IUCN LISTING Least Concern.

Weipa, QLD, Scott Eipper

Lake Moondarra, QLD, Scott Eipper

Duchess, QLD, Tie Eipper

Nullagine River, Marble Bar, WA, Brian Bush

WOMA *Aspidites ramsayi* (Macleay 1882)
(Sand Python, Ramsay's Python)

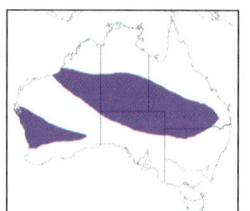

PRONUNCIATION As-pid-die-tees ram-zee-e.

ETYMOLOGY Ramsay's shield bearer; probably pertains to E. P. Ramsay, Australian zoologist.

TYPE LOCALITY Tyndarie, NSW.

APPEARANCE Large, robust python with indistinct head. Dorsal colouration reddish-orange to yellowish-brown or grey, with light to dark grey bands. Occasionally some individuals have purplish-black to black bands. Markings fade with age. Head and neck yellow to orange. Some individuals have dark marks over the eyes. Pupil elliptic; eye colouration usually dark. Labial pits covered by scales. Tongue and mouth lining dark. Ventral colouration orange to yellow, with or without dark markings. Adult female larger than male, reaching 220cm TL. **Scalation** MB 43–65 rows, 273–315 VENT, SUB 40–55 all single, occasionally a few divided posteriorly, and anal scale single. Scales matt in appearance and smooth.

RANGE Encountered in two distinct populations. One, Westmar, QLD, across Australia through western NSW, northern SA, south and central NT, to the south of Broome, WA. Second population is in WA, from Shark Bay to Perth, into the goldfields east of Kalgoorlie.

COMMENTS Nocturnal. Terrestrial; usually seen on sandy soils. Found in open woodland, brigalow, deserts and mulga woodland. Shelters in hollow logs or under leaf debris. Occasionally hunts in trees, but most prey items are caught in burrows. Southwestern population listed as rare and locally endangered by the state government. **Diet** Birds, mammals, lizards, snakes. **Reproduction** Oviparous. 5–15 per clutch. Neonates approximately 45cm TL and hatch in January–April.

Disposition Inoffensive, but will defensively bluff strike at a perceived threat.

BITE/VENOM HARMLESS

IUCN LISTING Least Concern.

5km N Sunrise Dam, WA, Brian Bush

50km E Port Hedland, WA, Brian Bush

Moomba, SA, Tie Eipper

Wallumbilla, QLD, Michael Payne

Water & Olive Pythons, genus *Liasis* Gray, 1840

The *Liasis* genus currently comprises five extant species and one subspecies, three of which come from Australia. Further taxonomic work may elevate Pilbara olive pythons to a full species and separate the distinct eastern and western clades within Water pythons. **Species-level identification difficulty** – 3.

ETYMOLOGY *Liasis*: possibly from 'lias', a type of blue limestone, or meaningless.

TYPE SPECIES *Liasis mackloti*.

Key to *Liasis*

1. Midbody rows more than 55 rows, white to pale yellow ventral colouration 2
   Midbody rows fewer than 55 rows, bright yellow ventral colouration .... *L. fuscus* (p. 40)

2. Midbody rows 58–63 rows, restricted to the Pilbara region .... *L. olivaceus barroni* (p. 41)
   Midbody rows 61–73 rows, not in the Pilbara region .... *L. o. olivaceus* (p. 42)

Liasis fuscus, Tully, QLD, Scott Eipper

Liasis olivaceus olivaceus, Lake Moondarra, QLD, Scott Eipper

WATER PYTHON *Liasis fuscus* W. C. H. Peters, 1873
(Rainbow Serpent, Brown Water Python, Yellow-bellied Python)

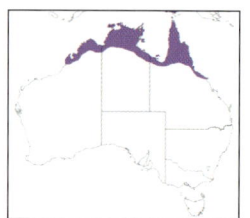

PRONUNCIATION *Lee-as-iss fus-cuss*.
ETYMOLOGY Dark/dusky.
TYPE LOCALITY Port Bowen, QLD.
APPEARANCE Large, slender python with indistinct head. Dorsal colouration greenish-brown to grey, and unpatterned. Lips usually peppered with grey or black over white. Pupil elliptic; eye colouration matches head. Tongue and mouth lining dark. Ventral colouration white beneath head, yellow to orange below body and grey under tail. Adult females larger than males, reaching 300cm TL. **Scalation** MB 40–50 rows, 270–300 VENT, SUB 60–90 divided, and anal scale single. Scales smooth and have an iridescent sheen that gives this species the name rainbow serpent. **Similar species** *L. olivaceus olivaceus* (p. 42).
RANGE Encountered from Conway, QLD, across northern Australia to Broome, WA.
COMMENTS Nocturnal but also basks during the day. Terrestrial to semi-aquatic; found in moist environments in tropical savannah and woodland, as well as in swamps and floodplains. Shelters in tree stumps and hollows, soil burrows and under fallen vegetation. Will take to the water when startled. **Diet** Rodents, mammals, birds', eggs and occasionally reptiles, including small crocodiles. Juveniles eat frogs, lizards and fish. **Reproduction** Oviparous. 9–20 per clutch. Neonates approximately 45cm TL and hatch in October–November. **Disposition** Generally defensive; will readily bite or musk on a perceived threat.
BITE/VENOM HARMLESS
IUCN LISTING Least Concern.

Evans Landing, QLD, Scott Eipper

Fogg Dam, NT, Adam Elliott

Tully, QLD, Scott Eipper

Tully, QLD, Scott Eipper

PILBARA OLIVE PYTHON *Liasis olivaceus barroni* Smith, 1981

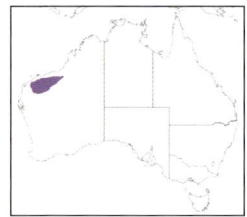

PRONUNCIATION *Lee-as-iss ol-ee-vee-cee-us ba-ron-ee.*
ETYMOLOGY Pertains to G. Barron of the WA Museum.
TYPE LOCALITY Tambrey, WA.
APPEARANCE Very large, robust python with indistinct head. Dorsal colouration dark brown to grey, and unpatterned. Upper labial scales edged with white to cream. Pupil elliptic; eye colouration matches head. Tongue and mouth lining dark. Ventral colouration cream to yellow. Adult females larger than males, reaching TL 450cm. **Scalation** MB 58–63 rows, 374–411 VENT, SUB 90–110 all divided, and anal scale single. Scales smooth and have an iridescent sheen. **Similar species** *L. olivaceus olivaceus* (p. 42).
RANGE Only found in the Pilbara region and Mt Augustus in WA.
COMMENTS Nocturnal. Terrestrial and semi-arboreal, living in open woodland, in rocky gorges usually with permanent water. Shelters in tree hollows and rock crevices. State-based assessments regard this subspecies as vulnerable. **Diet** Wallabies, rodents, flying foxes, birds. Young individuals eat rodents and lizards. **Reproduction** Oviparous. One recorded clutch of eight eggs cut from a dead wild specimen. Subspecies has never been bred in captivity, so neonate size is unknown but expected to be like that of the nominate form. **Disposition** Fairly inoffensive.
BITE/VENOM HARMFUL A bite from a large adult could cause a significant wound.
IUCN LISTING Least Concern.

Juvenile, Marillana Creek, WA, Brian Bush

Pilbara, WA, Scott Eipper

Pilbara, WA, Scott Eipper

Yarrie Station, WA, Brian Bush

OLIVE PYTHON *Liasis olivaceus olivaceus* Gray, 1842

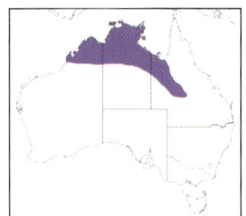

PRONUNCIATION *Lee-as-iss ol-ee-vee-cee-us ol-ee-vee-cee-us*.
ETYMOLOGY Olive coloured.
TYPE LOCALITY Port Essington, NT.
APPEARANCE Very large, robust python with indistinct head. Dorsal colouration pale to dark brown, olive to grey, and unpatterned. Upper labial scales white to cream. Pupil elliptic; eye colouration matches head. Tongue and mouth lining dark. Ventral colouration cream to pale yellow. Adult females larger than males, reaching TL 400cm. **Scalation** MB 60–80 rows, 321–377 VENT, SUB 96–119 all divided, and anal scale single. Scales smooth and have an iridescent sheen. **Similar species** *L. fuscus* (p. 40).

RANGE Encountered across northern Australia from along the Selwyn Range, QLD, across to the Kimberley, WA.

COMMENTS Nocturnal, occasionally basking by day, particularly during cooler months of the year. Terrestrial and semi-arboreal, living in open woodland, savannah, swamps, rocky hillsides and river edges. Occasionally encountered in houses. Shelters in tree hollows and rock crevices. **Diet** Mammals, reptiles and birds. **Reproduction** Oviparous. 5–21 eggs per clutch. Neonates approximately 68cm TL and hatch in January–February. **Disposition** Generally defensive, especially when young; usually calms with age.

BITE/VENOM HARMFUL A bite from a large adult could cause a significant wound.
IUCN LISTING Least Concern.

Kununurra, WA, Brian Bush

Katherine, NT, Tie Eipper

Pine Creek, NT, Scott Eipper

Sybella Creek, QLD, Scott Eipper

White-lipped Pythons, Genus *Leiopython* Hubrecht, 1879

Leiopython currently comprises two species. Historically up to six species have been suggested, based largely on morphological features and mitochondrial DNA. It is not clear if the species occur in Australia, and it is suggested that they may inhabit islands in the northern Torres Strait, such as Dauan Island, but no specimens are lodged in any institution from any Australian territory at the time of writing. The species that may occur in Australia have also been called *L. hoserae* and *L. meridonalis*. **Species-level identification difficulty** – 1.

ETYMOLOGY *Leiopython*: smooth python.

TYPE SPECIES *Leiopython gracilis*.

Leiopython fredparkeri, Merauke, Indonesia, Scott Eipper

SOUTHERN WHITE-LIPPED PYTHON *Leiopython fredparkeri* Schleip, 2008

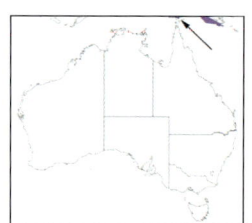

PRONUNCIATION *Lee-oh-pie-thon fred-par-ker-ree.*
ETYMOLOGY Pertains to F. Parker, Australian herpetologist.
TYPE LOCALITY Karimui, Chimbu District, Simbu Province, PNG.
APPEARANCE Large, robust python with slightly distinct head. Dorsal colouration dark brown to grey-black, becoming pale on lower flanks. Lips white with dark anterior edges. Pupil elliptic; eye colouration matches head. Tongue and mouth lining dark. Ventral colouration white. Adult females larger than males, reaching 310cm TL. **Scalation** MB 47–53 rows, 258–275 VENT, SUB 64–77 divided, and anal scale single. Scales smooth and have an iridescent sheen. **Similar species** *L. albertisii* (from PNG & Indonesia).

RANGE Found across southern NG on to the Bird's Head Peninsula region, where it is sympatric with *L. albertisii*. Suggested to occur on northern Torres Strait Islands such as Dauan. In NG, does not occur sympatrically with Water pythons *Liasis fuscus*. As the latter occur on both Saibai and Boigu islands, it has been suggested that *Leiopython* may not occur there.

COMMENTS Nocturnal but will also bask by day. Terrestrial, in dry savannah, eucalypt woodland, grassland and dense forest. Shelters in coconut husk piles and under fallen vegetation. **Diet** Mammals, birds. This species regurgitates 'furballs,' a trait unique to this genus and the closely related Ringed python *Bothrochilus boa*. **Reproduction** Oviparous. 5–13 per clutch. Neonates approximately 51cm TL and hatch in October–November. **Disposition** Strongly defensive; will readily bite or musk on a perceived threat.

BITE/VENOM HARMLESS
IUCN LISTING Least Concern.

Merauke, Indonesia, Scott Eipper

Papua, Indonesia, Lisa Farina

Sogeri, PNG, Scott Eipper

Merauke, Indonesia, Scott Eipper

Carpet, Green & Rough-scaled Pythons, genus *Morelia* Gray, 1842

Morelia currently comprises six extant species and three subspecies, six of which come from Australia. Further taxonomic work may elevate Inland carpet pythons to a full species. Many subspecies are recognized by other specialists based on morphology and ecology. Here, these differences are treated as intraspecific variations, as they are better regarded as races than subspecies based on current evidence. Three species are endemic to Australia. **Species-level identification difficulty** – 3.
ETYMOLOGY *Morelia*: probably meaningless.
TYPE SPECIES *Morelia variegata*.

Key to *Morelia*

1. Midbody rows smooth..2
   Midbody rows with keels...*M. carinata* (p. 47)

2. Body round in cross-section, adults not bright green.....................3
   Body vertically oval in cross-section, adults bright green,
   juveniles are bright yellow...*M. viridis* (p. 52)

3. Midbody rows 51 or less, not found in central Australia................4
   Midbody rows 52 or more, restricted to central Australia.........*M. bredli* (p. 46)

4. Scales rhomboid, found in northern and eastern Australia..............5
   Scales lanceolate, found in south-west
   Australia to the Eyre Peninsula...*M. imbricata* (p. 48)

5. Nasal scale usually with a posterior crease..............................*M. spilota* (p. 50)
   Nasal scale usually without a posterior crease.....................*M. s. metcalfei* (p. 49)

Morelia spilota spilota, Diamond race, Coomba Bay, NSW, Tie Eipper

Morelia spilota spilota, Coastal race, after eating a large meal, Joalah Falls, QLD, Scott Eipper

Morelia imbricata, Albany area, WA, Scott Eipper

CENTRALIAN CARPET PYTHON *Morelia bredli* (Gow, 1981)
(Bredl's Python)

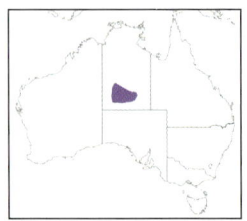

PRONUNCIATION *Mo-reel-e-ah bred-dul-e.*
ETYMOLOGY Pertains to J. Bredl sr, Australian zookeeper.
TYPE LOCALITY Pitchie Ritchie Park, Alice Springs, NT.
APPEARANCE Large, robust python with strongly distinct head. Dorsal colouration reddish-brown to dark brown, often darker on rear third, with white, yellow and black markings that transition into bands on rear third. Pupil elliptic; eye colouration matches head. Tongue and mouth lining dark. Ventral colouration white to cream, with orange, brown and black spotting. Adult females larger than males, reaching 300cm TL. **Scalation** MB 50–55 rows, 280–310 VENT, SUB 85–95 all divided, and anal scale single. Scales matt in appearance and smooth.
RANGE Encountered in NT around MacDonnell Ranges across to Hart's Range.
COMMENTS Nocturnal. Both terrestrial and arboreal, living along gorges, and in dry creek beds with old-growth eucalypts that have deep hollows. Shelters in tree hollows and rock crevices. Occasionally enters houses in Alice Springs. Sometimes treated as a subspecies of carpet python. **Diet** Mammals, birds. **Reproduction** Oviparous. 15–45 per clutch. Neonates approximately 30cm TL and hatch in November–March. **Disposition** Inoffensive but will bite if threatened.
BITE/VENOM HARMLESS
IUCN LISTING Least Concern.

Simpson's Gap, NT, Scott Eipper

Simpson's Gap, NT, Tie Eipper

Alice Springs, NT, Scott Eipper

Ellery Big hole, NT, Scott Eipper

ROUGH SCALED PYTHON *Morelia carinata* (L. A. Smith, 1981)

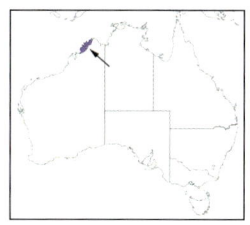

PRONUNCIATION *Mo-reel-e-ah ca-rin-nah-tah*.

ETYMOLOGY Keeled; pertaining to the distinctive midbody scales being keeled.

TYPE LOCALITY Mitchell River Falls, Admiralty Gulf on the NW coast of Kimberley, WA.

APPEARANCE Medium-sized, robust python with strongly distinct head. Dorsal colouration dark brown or grey, with inconsistent white bands and streaks. Noticeable colour change from day to night. Individuals that are brown by day become silver at night. Very long teeth, probably an adaptation for the species' prey items. Pupil elliptic; eye colouration matches head. Tongue dark; mouth lining pink. Ventral colouration white with brown peppering towards rear. Adult females larger than males, reaching 200cm TL. **Scalation** MB 45–65 rows, 298–292 VENT, SUB 83–89 that are mainly divided, and anal scale single. Scales matt in appearance and distinctly keeled. **Similar species** *Morelia spilota spilota* (p. 50).

RANGE Encountered in the Kimberley on the Mitchell River Plateau and adjacent areas; also on Bigge Island, WA.

COMMENTS Nocturnal. Predominantly arboreal. Found in vine thickets in deep sandstone gorges. Shelters in tree hollows and rock crevices. **Diet** Rodents, but will also eat birds. **Reproduction** Oviparous. 10–15 per clutch. Neonates approximately 47cm TL and hatch in December–January. **Disposition** Fairly inoffensive; will occasionally gape at a perceived threat.

BITE/VENOM HARMLESS

IUCN LISTING Least Concern.

Hunter River, WA, Scott Eipper

Hunter River, WA, Tie Eipper

Prince Regent, WA, Jake Meney

Prince Regent, WA, Matt Summerville

Western Carpet Python *Morelia imbricata* (L. A. Smith, 1981)
(Southwestern Carpet Python)

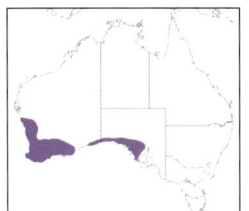

PRONUNCIATION *Mo-reel-e-ah im-bree-ca-tah*.

ETYMOLOGY Referencing the distinctive midbody scalation shape.

TYPE LOCALITY Jurien Bay, WA.

APPEARANCE Medium-sized, robust snake that is very variable in appearance with a distinct head. Dorsal colouration brown, black, grey or yellow, with white to cream or yellow markings that form irregular bands, stripes and blotches. Pupil elliptic; eye colouration matches head. Tongue and mouth lining dark. Ventral colouration white with grey or black flecks and spots. Adult females larger than males, reaching 270cm TL. **Scalation** MB 41–49 rows, 239–276 VENT, SUB 63–82 mainly divided, and anal scale single. Scales smooth and matt in appearance, with a distinctive lanceolate shape compared to rhomboid in other *Morelia*. **Similar species** *M. spilota metcalfei* (opposite).

RANGE Encountered from Kalbarri across the SW corner of WA, to Esperance. In SA on the western Eyre Peninsula, and on islands of the St Francis group.

COMMENTS Nocturnal, but often out basking by day. Both terrestrial and arboreal, living in forests, open woodland, rocky hills, grassland, mallee and heathland. Commonly found in houses. Shelters in rock crevices and tree hollows. Once thought to be a subspecies of carpet python, genetic and morphological evidence elevated it to a full species. **Diet** Mammals, birds, occasionally lizards. **Reproduction** Oviparous. 5–27 per clutch. Neonates approximately 42cm TL and hatch in March–April. Juveniles reddish with the same patterning as adults. **Disposition** Fairly inoffensive.

BITE/VENOM HARMLESS

IUCN LISTING Least Concern.

Dawesville, WA, Danny Melville

Cape Le Grand, WA, Danny Melville

Nullabor, SA, Shawn Scott

Nullabor, SA, Shawn Scott

INLAND CARPET PYTHON *Morelia spilota metcalfei* Wells & Wellington, 1985
(Murray-Darling Carpet Python)

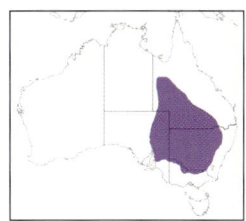

PRONUNCIATION *Mo-reel-e-ah spil-low-tah met-calf-ee.*
ETYMOLOGY Pertains to D. Metcalfe, Australian zoologist.
TYPE LOCALITY Warrumbungle Mountains, NSW.
APPEARANCE Medium-sized, very robust snake that is very variable in appearance, with a distinct head. Can acquire a heavy build as it ages. Dorsal colouration brown, black, grey or yellow, with white, cream, orange, yellow and pale brown markings that form irregular bands, stripes and blotches. Colouration shifts from dark in the south of the range to lighter with less black in the north. Juveniles reddish. Pupil elliptic; eye colouration matches head. Tongue and mouth lining dark. Ventral colouration white with grey or black flecks and spots. Adult females larger than males, reaching 220cm TL. **Scalation** MB 40–65 rows, 240–310 VENT, SUB 60–95 mainly divided, and anal scale single. Scales smooth and matt in appearance. **Similar species** *M. imbricata* (opposite), *M. spilota spilota* (p. 50).

RANGE Encountered along the Murray Darling basin, west of the GDR. From Pyramid Hill, Vic, west into eastern SA, and through western and central NSW, to Dajarra, QLD.

COMMENTS Nocturnal, but often basks during the day. Both terrestrial and arboreal, living in open woodland, rocky hills, grassland, savannah, swamps and river edges. Shelters in tree hollows and rock crevices. Occasionally found in houses. **Diet** Mammals, birds, periodically lizards as adults. Juveniles eat lizards. **Reproduction** Oviparous. 8–43 per clutch. Neonates approximately 45cm TL and hatch in March–April. **Disposition** Fairly inoffensive.

BITE/VENOM HARMLESS
IUCN LISTING Least Concern.

Warrumbungles, NSW, Scott Eipper

Baloone River, QLD, Tie Eipper

Cunnamulla, QLD, Anders Zimny

Mt Meg, VIC, Scott Eipper

Carpet Python *Morelia spilota spilota* (Lacépède, 1804)
(Diamond Python, Coastal Carpet Python, Jungle Carpet Python)

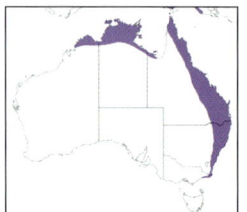

PRONUNCIATION *Mo-reel-e-ah spil-low-tah spil-low-tah.*
ETYMOLOGY Spotted.
TYPE LOCALITY Nouvelle-Hollande (Australia).
APPEARANCE Medium to large, robust snake that is very variable in appearance, with a distinct head. Species attains a heavy build with age. Dorsal colouration brown, black, grey or yellow, with white to cream or yellow markings that form irregular bands, stripes and blotches. Southern animals black with yellow flecking. Colouration shifts from dark in south of range to lighter with less black further north. Juveniles reddish. Pupil elliptic; eye colouration matches head. Tongue and mouth lining dark. Ventral colouration white with grey or black flecks and spots. In most populations, adult females larger than males, reaching up to 320cm TL. Usually, they reach a maximum size of 240cm. Regarded by some researchers as having three additional subspecies (*M. s. cheynei*, *M. s. macdowelli* and *M. s. variegata*). Here regarded as races on the basis that the forms have little genetic variation from each other, overlapping colour and pattern traits, and routinely hybridize, forming intergrades. **Scalation** MB 40–65 rows, 240–310 VENT, SUB 60–95 mainly divided, and anal scale single. Scales smooth and matt in appearance.
Similar species *M. s. metcalfei* (p. 49).

RANGE Encountered over much of eastern and northern Australia, from Marlo in Vic, to the Kimberley region of WA. Also found in PNG.

COMMENTS Nocturnal, but often found basking by day. Both terrestrial and arboreal, living in rainforests, forests, open woodland, rocky hills, grassland, savannah, swamps, mangroves and river edges. Shelters in tree hollows and rock crevices. Commonly found in houses and other properties. **Diet** Mammals, birds and occasionally lizards as adults. **Reproduction** Oviparous. 5–56 per clutch. Neonates approximately 42cm TL and hatch in December–April. **Disposition** Variable. Some populations inoffensive, while others are strongly defensive, readily biting and musking on a perceived threat.

BITE/VENOM HARMLESS
IUCN LISTING Least Concern.

Northern race, Howard Springs, NT, Scott Eipper

Coastal race, Chambers flat, QLD, Scott Eipper

Diamond race, Myall Lakes NP, NSW, Scott Eipper

Diamond race, Newcastle, NSW, Tie Eipper

Coastal race, Manorina, Mt Glorious, QLD, Scott Eipper

Jungle race, Julatten, QLD, Scott Eipper

Jungle race, Innisfail, QLD, Cody Eipper

Juvenile, coastal race, Glass House Mountains, QLD, Scott Eipper

Coastal race, Karawatha, QLD, Scott Eipper

Northern race, Black Mountain, QLD, Scott Eipper

Northern race, Humpty Doo, NT, Scott Eipper

Diamond race, Pearl Beach, NSW, Scott Eipper

SOUTHERN GREEN PYTHON *Morelia viridis* (Schlegel, 1872)
(Green Tree Python)

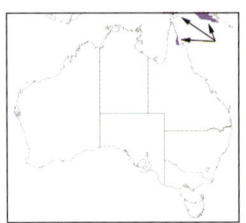

PRONUNCIATION *Mo-reel-e-ah vi-rid-diss.*
ETYMOLOGY Green.
TYPE LOCALITY Aru Islands, Indonesia.
APPEARANCE Medium-sized, relatively slender python with distinct head. Bright green, usually with a broken white vertebral stripe. Some individuals get a bluish flush, particularly when gravid. Juveniles yellow with brown vertebral stripe and brown and white flecks. Tail-tip usually a different colour from body, used for the caudal luring of prey. Pupil elliptic; eye colouration either yellowish or matches head. Tongue and mouth lining dark. Ventral colouration yellow or white, with green, blue and black markings. Adult females larger than males, reaching 150cm TL. **Scalation** MB 47–69 rows, 222–257 VENT, SUB 63–110 mainly or all divided, and anal scale single. Scales smooth and matt in appearance.

RANGE Encountered in QLD on Cape York Peninsula, in the Iron and McIlwraith Ranges. Also found in southern PNG.

COMMENTS Nocturnal, but often seen basking by day. Arboreal, living in rainforests and vine forests. Shelters in tree hollows and in epiphytes. **Diet** Mammals, birds and occasionally lizards. **Reproduction** Oviparous. 5–26 per clutch. Neonates approximately 32cm TL and hatch in October–March. **Disposition** Inoffensive but will bite if threatened.

BITE/VENOM HARMLESS

IUCN LISTING Least Concern.

Captive individual undergoing colour change, Scott Eipper

Iron Range, QLD, Scott Eipper

Iron Range, QLD, Shane Black

Juvenile, Iron Range NP, QLD, Jake Meney

Oenpelli Python, genus *Nyctophilopython* Wells & Wellington, 1985

This monotypic genus was until recently placed in both *Simalia* and *Morelia*. In 2020, it was placed in the genus *Narawan*, which provided genetic evidence supporting the initial morphology-based recognition of the genus, but this is a junior synonym of *Nyctophilopython*. **Species-level identification difficulty** – 2.

ETYMOLOGY Name means night-loving python, however, the author of the description has stated it refers to a python that prefers to be alone in the dark.

TYPE SPECIES *Python oenpelliensis*.

Nyctophilopython oenpelliensis, Noulangie Rock, NT, Scott Eipper

Nyctophilopython oenpelliensis, Noulangie Rock, NT, Dean Purcell

Oenpelli Python *Nyctophilopython oenpelliensis* (Gow, 1977)
(Narawan)

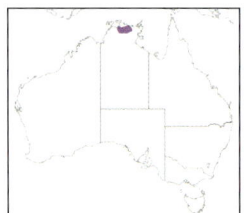

PRONUNCIATION *Nick-toe-fi-lo-pie-thon oh-en-pel-e-n-siss*.
ETYMOLOGY Oenpelli's Night-loving python; in reference to the town Oenpelli.
TYPE LOCALITY 6.5km SW of Oenpelli, NT.
APPEARANCE Very large, slender python with moderately distinct head. Dorsal colouration dark brown or grey, with irregular white bands and streaks that form blotches and bands. Marked shift in colour from day to night: individuals that are brown during the day become silver at night. Pupil elliptic; eye colouration bluish-grey. Tongue and mouth lining dark. Ventral colouration white with brown peppering towards rear. Adult females larger than males, reaching 600cm TL.
Scalation MB 65–75 rows, 420–450 VENT, SUB 150–170 mainly divided, and anal scale single. Scales smooth and matt in appearance.
RANGE Encountered in western Arnhem Land, NT.
COMMENTS Nocturnal. Mainly arboreal. Found around the Kombalgie sandstone gorges and outcrops, associated woodland and creek lines. Often seen around rock crevices and in trees in which it shelters. **Diet** Mammals, birds. **Reproduction** Oviparous. 10–15 per clutch. Neonates approximately 86cm TL and hatch in January–March. **Disposition** Inoffensive but will bite if threatened.
BITE/VENOM HARMFUL
IUCN LISTING Least Concern.

Nourlangie Rock, NT, Hal Cogger

Juvenile, Arnhemland, NT, Matt Summerville

Arnhemland, NT, Scott Eipper

Arnhem Escarpment, NT, Anders Zimny

Scrub Pythons, genus *Simalia* Gray, 1849

The genus *Simalia* currently comprises six species. Further taxonomic work on the genus is likely. Historically, the genus has been placed in *Morelia* and *Liasis*. Species were split primarily by genetic separation. Australia's largest pythons. **Species-level identification difficulty** – 4.

ETYMOLOGY Probably meaningless.

TYPE SPECIES *Boa amethistina*.

Key to *Simalia*

1. Infralabials 20–23, found south of Coen
   (see note in species account) .... *S. kinghorni* (p. 57)
   Infralabials 18–23, found north of Yarraden
   (see note in species account) .... *S. amethistina* (p. 56)

Simalia kinghorni, Lake Morris, QLD, Scott Eipper

Simalia kinghorni, Lake Morris, QLD, Scott Eipper

Southern Scrub Python *Simalia amethistina* (J. G. Schneider, 1801)
(Amethystine Python)

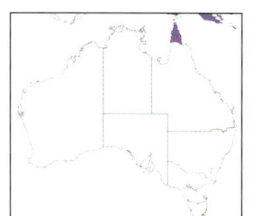

PRONUNCIATION *Si-mah-lee-ah am-e-this-tina*.
ETYMOLOGY Amethyst coloured.
TYPE LOCALITY Unknown.
APPEARANCE Very large, slender python with moderately distinct head. Dorsal colouration brown with yellowish and dark markings forming irregular bands and blotches. Colouration and patterning are more muted than in *S. kinghorni* (opposite). Pupil elliptic; eye colouration matches head. Tongue blue to purple with pale tines; mouth lining dark. Juveniles reddish. Ventral colouration white to cream without markings. Adult females larger than males, reaching 450cm TL. **Scalation** MB 35–50 rows, 270–340 VENT, SUB 80–120 mainly or all divided, and anal scale single. Scales smooth with pearlescent sheen caused by refraction. **Similar species** *S. kinghorni*. Animals found in overlap zone between Coen and Yarraden unable to be reliably split based on morphology.

RANGE Encountered in NE QLD, from Yarraden north, into PNG and Indonesia.

COMMENTS Nocturnal, occasionally basking during the day, particularly during cooler months of the year. Terrestrial but occasionally climbs trees. Lives in forests, savannah margins, swamps, mangroves and river edges. Shelters in tree hollows, in dense foliage and inside hollow logs. **Diet** Wallabies, brush-turkeys, bandicoots, possums, occasionally lizards. **Reproduction** Oviparous. 5–12 per clutch. Neonates approximately 65cm TL and hatch in November–March. **Disposition** Generally inoffensive. If handled, will readily bite and produce faecal matter as a deterrent to a predator. Very powerful snake.

BITE/VENOM DANGEROUS
IUCN LISTING Least Concern.

Dauan Island, QLD, Alexander Davies

Iron Range, QLD, Tie Eipper

Kerr Point, QLD, Scott Eipper

Weipa, QLD, Scott Eipper

Australian Scrub Python *Simalia kinghorni* (Stull, 1933)
(Amethystine Python)

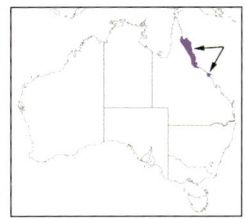

PRONUNCIATION *Si-mah-lee-ah king-hor-nee.*
ETYMOLOGY Pertains to J. R Kinghorn of the Australian Museum.
TYPE LOCALITY Lake Barrine, QLD.
APPEARANCE Very large, slender python with moderately distinct head. Ventral colouration brown, with yellowish and black markings forming irregular bands and blotches. Colouration and patterning are more distinct than in *S. amethistina* (opposite). Pupil elliptic; eye colouration matches head. Tongue blue to purple with pale tines; mouth lining dark. Juveniles reddish. Ventral colouration white to cream without markings. Adult females larger than males, reaching 450cm TL. Exceptional individuals reach more than 600cm, with an unverified report of 850cm. **Scalation** MB 35–50 rows, 270–340 VENT, SUB 80–120 mainly or all divided, and anal scale single. Scales smooth with pearlescent sheen caused by refraction. **Similar species** *S. amethistina*. Animals found in overlap zone between Coen and Yarraden unable to be reliably split based on morphology.

RANGE Encountered in NE QLD, between Coen and Conway Range.

COMMENTS Nocturnal, occasionally basking by day, particularly during cooler months of the year. Terrestrial but occasionally climbs trees. Lives in forests, savannah margins, swamps, mangroves and river edges. Shelters in tree hollows, in dense foliage and inside hollow logs. Commonly encountered in houses throughout its range. **Diet** Wallabies, brush-turkeys, bandicoots, possums, occasionally lizards. **Reproduction** Oviparous. 5–12 per clutch. Neonates approximately 70cm TL and hatch in December–April. **Disposition** Generally inoffensive. If handled, will readily bite and produce faecal matter as a deterrent to a predator. Underestimating the ability of this species has resulted in the death of a human and numerous near misses. A very powerful snake.

BITE/VENOM DANGEROUS
IUCN LISTING Least Concern.

Conway Range, QLD, Scott Eipper

Dinden NP, QLD, Scott Eipper

Juvenile, Julatten, QLD. Matt Summerville

Paluma, QLD, Scott Eipper

Family Colubridae (Colubrid Snakes)

Colubrids were thought to be the most diverse group of advanced snakes that were not vipers or elapids. Subsequent revisions have broken this family into many smaller families and subfamilies.

The Australian colubrids appear to have recently arrived from SE Asia via land bridges that were exposed during recent ice ages or that they have rafted across narrow straits. This appears to have occurred multiple times in the last 20 million years. This may explain the restriction of the colubrids to northern and eastern Australia, and why each of the genera found in Australia are further diversified in other parts of SE Asia.

In Australia, the colubrids have been split into three families: Colubridae, Natricidae and Homalopsidae. However, in earlier publications, these were all referred to as colubrids.

All solid-toothed colubrids were previously thought to be non-venomous, but subsequent research has shown that all species have rudimentary toxins within the saliva. This, coupled with elongated teeth, means that they are technically venomous, but they are essentially harmless outside of an allergic reaction. Outside Australia, some colubrid species have killed healthy adults. These species were at one point thought to be harmless, before unfortunately proving otherwise. Two species are present in Australian territories due to human introduction. All Australian species lay eggs.

Key to Australian Colubrid Snakes

1. Anal scale single 2
   Anal scale divided..3

2. Iris black,; ventrals fewer than 225, 17 or fewer
   midbody scale rows ....*Stegonotus australis* (p. 71)
   Iris grey, yellow to orange; ventrals more than 225,
   19 or more midbody scale rows..*Boiga irregularis* (p. 60)

3. Fewer than 19 midbody scale rows 4
   More than 19 midbody scale rows ....*Pantherophis guttatus* (p. 69)

4. 13–15 midbody scale rows 5
   17 midbody scale rows; restricted to Christmas Island................*Lycodon capucinus* (p. 67)

5. Black stripe from rostral through eye on to nape............................*Dendrelaphis calligastra* (p. 63)
   No black stripe from rostral through eye on to nape....................*Dendrelaphis punctulatus* (p. 64)

Dendrelaphis punctulatus, Brassall, QLD, Scott Eipper.

Cat Snakes, genus *Boiga* Fitzinger, 1849

Boiga currently comprises 37 species worldwide, and is represented in Australia by a single species. The night tiger form that is distinctly banded has been suggested to be a different species. **VENOM** Toxicity unknown; no evidence of severe envenomations in Australia other than allergic reactions. Locally, bites usually result in stinging, mild pain, itchiness and swelling. **Species-level identification difficulty** (within Australia) – 1.

ETYMOLOGY Boa.

TYPE SPECIES *Coluber irregularis*.

Boiga irregularis, Mt Carbine, QLD, Shane Black

Boiga irregularis, Springbrook NP, QLD, Scott Eipper

Brown Tree Snake *Boiga irregularis* (Merrem *in* Bechstein, 1802)
(Night Tiger, Doll's Eye Snake, Banded Cat Snake)

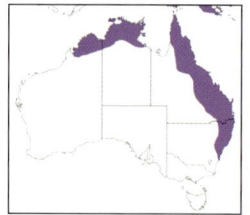

PRONUNCIATION *Boy-garr e-reg-u-lar-riss.*
ETYMOLOGY Irregular boa, pertaining to irregular pattern on dorsum.
TYPE LOCALITY Unknown.
APPEARANCE Long, slender snake with laterally compressed body, distinct head and lengthy prehensile tail. Dorsal colouration reddish-brown to dark brown, with darker reticulations; alternatively reddish-orange with contrasting white bands. Pupil elliptic; eye colouration usually matches head, but can be bright orange-yellow or grey. Tongue pale; mouth lining pink. Ventral colouration white, yellow or orange. Adult males larger than females, reaching 200cm TL. **Scalation** MB 19–23 rows, 225–265 VENT, SUB 85–130 all divided, and anal scale single. Distinct row of enlarged vertebral midbody scales. Scales matt in appearance and smooth.
RANGE Encountered through northern and eastern Australia, from the Kimberley, WA, to Wollongong, NSW. Also encountered through PNG and Indonesia, and introduced to Guam.
COMMENTS Nocturnal but basks throughout the day. Arboreal, inhabiting forests, gorges, savannah, swamps and mangroves. Shelters in caves, crevices and tree hollows. **Diet** Skinks, dragon lizards, geckos, small mammals, bats, rodents and birds. **Reproduction** Oviparous. 3–11 per clutch. Neonates approximately 45cm TL and hatch in September–December. **Disposition** Generally defensive, and will readily bite or musk on a perceived threat.
BITE/VENOM HARMFUL
IUCN LISTING Least Concern.

Fogg Dam, NT, Shane Black

Lake Morris, QLD, Scott Eipper

Cutta Cutta Caves, NT, Scott Eipper

White Cedar, QLD, Scott Eipper

Evans Landing, QLD, Scott Eipper

Dinden NP, QLD, Scott Eipper

Bronzebacks, Tree Snakes, genus *Dendrelaphis* Boulenger, 1890

Dendrelaphis currently comprises 48 species worldwide, and is represented in Australia by two species. One further species – *D. macrops* – is found on Papuan islands neighbouring Australia. *D. macrops* is similar to *D. punctulatus*, but differs by having a larger eye and having greater than 140 subcaudals. *D. calligastra* is also known in literature as *D. calligaster*. In this case it was incorrect to change the gender of the species name within the rules of nomenclature. **VENOM** Toxicity unknown; no evidence of severe envenomations recorded. Locally, bites may result in mild stinging, itchiness and mild swelling. **Species-level identification difficulty** (within Australia) – 3.

ETYMOLOGY Tree snake.

TYPE SPECIES *Ahaetula caudolineata*.

Dendrelaphis calligastra, Julatten, QLD, Shane Black

Dendrelaphis punctulatus, Airlie Beach, QLD, Scott Eipper

Northern Tree Snake *Dendrelaphis calligastra* (Günther, 1867)
(Coconut Tree Snake)

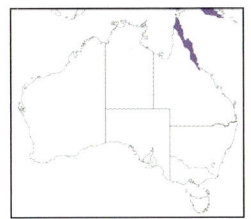

PRONUNCIATION *Den-dre-lay-fiss cal-e-gas-trah*.
ETYMOLOGY Beautiful-bellied tree snake.
TYPE LOCALITY Cape York, QLD.
APPEARANCE Medium-sized, very slender snake with head slightly distinct from body. Dorsal colouration grey to dark brown, occasionally with black speckling. Anterior edges of most midbody scales pale, only becoming visible when the snake inflates its body in response to a threat. Side of head has bright yellow to white streak, with upper margin edged with black and extending from snout, beneath eye and on to neck. Pupil round; eye colouration matches head. Tongue dark; mouth lining pale. Ventral colouration greyish and mottled with dark spotting. Females larger than males, reaching 120cm TL. **Scalation** MB 13–15 rows, 180–230 VENT, SUB 90–150 all divided, and anal scale divided. Scales smooth with matt finish. Distinctive keel on either side of ventrals, presumably to aid in climbing. **Similar species** *Dendrelaphis punctulatus* (p. 64).

RANGE Encountered in NE QLD, from Mt Elliot north to the Torres Strait Islands. Also found in PNG and Indonesia.

COMMENTS Diurnal. Arboreal, living in forests, savannah, swamps and mangroves. Shelters in tree hollows and dense vegetation. Occasionally encountered sleeping on vegetation at night. **Diet** Mainly frogs and lizards. **Reproduction** Oviparous. 5–12 per clutch. Neonates 22cm TL hatching in January–March. **Disposition** Generally inoffensive, quick to flee, and will readily musk and bluff strike at a perceived threat.

BITE/VENOM HARMLESS
IUCN LISTING Least Concern.

Dauan Island, QLD, Alexander Davies

Iron Range NP, QLD, Reid Newell

Kuranda, QLD, Paul Horner

Mt Lewis, QLD, Shane Black

Common Tree Snake *Dendrelaphis punctulatus* (Gray, 1826)
(Green Tree Snake, Yellow-bellied Black Snake, Common Bronzeback)

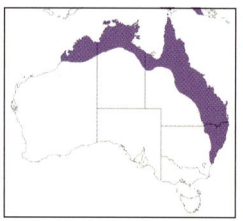

PRONUNCIATION *Den-dre-lay-fiss punkt-u-la-tuss.*
ETYMOLOGY Fine spotted tree snake.
TYPE LOCALITY Careening Bay, WA.
APPEARANCE Medium-sized, slender snake with head slightly distinct from body. Colouration very variable, and depends on location. There are several colour forms, including golden-yellow with a grey or white head, golden-brown with a darker head, black with a lighter underside, black with an electric-yellow underside, and blue to blue-grey, or green with a yellow flush to the head. Anterior edge of most midbody scales pale white to pale blue, only becoming visible when the snake inflates its body in response to a threat. Pupil round; eye colouration matches head. Tongue dark; mouth lining pale. Ventral colouration whitish, blue, green, black or yellow. Females larger than males, reaching 200cm TL. **Scalation** MB 13–15 rows, 180–230 VENT, SUB 100–150 all divided, and anal scale divided. Scales smooth with matt finish. Distinctive keel on either side of ventrals, presumably to aid in climbing. **Similar species** *Dendrelaphis calligastra* (p. 63), *D. macrops* (from PNG and Indonesia).

RANGE Encountered in eastern NSW, from Batemans Bay, through eastern QLD, and across the top end of the NT and Kimberley region of WA. Also found in PNG and Indonesia.

COMMENTS Diurnal. Arboreal, living in a wide variety of habitats, including forests, rainforest edges, savannah, swamps and mangroves. Very common in some urban areas that are well vegetated, and near creeks or streams. Shelters in tree hollows, rock crevices and dense vegetation. Occasionally encountered sleeping on vegetation at night. Very commonly found in urban environments. **Diet** Mainly frogs and small lizards, but will take small fish or mammals. **Reproduction** Oviparous. 3–14 per clutch. Neonates approximately 32cm TL and hatch in January– March. **Disposition** Generally inoffensive and quick to flee, but will readily musk and bluff strike at a perceived threat. If pressed further some individuals will bite.

BITE/VENOM HARMLESS
IUCN LISTING Least Concern.

Cape Pallarenda, QLD, Scott Eipper

Darwin, NT, Phill Mangion

Juvenile, Marsden, QLD, Scott Eipper

Mt Molloy, QLD, Shane Black

COLUBRIDS 65

Parklands, QLD, Scott Eipper

Kroombit Creek, QLD, Scott Eipper

Cornubia, QLD, Scott Eipper

Wolf Snakes, genus *Lycodon* Boie, in Fitzinger, 1826

Lycodon currently comprises 72 species worldwide. It is represented in Australia by a single introduced species. The accidental introduction into Christmas Island first recorded in 1987 has been detrimental to the native lizard fauna. Along with the Yellow crazy ant *Anoplolepis gracilipes*, Giant centipede *Scolopendra subspinipes* and Feral cats *Felis cattus*, introduced species have been implicated in the local extinction in the wild of four species and in threatening two others. **VENOM** Toxicity unknown, and there is no recorded evidence of severe envenomations. May produce swelling and itchiness. **Species-level identification difficulty** (within Australia) – 1.
ETYMOLOGY Wolf tooth in reference to fang-like anterior dentition.
TYPE SPECIES *Coluber aulicus*.

Lycodon capucinus, Medewei, Bali, Indonesia, Scott Eipper

Lycodon capucinus, Christmas Island, Hal Cogger

Common Wolf Snake *Lycodon capucinus* Boie, 1827

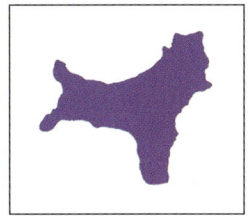

PRONUNCIATION Lie-co-don cap-u-sin-uss.
ETYMOLOGY Hooded wolf tooth; in reference to dark head marking.
TYPE LOCALITY Tjihandjawar, Java, Indonesia.
APPEARANCE Small, slender snake with distinct head. Dorsal colouration grey to dark brown, with white to yellow reticulations that form irregular cross-bands. A lower lateral zone paler than dorsum. Crown dark coloured, formed by pale markings on sides of head and pale nuchal band. Pupil elliptic; eye colouration black-brown. Tongue pink; mouth lining pale. Ventral colouration white to pale grey. Adult females larger than males, reaching 80cm TL. **Scalation** MB 17 rows, 170–225 VENT, SUB 50–80 all divided, and anal scale entire. Scales glossy in appearance and smooth.

RANGE Introduced into Australia on Christmas Island. Encountered throughout southern Asia.

COMMENTS Nocturnal. Terrestrial and arboreal, living in forests, savannah, swamps and mangroves. Commonly enters houses on Christmas Island. **Diet** Lizards. **Reproduction** Oviparous. 3–11 per clutch. Neonates approximately 20cm TL and hatch in October–March, although may produce multiple clutches per year. **Disposition** Generally inoffensive, quick to flee, and will readily musk and bluff strike at a perceived threat. If pressed further some individuals will bite.

BITE/VENOM HARMFUL

IUCN LISTING Least Concern.

Christmas Island, Angus McNab

Medewei, Bali, Indonesia, Scott Eipper

Medewei, Bali, Indonesia, Tie Eipper

Medewei, Bali, Indonesia, Tie Eipper

Rat Snakes, genus *Pantherophis* Leopold, in Fitzinger, 1843

Pantherophis currently contains eight species worldwide. The taxonomy of the genus has been unstable, with up to 10 species recognized. A single species has been introduced to Australia. Corn snakes have become established in urban centres by the illegal pet trade via accidental escapes and deliberate releases. The impact on native wildlife is unknown at this stage. If seen, authorities should be notified. **VENOM** Toxicity unknown, and there is no recorded evidence of severe envenomations. **Species-level identification difficulty** (within Australia) – 1.

ETYMOLOGY Predator of all snakes.

TYPE SPECIES *Coluber guttatus*.

Pantherophis guttatus, (albino), Nerang, QLD, Scott Eipper

Pantherophis guttatus, captive bred, Scott Eipper

Corn Snake *Pantherophis guttatus* (Linnaeus, 1766)
(Red Rat Snake, Chicken Snake)

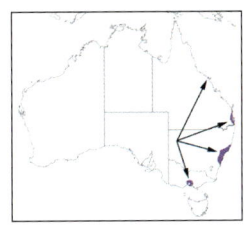

PRONUNCIATION *Pan-ther-roe-fis goo-tah-tus.*
ETYMOLOGY Spotted predator of all snakes.
TYPE LOCALITY Charleston, South Carolina, USA.
APPEARANCE Large, moderately robust snake with slightly distinct head. Dorsal colouration orange-brown, with black-edged red, rhomboid-shaped vertebral blotches. Lower flanks often yellowish. Albino individuals commonly encountered, due to captive lineage origins. Pupil round, with eye colouration matching that of head. Tongue pink; mouth lining pale. Striking undersides have a black-and-white chequered pattern. Adult females larger than males, reaching 180cm TL. **Scalation** MB 27–29 rows, 215–240 VENT, SUB 61–79 mostly divided, and anal scale divided. Scales glossy in appearance and smooth.

RANGE Introduced. Most individuals have been found around Sydney, NSW, with additional ones being located around Melbourne, Vic, and Brisbane, QLD. Native to southeastern USA.

COMMENTS Nocturnal but basks by day. Terrestrial but very adaptable, climbing trees while hunting. Lives in open forests, swamps and urban areas. The large numbers of adults and juveniles that have been found suggest that this species has become established in the wild. **Diet** Mammals, birds, bird eggs. Juveniles eat lizards and frogs. **Reproduction** Oviparous. 3–40 per clutch. Neonates approximately 13cm TL, hatching in summer and early autumn in the USA. Can have multiple clutches per year. **Disposition** Generally inoffensive and quick to flee, but will readily musk and bluff strike at a perceived threat. If pressed further some individuals will bite.

BITE/VENOM HARMLESS
IUCN LISTING Least Concern.

Frankston, (albino), VIC, Scott Eipper

Campbelltown, (albino), NSW, Scott Eipper

Nerang, (albino), QLD, Scott Eipper

Captive bred, Scott Eipper

Genus *Stegonotus* Duméril, Bibron & Duméril, 1854

Stegonotus currently comprises 25 species, one of which occurs in Australia. The genus has been the subject of numerous taxonomic reviews. Historically, *S. cucullatus* and *S. parvus* occurred in Australia; subsequent revisions restricted both of these species to Indonesia. **VENOM** Toxicity unknown, and no recorded evidence of severe envenomations. The large teeth inflict deep cuts that sting intensely and take significant time to form a blood clot, indicating that the saliva is toxic. **Species-level identification difficulty** (within Australia) – 1.

ETYMOLOGY Covered back.

TYPE SPECIES *Stegonotus mulleri*.

Stegonotus australis, Darwin, NT, Scott Eipper

Stegonotus australis, Malanda, QLD, Scott Eipper

Australian Slaty Grey Snake *Stegonotus australis* (Günther, 1872)

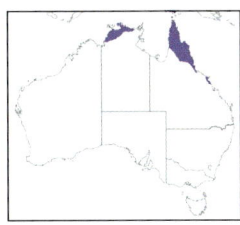

PRONUNCIATION *Steg-o-no-tuss os-trah-liss*.
ETYMOLOGY Southern covered back.
TYPE LOCALITY Cape York, QLD.
APPEARANCE Large, moderately robust snake with slightly distinct head. Dorsal colouration black, dark grey or dark brown; lower flanks pale grey. Enlarged teeth for slicing open reptile eggs. Pupil round; eye colouration black-brown. Tongue pink; mouth lining pale. Ventral colouration white to cream, occasionally pinkish. Adult females larger than males, reaching 180cm TL. **Scalation** MB 17 or (rarely 19) rows, 170–225 VENT, SUB 65–105 all divided, and anal scale single. Scales glossy in appearance and smooth. **Similar species** *Cryptophis nigrescens* (p. 112) differs by having loreal scale.

RANGE Encountered across northern Australia from Wadeye, NT, across top end region of the NT, on to Cape York, and south to Sarina, QLD. Also found in PNG and parts of Indonesia.

COMMENTS Nocturnal. Terrestrial, living in forests, along waterways in savannah, swamps and floodplains. Commonly enters houses. Shelters beneath tree roots and debris, and in soil cracks. **Diet** Mainly reptile eggs, frogs, mammals and fish. **Reproduction** Oviparous. 7–16 per clutch. Neonates approximately 32cm TL and born in December–April. **Disposition** Generally defensive; will readily bite or musk on a perceived threat.

BITE/VENOM HARMFUL
IUCN LISTING Least Concern.

Humpty Doo, NT, Scott Eipper

Jum Rum Creek, QLD, Scott Eipper

Barron Gorge, QLD, Scott Eipper

Jum Rum Creek, QLD, Scott Eipper

Family Natricidae (Water Snakes)

The water snakes are a group of both solid-toothed and rear-fanged snakes that are mildly venomous. For many years this was thought to be a subfamily of Colubridae. Natricidae are much more diverse in SE Asia, with only one of the approximately 240 species in the family occurring in northern Australia.

Keelbacks, genus *Tropidonophis* Gray, 1841

Tropidonophis currently comprises 20 species worldwide. One species occurs in Australia. Keelbacks have strong, muscular bodies that are easily depressed horizontally to aid in movement and basking. The keeled scales assist with directional stability and reduce friction when moving. Like lizards, *T. mairii* can drop their tails when in distress. **VENOM** The enlarged rear teeth inflict deep cuts but are harmless. The toxicity is unknown, and no evidence of severe envenomations has been recorded. **Species-level identification difficulty** (within Australia) – 2.

ETYMOLOGY Keeled-snake.

TYPE SPECIES *Tropidonophis picturatus*.

Tropidonophis mairii, Mission River, QLD, Scott Eipper

Tropidonophis mairii, Noulangie, NT, Scott Eipper

Tropidonophis mairii, Cudgen, QLD, Scott Eipper

KEELBACK *Tropidonophis mairii* (Gray, 1841)
(Fresh-water Snake)

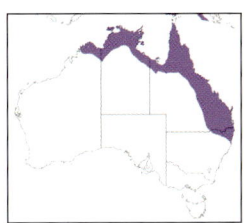

PRONUNCIATION *Trop id-o-no-fisss mare-ree*.

ETYMOLOGY Mair's keeled snake; pertains to A. Mair, who donated the type specimen.

TYPE LOCALITY Port Essington, NT.

APPEARANCE Medium-sized, slender snake with slightly distinct head. Quite variable in colour. Dorsal colouration ranges from grey, olive, yellowish, all shades of brown, to black, with darker flecks and spots. Round pupil; iris the same colour as head. Mouth lining pink; tongue generally dark at the tines and reddish-pink, but can vary in colour. Ventral colouration white, yellow, orange to greenish, with darker spotting. Adult females larger than males, reaching 120cm TL. **Scalation** MB 15 or (rarely 17) rows, 130–165 VENT, SUB 50–85 all divided, and anal scale single. Scales strongly keeled. **Similar species** *Tropidechis carinatus* (p. 222); distinguished by having loreal scale and corners of mouth upturned, forming a 'smile'.

RANGE Encountered across northern and eastern Australia, from the Kimberley, WA, to Grafton, NSW. Also encountered in PNG and parts of Indonesia.

COMMENTS Cathemeral. Terrestrial, living in forests, along waterways, in savannah, swamps and floodplains. Commonly enters gardens. Shelters under bushes, fallen logs, surface debris, and animal burrows and mud holes. **Diet** Frogs, lizards, fish. Known to eat small Cane toads; however, it takes a significant toll on a snake's physiology to metabolize this toad's toxins. **Reproduction** Oviparous. Females can be gravid throughout the year, laying 3–18 eggs per clutch. Neonates approximately 14cm TL. **Disposition** Inoffensive, but once disturbed readily bites or musks on a perceived threat.

BITE/VENOM HARMLESS

IUCN LISTING Least Concern.

Mt Molloy, QLD, Scott Eipper

Weipa, QLD, Scott Eipper

Slacks Creek, QLD, Scott Eipper

Spring Gully, QLD, Scott Eipper

Family Elapidae (Front-fanged Venomous Snakes)

Elapids are found on all continents except Antarctica, and include snakes that strike fear in some people and induce wonder in others. All sea snakes are now regarded as elapids. Previously, sea snakes were placed in their own families, Hydrophiidae and Laticaudidae. Both the Australian terrestrial elapids and the sea snakes and sea kraits are in their subfamily Hydrophiinae within the Elapidae. Elapids are the most species-rich group of snakes in Australia, with 99 terrestrial species being endemic. These venomous snakes give birth to live young (sea snakes and some terrestrial elapids) or lay eggs (most terrestrial elapids). The family contains some of the world's best-known venomous snakes, such as Black mambas *Dendroaspis polylepis*, taipans *Oxyuranus*, tiger snakes *Notechis*, Eastern coral snakes *Micrurus fulvius*, Indian cobras *Naja naja* and Common kraits *Bungarus caerulescens*.

Some species have amazing toxicity, including the Inland taipan *Oxyuranus microlepidotus*, the world's most toxic snake to mice. Bites on humans typically only occur when they try to pick up, provoke or kill the snake, or unwittingly stand on it when walking through long grass or dense ground litter. There are several species and genera that are poorly understood, and taxonomic changes will be unavoidable.

Key to Australian Terrestrial Elapids

1 Tail does not terminate in soft spiny lure, no subocular scales.................. 2
  Tail terminates in soft spiny lure, with subocular scales present............... *Acanthophis* (p. 76)

2 At least some or all subcaudals divided.. 3
  Subcaudals undivided... 15

3 Anal scale normally divided; 21 or usually fewer midbody scale rows.. 4
  Anal scale single; 21–23 midbody scale rows...................................... *Oxyuranus* (p. 169)

4 Subcaudals less than 35.. 5
  Subcaudals more than 35.. 10

5 Colour pattern not consisting of alternate black and white bands........ 6
  Colour pattern consisting of black and white bands
  from head to tail.. *Vermicella* (p. 223)

6 Body with or without cross-bands; if unbanded the
  ventral surface is immaculate white or cream..................................... 7
  Body without cross-bands; ventral surface coloured or patterned....... *Cacophis* (p. 104) (part)

7 17 midbody scale rows... 8
  15 midbody scale rows... 9

8 Rostral scale rounded... *Simoselaps* (p. 204)
  Rostral scale shovel-shaped... *Brachyurophis* (p. 94)

9 Restricted to QLD; medium to robust build;
  multiple maxillary teeth following the fang.. *Antaioserpens* (p. 87)
  Found in WA or SA; very slender;
  one maxillary tooth following the fang... 28

10 Nasal and preocular scales widely separated..................................... 11
   Nasal and preocular scales in contact... 12

11 19–21 midbody scale rows.. *Glyphodon* (p. 147)
   15–17 midbody scale rows.. *Furina* (p. 143)

12	17 or more midbody scale rows	**13**
	15 midbody scale rows	**14**

13 All or most subcaudals are divided, usually one anterior
temporal scale between the 5th supralabial and parietal scale ....*Pseudonaja* (p. 187)
Usually, the anterior subcaudals entire, the remaining
divided, usually two anterior temporal scales between
the 5th supralabial and parietal scale ....*Pseudechis* (p. 177) (part)

14 Diameter of the eye noticeably greater than
its distance from the mouth ....*Demansia* (p. 115)
Diameter of the eye is equal to or less than
its distance from the mouth ....*Cacophis* (p. 104) (part)

15 Dorsal scales smooth **16**
Dorsal scales strongly keeled ....*Tropidechis carinatus* (p. 222)

16 Anal scale normally single **17**
Anal scale normally divided ....*Hemiaspis* (p. 150)

17 Ventral scales smooth **18**
Ventral scales keeled ....*Hoplocephalus* (p. 153)

18 Frontal scale is longer than wide,
midbody scales short and rounded **19**
Frontal scale is roughly equal in length and width,
midbody scales long and oblique ....*Notechis* (p. 161)

19 Frontal scale less than one and a half as wide, as supraocular **20**
Frontal scale more than one and a half as wide, as supraocular **24**

20 Ventral scales 175 or more **21**
Ventral scales 174 or less **22**

21 Head black, body grey to brown, lips barred with white ....*Paroplocephalus atriceps* (p. 176)
The head is the same colour as the body,
lips not barred with white ....*Pseudechis* (p. 177) (part)

22 Usually 19 midbody scale rows (rarely 17–21); colouration uniform ....*Echiopsis curta* (p. 139)
17 midbody scale rows or less **23**

23 Lips barred with white or cream; lower lateral midbody
scales abutting the ventrals usually lighter than the body ....*Austrelaps* (p. 90)
Lips not barred with white, lower lateral
scales same colour as the body ....*Drysdalia* (p. 134)

24 15–21 midbody scale rows; supralabials uniform or
paler towards mouth **25**
17 midbody scale rows; supralabials prominently barred ....*Denisonia* (p. 131)

25 15–21 midbody scale rows; ventral scales plain
coloured or with dark flecking not in a crescent shape **26**
15 midbody scale rows; centre of ventral scales
with a dark pigment in a crescent shape ....*Elapognathus* (p. 140)

26 15–21 midbody scale rows; head with prominent spots, flecks or a
pigment forming a hood, white or cream coloured ventral surface .... *Suta* (p. 209)
15 midbody scale rows; head usually the same colour
as body or with prominent pink sides of the head and body **27**

27 Internasal scales absent .... *Rhinoplocephalus bicolor*
(p. 203)
Internasal scales present .... *Cryptophis* (p. 109)

28 More than 170 ventral scales; no black vertebral stripe .... *Narophis bimaculatus*
(p. 158)
Less than 150 ventral scales; usually with or
an indication of black vertebral stripe .... *Neelaps calonotos* (p. 160)

Brachyurophis australis, Comet, QLD, Angus McNab

Death Adders, genus *Acanthophis* Daudin, 1803

This genus currently contains nine species, with eight found in Australia. Phylogenetic relationships within the genus suggest that there are several species complexes that require further research to define whether they are distinct species, subspecies or just geographical variants.

Unlike most Australian snakes, death adders often do not move away at the approach of a predator, relying on their superb camouflage. They settle down into leaf litter and loose soil to ambush a prospective meal. Their tail is usually placed near the head and wiggled to give the appearance of a grub or worm, luring the prospective meal. Six species are endemic to Australia. **VENOM** Predominantly neurotoxic, weakly haemolytic, with anticoagulants. Some species are also myotoxic. Envenomations from this genus can lead to symptoms such as headache, nausea, vomiting, abdominal pain, diarrhoea, dizziness, collapse and convulsions. Bites have resulted in, or could cause fatalities. Death adder or Polyvalent antivenom is used to treat envenomations. **Species-level identification difficulty** (within Australia) – 5.

TYPE SPECIES *Acanthophis cerastinus*.

ETYMOLOGY Spine-snake.

Acanthophis antarticus, Umina, NSW, Tie Eipper

Acanthophis praelongus, Magnetic Island, QLD, Scott Eipper

Key to *Acanthophis*

1 Midbody scales usually in 19 rows
   (rarely 17 or 21); restricted to WA ....*A. wellsei* (p. 86)
   Midbody scales in 21–23 rows 2

2 Posterior third of body smooth to weakly keeled 3
   The posterior third of body strongly keeled;
   reddish-orange with yellow cross-bands ....*A. pyrrhus* (p. 83)

3 Anterior dorsal scales smooth to weakly keeled 4
   Anterior dorsal scales strongly keeled 7

4 Supraocular scales strongly raised (figure A) 5
   Supraocular scales moderately raised (figure B) 6

5 Supralabial scales usually heavily pigmented;
   lacks well-defined black spots on supralabials 5, 6 & 7 ....*A. praelongus* (p. 82)
   Supralabial scales plain or with moderate stippling,
   often with well-defined black spots on supralabial scales 5, 6 & 7 ....*A. laevis* (p. 81)

6 Relatively ill-defined darker cross-bands unless the snake
   is flattened in response to a threat; dark pigment on
   supralabial scales is concentrated high on the scale,
   forming a pale lower edge; head dorsally depressed ....*A. hawkei* (p. 80)
   Well-defined darker cross-bands; pigment evenly distributed
   on supralabial scales; head not dorsally depressed ....*A. antarcticus* (p. 78)

7 Posterior edge of frontal scale does not extend
   beyond posterior edge of supraocular scales; lower
   secondary temporal equal to or smaller than sixth
   supralabial scale; head dorsally depressed ....*A. cryptamydros* (p. 79)
   Posterior edge of the frontal scale extends
   beyond posterior edge of supraocular scales;
   Head is not dorsally depressed ....*A. rugosus* (p. 84)

figure A   figure B

COMMON DEATH ADDER *Acanthophis antarcticus* (Shaw & Nodder, 1802)
(Southern Death Adder, Death Adder)

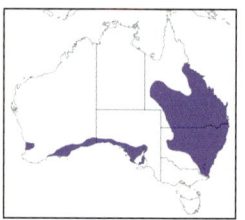

PRONUNCIATION *Ahh-can-tho-fis an-tar-tic-cus*.
ETYMOLOGY Southern spine-snake.
TYPE LOCALITY Australasia.
APPEARANCE Medium-sized, robust snake with head distinct from body. Dorsal colouration reddish-brown to charcoal-grey with lighter cross-bands. Lips white with black or dark grey markings. Tail terminates in soft spine that is white to yellow, or occasionally black. Pupil vertically elliptic; iris matches head colouration. Tongue dark; buccal cavity pink. Ventral colouration similar to that of lighter bands with darker flecking. Adult females larger than males, reaching 75cm TL. Exceptionally large females have measured up to 115cm. **Scalation** MB 21–23 rows, 110–135 VENT, SUB 35–60 single but posteriorly divided, and anal scale single. Scales matt in appearance and weakly keeled. **Similar species** *A. hawkei* (p. 80), *A. rugosus* (p. 84), *A. praelongus* (p. 82).

RANGE Encountered from the NSW/Vic border in SE, through most of NSW into QLD, north near Townsville and west to near Mitchell. Also occurs along southern coast of Australia from near Ardrossan, SA, to Perth in WA.

COMMENTS Nocturnal. Terrestrial. Found in dry forests, rainforest margins, mallee, heath, brigalow and grassland. Shelters beneath leaf litter or overhanging vegetation. **Diet** Skinks, dragons, geckos, frogs, birds, small mammals. **Reproduction** Viviparous. 2–32 per litter. Neonates approximately 17.5cm TL and born in December–March. **Disposition** Generally inoffensive unless threatened.

BITE/VENOM DANGEROUSLY VENOMOUS
IUCN LISTING Least Concern.

Gosford, NSW, Tie Eipper

Araluen, WA, Brian Bush

Mungo, NSW Scott Eipper

West Head, NSW, Scott Eipper

KIMBERLEY DEATH ADDER *Acanthophis cryptamydros*
Maddock, Ellis, Doughty, Smith & Wüster, 2015

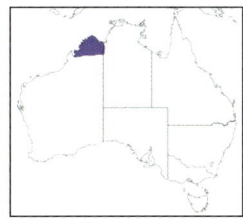

PRONUNCIATION *Ahh-can-tho-fis crypt-a-mid-ross.*
ETYMOLOGY Hidden, indistinct spine-snake.
TYPE LOCALITY 1km NW of Theda Station homestead, WA.
APPEARANCE Medium-sized, robust snake with head distinct from body. Dorsal colouration reddish-brown to orange or greyish with lighter cross-bands. Lips cream with brown to grey markings. Tail terminates in soft spine that is white to black. Pupil vertically elliptic; iris matches head colouration. Tongue dark; buccal cavity pink. Ventral colouration cream to yellow. Adult females larger than males, reaching 55cm TL. **Scalation** MB 22–23 rows, 125–139 VENT, SUB 46–56 single but posteriorly divided, and anal scale divided. Scales matt in appearance and strongly keeled. **Similar species** *A. rugosus* (p. 84).

RANGE Encountered in Kimberley region of WA and extending into neighbouring areas of the NT.

COMMENTS Nocturnal. Terrestrial. Lives in savannah woodland in close association with rocky outcrops. Shelters beneath leaf litter or overhanging vegetation, or inside spinifex clumps. **Diet** Agamids, skinks, geckos, small mammals, birds. **Reproduction** Viviparous. Up to 27 per litter. Neonates approximately 18cm TL and born in February–April. **Disposition** Generally inoffensive unless threatened.

BITE/VENOM DANGEROUSLY VENOMOUS
IUCN Vulnerable.

Charnley River Station, WA, Ian 'Bushrat' Bool

Lake Argyle, WA, Rob Valentic

Doongan Station, WA, Anders Zimny

Yampi Peninsula, WA, Ian 'Bushrat' Bool

Floodplain Death Adder Acanthophis hawkei Wells & Wellington, 1985

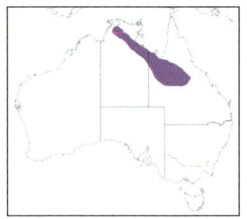

PRONUNCIATION *Ahh-can-tho-fis hawk-e.*

ETYMOLOGY Hawke's spine-snake. Pertains to Robert Hawke, former Prime Minister of Australia.

TYPE LOCALITY 1.5m SW of Brunette Downs Station Homestead, Barkly Tablelands, NT.

APPEARANCE Large, robust snake with head distinct from body. Dorsal colouration sandy-yellow to charcoal-grey with lighter cross-bands. Edges of scales in cross-bands are much brighter than other parts, so when the snake flattens out indicating that it is threatened, bright colours appear, startling a predator. Lips white with smudged darker markings. Tail terminates in soft spine that is white to orange or occasionally black. Pupil vertically elliptic; iris matches head colouration. Tongue dark; buccal cavity pink. Ventral colouration similar to that of lighter bands with darker flecking. Adult females larger than males, reaching 80cm TL. Exceptionally large females have measured up to 110cm. **Scalation** MB 23 rows, 110–155 VENT, SUB 35–60 single but posteriorly divided, and anal scale single. Scales matt in appearance and weakly keeled. **Similar species** *A. antarcticus* (p. 78), *A. rugosus* (p. 84).

RANGE Encountered south of Darwin, NT, to Longreach, QLD, via the Gulf of Carpentaria.

COMMENTS Predominantly nocturnal. Terrestrial. Found on black soil plains, grassland, floodplains and swamps. Shelters beneath leaf litter or overhanging vegetation. **Diet** Birds, lizards, frogs, mammals. **Reproduction** Viviparous. 8–27 per litter. Neonates approximately 21cm TL and born in February–May. **Disposition** Generally inoffensive unless threatened.

BITE/VENOM DANGEROUSLY VENOMOUS

IUCN Vulnerable.

Anthony's Lagoon, NT, Tie Eipper

Fogg Dam, NT, Phill Mangion

Anthony's Lagoon, NT, Scott Eipper

Fogg Dam, NT, Shane Black

Smooth-scaled Death Adder *Acanthophis laevis* Macleay, 1877

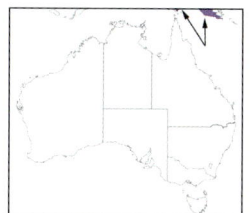

PRONUNCIATION *Ahh-can-tho-fis lay-viss.*
ETYMOLOGY Smooth spine-snake.
Type Locality Katow, PNG.
APPEARANCE Medium-sized, robust snake with head distinct from body. Dorsal colouration yellow to reddish-brown to dark grey, with lighter cross-bands often edged with black spotting posteriorly. Lips usually plain or with moderate stippling. Usually conspicuous black or dark grey markings behind eye on to lower temporal region. Pupil vertically elliptic; iris matches head colouration. Tongue dark; buccal cavity pink. Ventral colouration similar to that of lighter bands dorsally, with darker flecking that can coalesce to form bands. Adult females larger than males, reaching 50cm TL. Exceptionally large females have measured up to 100cm. **Scalation** MB 21–23 rows, 110–135 VENT, SUB 35–57 single but posteriorly divided, and anal scale single. Scales matt in appearance and smooth. **Similar species** *A. rugosus* (p. 84), *A. ceramensis* (Indonesia and PNG).
RANGE Restricted in Australia to Dauan Island in the Torres Strait. Widespread in NG.
COMMENTS Predominantly nocturnal. Terrestrial, living in rainforests, grassland and woodland. Shelters beneath leaf litter or overhanging vegetation. **Diet** Lizards, frogs, small mammals. **Reproduction** Viviparous. 8–14 per litter, born in July–August. **Disposition** Generally inoffensive unless threatened.
BITE/VENOM DANGEROUSLY VENOMOUS
IUCN LISTING Least Concern.

Dauan Island, QLD, Alexander Davies

Edebu, Central Province, PNG, Mark O'Shea

Gulf Province, PNG, Wolfgang Wüster

Kikori, Gulf Province, PNG, David Williams

NORTHERN DEATH ADDER *Acanthophis praelongus* Ramsay, 1877

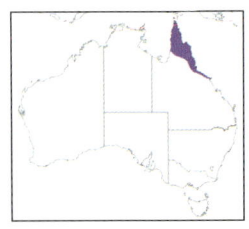

PRONUNCIATION *Ahh-can-tho-fis prey-lon-gus.*
ETYMOLOGY Very long spine-snake, referring to the tail length.
TYPE LOCALITY Cape York, North Australia.
APPEARANCE Small to medium-sized, robust snake with head distinct from body. Dorsal colouration yellow to reddish-brown, to dark grey with lighter cross-bands. Lips white with black or dark grey markings. Tail terminates in soft spine that is white to orange or occasionally black. Pupil vertically elliptic; iris matches head colouration. Tongue dark; buccal cavity pink. Ventral colouration similar to that of the lighter bands with darker flecking. Adult females larger than males, reaching 50cm TL. **Scalation** MB 19–21 rows, 110–135 VENT, SUB 35–60 single but posteriorly divided, and anal scale single. Scales matt in appearance and weakly keeled.
Similar species *A. antarcticus* (p. 78).
RANGE NE QLD, from the Torres Strait Islands to the Whitsunday region.
COMMENTS Predominantly nocturnal. Terrestrial, living in rainforests, grassland and woodland. Shelters beneath leaf litter or overhanging vegetation. **Diet** Lizards, frogs, small mammals. **Reproduction** Viviparous. 6–17 per litter. Neonates approximately 14cm TL and born in December–March. **Disposition** Generally inoffensive unless threatened.
BITE/VENOM DANGEROUSLY VENOMOUS
IUCN LISTING Least Concern.

Black Mountain, QLD, Shane Black

Yungaburra, QLD, Scott Eipper

Badu Island, QLD, Alexander Davies

Wenlock River, QLD, Anders Zimny

Desert Death Adder *Acanthophis pyrrhus* Boulenger, 1898

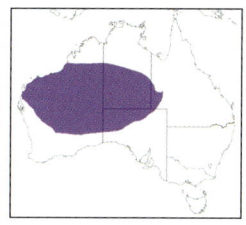

PRONUNCIATION *Ahh-can-tho-fis pier-russ.*

ETYMOLOGY Flame-coloured spine-snake.

TYPE LOCALITY Station Point, NT.

APPEARANCE Medium-sized, moderately robust snake with head distinct from body. Dorsal colouration orange to reddish-brown with lighter cross-bands. Lips peppered orange and brown, with lower margins white. Tail terminates in soft spine, which is white or occasionally black. Pupil vertically elliptic; iris matches head colouration. Tongue dark; buccal cavity pink. Ventral colouration white to cream. Adult females larger than males, reaching 60cm TL. Exceptionally large females have measured up to 90cm. **Scalation** MB 19–21 rows, 120–162 VENT, SUB 45–67 single but last few are divided, and anal scale single. Scales matt in appearance and strongly keeled. **Similar species** *A. wellsei* (p. 86).

RANGE Encountered in western and central Australia.

COMMENTS Predominantly nocturnal. Terrestrial. Lives in both sand ridge and rocky deserts and associated adjoining habitats, with a strong preference for spinifex. Shelters beneath leaf litter and overhanging vegetation, or inside spinifex clumps. **Diet** Mainly skinks, occasionally mammals. **Reproduction** Viviparous. 9–14 per litter. Neonates approximately 15cm TL and born in February–April. **Disposition** Generally inoffensive unless threatened.

BITE/VENOM DANGEROUSLY VENOMOUS

IUCN LISTING Least Concern.

West Macdonnell Ranges, NT, Scott Eipper

Port Hedland, WA, Scott Eipper

West Macdonnell Ranges, NT, Tie Eipper

Alice Springs, NT, Jesse Campbell

Rough-scaled Death Adder *Acanthophis rugosus* Loveridge, 1948
(Woodland Death Adder)

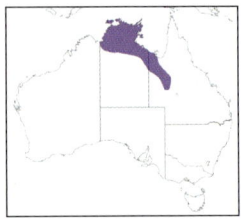

PRONUNCIATION *Ahh-can-tho-fis roo-go-suss.*

ETYMOLOGY Rough spine-snake.

TYPE LOCALITY Merauke, SW Dutch New Guinea.

APPEARANCE Medium-sized, robust snake with head distinct from body. Dorsal colouration yellow to charcoal-grey with lighter cross-bands. Edges of scales in these cross-bands much brighter, so that, when the snake flattens out, indicating that it is scared or threatened, bright colours appear, startling a predator. Lips white with black or dark grey markings. Raised scales above eyes and keeled head shields. Tail has soft spine that is whitish to orange, and occasionally black. Pupil vertically elliptic; iris matches head colouration. Tongue dark; buccal cavity pink. Ventral colouration similar to that of the lighter bands with darker flecking. Adult females larger than males, reaching 70cm TL. Exceptionally large females have measured up to 95cm. **Scalation** MB 23 rows, 115–165 VENT, SUB 53 single but posteriorly divided, and anal scale single. Scales matt in appearance and strongly keeled. **Similar species** *A. antarcticus* (p. 78), *A. cryptamydros* (p. 79), *A. hawkei* (p. 80), *A. laevis* (p. 81).

RANGE Encountered across northern Australia, from the Kimberley region across into western QLD. Also found in PNG and Indonesia.

COMMENTS Nocturnal. Terrestrial. Usually encountered in ambush position. Lives in rocky woodland, tropical savannah, grassland and breakaways. Shelters beneath leaf litter or overhanging vegetation. **Diet** Skinks, dragons, occasionally mammals. **Reproduction** Viviparous. 6–24 per litter. Neonates approximately 17.5cm TL. Selwyn Range (QLD) animals give birth in March–August. Those in NT give birth in April–September. **Disposition** Generally inoffensive unless threatened.

BITE/VENOM DANGEROUSLY VENOMOUS

IUCN LISTING Least Concern.

Cape Crawford, NT, Scott Eipper

Juvenile, Douglas Daly, NT, Scott Eipper

Selwyn Range, QLD, Scott Eipper

Juvenile, Cape Crawford, NT, Scott Eipper

Cape Crawford, NT, Scott Eipper

Deception Bay, Groote Eylant, NT, Rob Valentic

Mica Creek, QLD, Scott Eipper

Rum Jungle, NT, Scott Eipper

Tolmer Falls, NT, Scott Eipper

Juvenile, Douglas Daly, NT, Scott Eipper

Selwyn Range, QLD, Tie Eipper

Dajarra, QLD, Tie Eipper

PILBARA DEATH ADDER *Acanthophis wellsei* Hoser, 1998

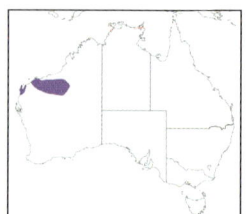

PRONUNCIATION *Ahh-can-tho-fis wells-e.*

ETYMOLOGY Wells' spine-snake. Pertains to R. Wells, Australian herpetologist.

TYPE LOCALITY Wittenoom Gorge, WA.

APPEARANCE Small to medium-sized, robust snake with head distinct from body. Dorsal colouration pale brown to orange with lighter crossbands; another colour phase is red with black bands and black head. Tail terminates in soft spine that is white or occasionally black. Pupil vertically elliptic; iris matches head colour. Tongue dark; buccal cavity pink. Ventral colouration similar to that of lighter bands with darker flecking. Adult females larger than males, reaching 45cm TL.

Scalation MB 17–21 rows, 119–143 VENT, SUB 41–64 single but posteriorly divided, and anal scale single. Scales matt in appearance and strongly keeled. **Similar species** *A. pyrrhus* (p. 83).

RANGE Encountered in the Pilbara region and Cape Range area, WA.

COMMENTS Nocturnal. Terrestrial. Occurs in rocky desert and gorges, and associated adjoining habitats, with a strong preference for spinifex. Shelters beneath leaf litter and overhanging vegetation, or inside spinifex clumps. **Diet** Mainly skinks, occasionally small mammals. **Reproduction** Viviparous. 9–20 per litter. Neonates approximately 14cm TL and born in December–March. **Disposition** Generally inoffensive unless threatened.

BITE/VENOM DANGEROUSLY VENOMOUS

IUCN LISTING Least Concern.

Exmouth, WA, Scott Eipper

Karijini NP, WA, Matt Summerville

Exmouth, WA, Scott Eipper

Pilbara, WA, Scott Eipper,

Warros, genus *Antaioserpens* Wells & Wellington, 1984

This genus currently contains two species, both endemic to Australia. They were historically placed in the genus *Simoselaps*. Nocturnal snakes, they have been found hunting sleeping skinks in leaf litter. **VENOM** Toxicity unknown, and there has been no evidence of envenomations. **Species-level identification difficulty** – 2.

TYPE SPECIES *Simoselaps warro*.

ETYMOLOGY For *Antaios* – an ancient Greek wrestler – pertaining to the strength of these snakes.

Key to *Antaioserpens*

1 Body grey, brown to reddish-orange
  without prominent speckling..*A. albiceps* (p. 88)
  Body grey, brown to reddish-orange
  with prominent speckling and blotches..*A. warro* (p. 89)

Antaioserpens albiceps, Kuranda, QLD, Hal Cogger

Antaioserpens albiceps, Archer River, QLD, Anders Zimny

Antaioserpens warro, Mitchell, QLD, Jamie Gover

NORTH-EASTERN PLAIN-NOSED BURROWING SNAKE,
Antaioserpens albiceps (Boulenger, 1898)
(Warro)

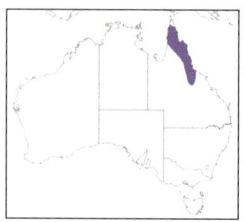

PRONUNCIATION *Ann-tay-o-ser-pens al-bee-ceps.*
ETYMOLOGY White-headed wrestling serpent.
TYPE LOCALITY Port Douglas, QLD.
APPEARANCE Small, moderately robust snake with head indistinct from body. Dorsal colouration grey-brown to reddish-orange, with lighter yellowish centres on each midbody scale. Head and lips whitish and strongly peppered with dark grey markings. Back of head has creamy-yellow bar with broad black band on nape. Juveniles have much broader white stripe on nape that covers much more of head. Pupil round; iris pale. Tongue and buccal cavity pinkish. Ventral colouration creamish-white. Adult females larger than males, reaching 42cm TL. **Scalation** MB 15 rows, 135–165 VENT, SUB 15–25 and anal scale divided. Scales glossy in appearance and smooth. **Similar species** *A. warro* (opposite).
RANGE Encountered in NE QLD, from Clermont to the northern Cape York Peninsula.
COMMENTS Predominantly nocturnal. Terrestrial. Encountered in grassland, open savannah and woodland. **Diet** Lizards. **Reproduction** Oviparous. 3–5 eggs per clutch. Neonates approximately 16cm TL hatching in February. **Disposition** Inoffensive but will bite if threatened.
BITE/VENOM HARMFUL
IUCN LISTING Least Concern.

Archer River, QLD, Anders Zimny

Juvenile, Aurukun, QLD, Anders Zimny

Torrens Creek, QLD, Hal Cogger

Warrego Burrowing Snake Antaioserpens warro (De Vis, 1884)

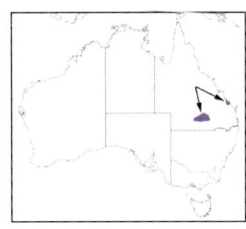

PRONUNCIATION *Ann-tay-o-ser-pens wah-row*.
ETYMOLOGY Warro wrestling serpent – in reference to the type locality.
TYPE LOCALITY Warro Station, Port Curtis, eastern QLD.
APPEARANCE Small, moderately robust snake with head indistinct from body. Dorsal colouration grey-brown to reddish-orange, with random dark brown groupings of scales. Head and lips whitish with strongly peppered dark grey markings. Back of head has a faded orange bar with broad black band on nape. Pupil round; iris pale. Tongue pale. Juveniles probably brighter in colour than adults. Ventral colouration creamish-white. Adult females larger than males, reaching 44cm TL. **Scalation** MB 15 rows, 139–150 VENT, SUB 15–17 and anal scale divided. Scales smooth and glossy in appearance. **Similar species** *A. albiceps* (opposite).
RANGE Central and eastern QLD, from Mitchell to Morven and Charleville. Also recorded from Blackdown Tablelands and around Port Curtis.
COMMENTS Predominantly nocturnal. Terrestrial. Encountered in woodland, mulga and grassland. **Diet** Lizards. **Reproduction** Thought to lay eggs. **Disposition** Inoffensive but will bite if threatened.
BITE/VENOM HARMFUL
IUCN LISTING Least Concern.

Mitchell, QLD, specimen courtesy of Jamie Williams, Rob Valentic

Copperheads, genus *Austrelaps* Worrell, 1963

This genus currently contains three species, all endemic to Australia. They are active hunters with excellent vision, and are found in southeastern Australia, usually around water, in forests or grassland. They are occasionally active after dark. Male combat has been recorded in the genus.
VENOM The venom is neurotoxic, with haemolysins, myotoxins, anticoagulants and cytotoxins. Bites have resulted in, or could cause fatalities. Tiger Snake or Polyvalent antivenom is used to treat envenomations. **Species-level identification difficulty** – 5.

TYPE SPECIES *Hoplocephalus superbus*.

ETYMOLOGY Southern-Elaps, pertaining to the genus of African garter-snakes *Elapsoidea* – the type genus for the family Elapidae.

Key to *Austrelaps*

1   Supralabial scales are usually strongly barred with grey, brown or black; no or point contact between postocular scale and lower anterior temporal scale...................**2**
    Supralabial scales are usually weakly barred with grey, brown or black; usually broad contact between postocular scale and lower anterior temporal scale..*A. superbus* (p. 93)

2   Ventral scales 155 or fewer; only found in South Australia..........................*A. labialis* (opposite)
    Ventral scales 150 or more; not found in South Australia............................*A. ramsayi* (p. 92)

Austrelaps superbus, Tunbridge, TAS, Ryan Francis

Austrelaps ramsayi, Mansfield, VIC, Scott Eipper

Pygmy Copperhead *Australaps labialis* (Jan, 1859)

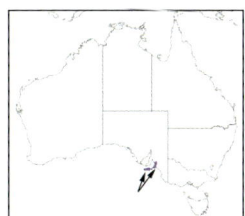

PRONUNCIATION *Os-tree-laps lay-bee-ahh-liss.*
ETYMOLOGY Lipped southern-Elaps.
TYPE LOCALITY Islet 477, Pelican Lagoon, Kangaroo Island.
APPEARANCE Medium-sized, robust snake with indistinct head. Dorsal colouration grey or dark brown. Lips boldly marked with black or dark grey on white. Large eyes with round pupils; iris reddish to orange. Tongue and inside of mouth pink. Ventral colouration cream to yellow with occasional orange flecks. Adult males larger than females, reaching 75cm TL. **Scalation** MB 15 (rarely 17) rows, 133–155 VENT, SUB 35–58 and anal scale single. Scales smooth and matt in appearance.
RANGE Encountered in SE SA in the Adelaide Hills and Kangaroo Island.
COMMENTS Diurnal, but crepuscular in hot weather. Terrestrial. Usually found under cover or among grass tussocks in swamps, grassland and moist forests. **Diet** Mainly frogs and lizards, but will also eat other snakes and occasionally mammals. **Reproduction** Viviparous. 3–32 per litter. Neonates approximately 16cm TL and born in December–March. **Disposition** Generally inoffensive unless threatened.
BITE/VENOM DANGEROUSLY VENOMOUS
IUCN LISTING Least Concern.

Parndara, SA, Shawn Scott

Adelaide Hills, SA, Shawn Scott

Adelaide Hills, SA, Shawn Scott

HIGHLAND COPPERHEAD *Austrelaps ramsayi* (Krefft, 1864)
(Alpine Copperhead)

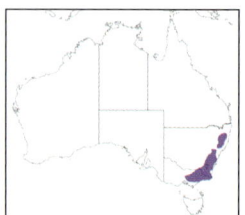

PRONUNCIATION *Os-tree-laps ram-say-e.*

ETYMOLOGY Ramsay's southern-Elaps. Pertains to E. P. Ramsay, a curator of the Australian Museum.

TYPE LOCALITY Moss Vale, NSW.

APPEARANCE Large, robust snake with indistinct head. Dorsal colouration light brown to charcoal-grey; thin vertebral stripe in some individuals. Lips boldly marked with black or dark grey on white. Pupil round; iris coppery-gold. Tongue dark to pinkish. Ventral colouration cream to yellow with occasional orange flecks. Adult males larger than females, reaching 130cm TL. **Scalation** MB 15 (rarely 17) rows, 150–171 VENT, SUB 35–58 and anal scale single. Scales smooth and matt in appearance. **Similar species** *A. superbus* (opposite).

RANGE Encountered in southeastern Australia from eastern Vic, up the eastern seaboard to the QLD border. The distribution becomes patchier in the north of the range.

COMMENTS Diurnal; occasionally nocturnal in warm weather. Terrestrial. Found in forests, grassland, heaths, swamps and riverine systems. Also readily exploits human-disturbed environments, such as rural areas and suburban gardens. Shelters under rocks, ground debris and logs. **Diet** Lizards, frogs, small mammals. **Reproduction** Viviparous. 5–32 per litter. Neonates approximately 17cm TL and born in February–May. **Disposition** Generally inoffensive unless threatened.

BITE/VENOM DANGEROUSLY VENOMOUS

IUCN LISTING Least Concern.

Juvenile, Orange, NSW, Scott Eipper

Blue Mountains, NSW, Scott Eipper

Guyra, NSW, Scott Eipper

Mansfield, VIC, Scott Eipper

LOWLAND COPPERHEAD *Austrelaps superbus* (Günther, 1858)
(Superb Snake)

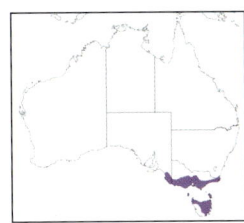

PRONUNCIATION *Os-tree-laps sup-purr-bus*.

ETYMOLOGY Superb southern-Elaps.

TYPE LOCALITY Tasmania.

APPEARANCE Large, robust snake with indistinct head. Dorsal colouration light orange-brown to charcoal-grey; lateral stripe invariably present that is usually yellow to copper coloured, and often extends up on to nape, giving the species the name copperhead. Some individuals also have spots dotted along body. Lips have dark markings on a white background. Usually, the markings do not have a sharp delineation like those of pygmy and highland copperheads; in juveniles, they are more prominent. Pupil round; iris cream to reddish-tan. Tongue dark in some individuals, pale in others; pink inside mouth. Ventral colouration cream to yellow with occasional orange flecks. Adult males larger than females, reaching 130cm TL. **Scalation** MB 15 (rarely 17) rows, 140–165 VENT, SUB 35–58 and anal scale single. Scales smooth and matt in appearance. **Similar species** *A. ramsayi* (opposite).

RANGE Encountered in SA, across Vic to Tas, including some offshore islands, and into southern NSW.

COMMENTS Diurnal; occasionally nocturnal in warm weather. Found in forests, grassland, heaths, swamps and riverine systems. Also uses urban environments. Shelters under logs, rocks and ground debris. Also utilizes burrows and curls up in or around bases of grass tussocks. **Diet** Lizards, frogs, small mammals. **Reproduction** Viviparous. 2–32 per litter. Neonates approximately 20cm TL and born in January–April. **Disposition** Generally inoffensive unless threatened.

BITE/VENOM DANGEROUSLY VENOMOUS

IUCN LISTING Least Concern.

Juvenile, Reedy Marsh, TAS, Ryan Francis

Stoney Rises, VIC, Adam Elliott

French Island, VIC, Scott Eipper

Dandenong, VIC, Scott Eipper

Coral Snakes, genus *Brachyurophis* Gunther, 1863

This genus currently contains eight species restricted to Australia. Phylogenetic relationships within the genus suggest that there is further work required to determine if there are unresolved taxa. These are nocturnal snakes that are usually encountered crossing roads at night. They specialize in eating reptile eggs. The snout is blunt and upturned, an adaptation to fossorial habits. **VENOM** Toxicity is unknown, and there is no evidence of envenomations. **Species-level identification difficulty** – 5.

TYPE SPECIES: *Brachyurophis semifasciatus*.
ETYMOLOGY Short-tailed snake.

Key to *Brachyurophis*

1. Body with bands or pattern..3
   Body patternless..2

2. Midbody scales rows 17; ventral scales 140–160..............................*B. incinctus* (p. 100)
   Midbody scales rows 15; ventral scales 135–145..............................*B. morrisi* (p. 101)

3. Nasal scale separated from preocular scale..4
   Nasal scale in contact with preocular scale..5

4. Dark nuchal bar is 3–6 scales wide, dark body bands are 1.25–2 scales wide, found at Plumridge Lakes, WA and further west..................*B. fasciolatus fasciolatus* (p. 99)
   Dark nuchal bar is 7–15 scales wide; dark body bands are 0.5–1 scale wide; found east of Cosmo Newbery, WA.......................*B. f. fasciata* (p. 98)

5. Frontal scale as wide as it is long, triple the width of the supraocular scales..6
   Frontal scale longer than it is wide, double the width of the supraocular scales..*B. australis* (p. 96)

6. Dark bands are no more than triple the width of pale cross bands......7
   Dark bands are usually 5 times broader than pale cross bands..............*B. approximans* (opposite)

7. Up to 59 dark bands on the body and tail..8
   60 or more dark bands on the body and tail....................................*B. semifasciatus* (p. 103)

8. Dark bands are usually 2–3 scales wide..*B. campbelli* (p. 97)
   Dark bands are usually 4–5 scales wide..*B. roperi* (p. 102)

Brachyurophis approximans, Auski Roadhouse, WA, Brian Bush

Brachyurophis australis, Longreach, QLD, Michael Payne

NORTH-WEST SHOVEL-NOSED SNAKE
Brachyurophis approximans (Glauert, 1954)

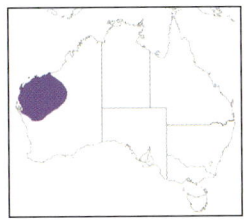

PRONUNCIATION *Brah-key-u-roe-fiss ah prox-e-mans*.

ETYMOLOGY Approaching short-tailed snake – pertaining to the similarity with *B. campbelli*.

TYPE LOCALITY Muccan Station, NW Australia.

APPEARANCE Small, robust snake with head that is indistinct from body. Dorsal colouration dark brown to charcoal with thin, one-scale-wide, cream to pale grey irregular cross-bands. Pupil round; iris dark. Tongue pale; pink inside mouth. Ventral colouration creamish-white. Both sexes reach 37cm TL. **Scalation** MB 17 rows, 151–181 VENT, SUB 19–27 and anal scale divided. Scales smooth. **Similar species** *B. roperi* (p. 102), *B semifasciatus* (p. 103).

RANGE Encountered in WA, from the Pilbara region, south to Yalgoo in the goldfields region.

COMMENTS Nocturnal. Terrestrial. Lives on heavy clay to rocky soils. Shelters beneath timber in mulga woodland and heath associations. **Diet** Reptile eggs. **Reproduction** Oviparous. 2–4 per clutch. Neonates approximately 11cm TL and hatch in January–March. **Disposition** Generally inoffensive unless threatened.

BITE/VENOM HARMFUL

IUCN LISTING Least Concern.

North West Cape, WA, Adam Elliott

Maud Hill, Coral Bay, WA, Hal Cogger

North West Cape, WA, Adam Elliott

Marillana Creek, WA, Brian Bush

Australian Coral Snake *Brachyurophis australis* (Krefft, 1864)
(Eastern Shovel-nosed Snake)

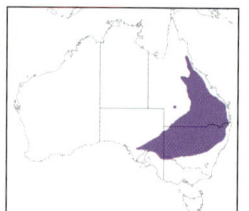

PRONUNCIATION *Brah-key-u-roe-fiss os-trah-liss.*
ETYMOLOGY Southern short-tailed snake.
TYPE LOCALITY Neighbourhood of Port Curtis and Clarence River, QLD.
APPEARANCE Small, robust snake with head that is indistinct from body. Dorsal colouration reddish-brown, pink to orange, with narrow, darkened cross-bands. Conspicuous dark nape blotch or band that is wider than the following bands on forebody. Pupil vertically oval; iris dark reddish-brown. Tongue has pale tines with a darker base; pink inside mouth. Ventral colouration creamish-white. Both sexes reach 50cm TL. **Scalation** MB 17 rows, 140–170 VENT, SUB 15–31 and anal scale divided. Scales smooth.
RANGE Encountered from Mingela, QLD, south through QLD mainly west of the GDR, through NSW, NW Vic, across to Port Pirie, SA.
COMMENTS Nocturnal. Terrestrial. Encountered on heavy clay to rocky or sandy soils in mulga, brigalow, open woodland and mallee associations. Shelters under rocks and fallen timber, and beneath leaf litter. **Diet** Lizards, reptile eggs. **Reproduction** Oviparous. 4–6 per clutch. Neonates approximately 11cm TL and hatch in October–March. **Disposition** Inoffensive but will bite if threatened.
BITE/VENOM HARMFUL
IUCN LISTING Least Concern.

Juvenile, Clermont, QLD, Rob Valentic

Herveys Range, QLD, Anders Zimny

Belyando Crossing QLD, Scott Eipper

Belyando Crossing QLD, Scott Eipper

Einasleigh Shovel-nosed Snake *Brachyurophis campbelli* (Kinghorn, 1929)

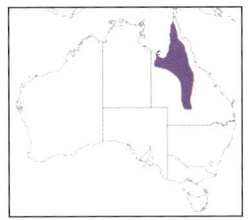

PRONUNCIATION *Brah-key-u-roe-fiss cam-bell-ee.*

ETYMOLOGY Campbell's short-tailed snake, after collector of the holotype, W. Campbell.

TYPE LOCALITY Almaden, QLD.

APPEARANCE Small, robust snake with head indistinct from body. Dorsal colouration pale grey to yellowish, pink to orange, with dark brown to charcoal-coloured, broad irregular cross-bands. Conspicuous dark nape blotch or mark that is broader than the following bands. Pupil round; iris dark. Tongue pale; pink inside mouth. Ventral colouration creamish-white. Both sexes reach 40cm TL. **Scalation** MB 15–17 rows, 140–190 VENT, SUB 14–30 and anal scale divided. Scales smooth. **Similar species** *B. roperi* (p. 102).

RANGE Encountered in QLD, from the western Cape York Peninsula, south to Townsville, and west to Camooweal.

COMMENTS Nocturnal. Terrestrial. Shelters beneath timber and rocks in open tropical woodland and on rocky hillsides. The population with 15 midbody scales, found in eastern part of its range, has the name *woodjonesii* available. **Diet** Reptile eggs. **Reproduction** Oviparous, laying six eggs per clutch in October–January. **Disposition** Generally inoffensive unless threatened.

BITE/VENOM HARMFUL

IUCN LISTING Least Concern.

Blue Lagoon, QLD, Angus McNab

Almaden, QLD, Scott Eipper

Almaden, QLD, Scott Eipper

Eastern Narrow-banded Shovel-nosed Snake
Brachyurophis fasciolatus fasciata (Stirling & Zietz, 1893)

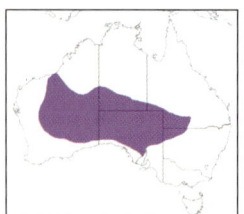

PRONUNCIATION *Brah-key-u-roe-fiss fas-cee-o-lah-tus fas-cee-ah-tah*.
ETYMOLOGY Small-banded, short-tailed snake.
TYPE LOCALITY Near the Barrow Range, WA.
APPEARANCE Small, robust snake with head that is indistinct from body. Dorsal colouration white to cream, with pale pink to orange flecks and regular thin, dark brown to charcoal, ragged-edged crossbands. Conspicuous dark nape blotch or band that is wider than the following bands. Pupil round; iris dark. Tongue pale; pink inside mouth. Ventral colouration creamish-white. Both sexes reach 41cm TL. **Scalation** MB 17 rows, 140–171 VENT, SUB 19–27 and anal scale divided. Scales smooth. **Similar species** *B. f. fasciolatus* (opposite).
RANGE Encountered in eastern WA, through central Australia, to western NSW and QLD. An isolated population in the Little and Great Sandy Deserts, WA, may represent an undescribed taxon.
COMMENTS Nocturnal. Terrestrial. Lives on sandy soils. Shelters beneath cover and leaf litter in mulga woodland, heaths, sand ridge deserts and mallee. **Diet** Feeds exclusively on reptile eggs. **Reproduction** Oviparous. 4–7 per clutch. Neonates approximately 13cm long and recorded as hatching in February. **Disposition** Generally inoffensive unless threatened.
BITE/VENOM HARMFUL
IUCN LISTING Least Concern.

Ilkurlka, WA, Brian Bush

Innaminka, SA, Hal Cogger

George Gill Ranges, NT, Paul Horner

Western Narrow-banded Shovel-nosed Snake

Brachyurophis fasciolatus fasciolatus (Günther, 1872)

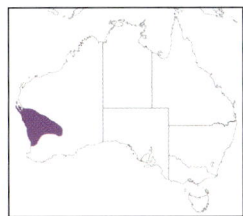

PRONUNCIATION *Brah-key-u-roe-fiss fas-cee-o-lah-tus.*

ETYMOLOGY Small-banded, short-tailed snake.

TYPE LOCALITY Perth, WA.

APPEARANCE Small, robust snake with head that is indistinct from body. Dorsal colouration white to cream, with pale pink to orange flecks and regular, thick, dark brown to charcoal, ragged-edged crossbands. Conspicuous dark nape blotch or band that is wider than the following bands. Pupil round; iris dark. Tongue pale; pink inside mouth. Ventral colouration creamish-white. Adult females larger than males, reaching 37cm TL. **Scalation** MB 17 rows, 140–175 VENT, SUB 15–30 and anal scale divided. Scales smooth. **Similar species** *B. f. fasciata* (opposite).

RANGE Encountered in WA, from Perth east to Laverton in the goldfields region, and north to Shark Bay.

COMMENTS Nocturnal. Terrestrial. Lives on sandy soils. Shelters beneath cover and leaf litter in mulga and mallee woodland, heaths and inside stick-ant nests. **Diet** Lizards and reptile eggs. **Reproduction** Oviparous. 2–5 per clutch. Neonates approximately 13cm long and hatch in January–March. **Disposition** Generally inoffensive unless threatened.

BITE/VENOM HARMFUL

IUCN LISTING Least Concern.

Jurien Bay, WA, Jake Meney

Golden Grove, SA, Brian Bush

Pinjar, WA, David Robinson

Pinjar, WA, David Robinson

Unbanded Shovel-nosed Snake *Brachyurophis incinctus* (Storr, 1968)

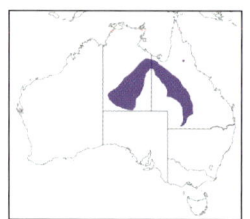

PRONUNCIATION *Brah-key-u-roe-fiss in-sink-tuss.*
ETYMOLOGY Unbanded short-tailed snake.
TYPE LOCALITY Near Alice Springs, NT.
APPEARANCE Small, robust snake with head that is indistinct from body. Dorsal colouration reddish-brown, pink to orange, without the characteristic bands of other genus members. Conspicuous dark nape blotch or band, usually with an additional dark band over eyes. Pupil round; iris dark. Tongue pale; pink inside mouth. Ventral colouration creamish-white. Both sexes reach 32cm TL. **Scalation** MB 17 rows, 140–165 VENT, SUB 18–31 and anal scale divided. Scales smooth.

RANGE Encountered from Quilpie, QLD, north to Mt Isa and SW across the Barkly Tableland to Alice Springs, NT.

COMMENTS Nocturnal. Terrestrial. Lives on heavy clay and sandy soils in mulga, open woodland and sand ridge deserts. Found under rocks and fallen timber, and beneath leaf litter. **Diet** Reptile eggs. **Reproduction** Oviparous. 3–5 per clutch. Neonates approximately 12cm TL, hatching in December–February. **Disposition** Generally inoffensive unless threatened.

BITE/VENOM HARMFUL

IUCN LISTING Least Concern.

Glen Helen, NT, Hal Cogger

Lake Moondarra, QLD, Scott Eipper

Lake Moondarra QLD, Scott Eipper

Arnhem Shovel-nosed Snake *Brachyurophis morrisi* (Horner, 1998)

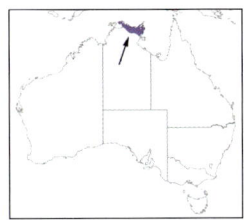

PRONUNCIATION *Brah-key-u-roe-fiss mor-riss-e.*

ETYMOLOGY Morris's short-tailed snake. Pertains to I. Morris, Australian ecologist.

TYPE LOCALITY Southern end of Elcho Island, NT.

APPEARANCE Small, robust snake with head indistinct from body. Dorsal colouration reddish-brown to brown, without the characteristic bands of other genus members. Conspicuous dark nape blotch or band; additional dark band usually present over eyes. Usually dark orange between bands. Pupil round; iris dark. Tongue pale; pink inside mouth. Ventral colouration creamish-white. Both sexes reach 33cm TL. **Scalation** MB 15 rows, 135–145 VENT, SUB 15–25 and anal scale divided. Scales smooth.

RANGE Encountered in the northern NT, from Elcho Island and the Cobourg Peninsula, to Nabarlek.

COMMENTS Nocturnal. Terrestrial. Lives on heavy clay and sandy soils in tropical woodland. Shelters under rocks and fallen timber, and beneath leaf litter. **Diet** Expected to feed on reptile eggs. **Reproduction** Thought to lay eggs. **Disposition** Generally inoffensive unless threatened.

BITE/VENOM HARMFUL

IUCN LISTING Least Concern.

Elcho Island, NT, Paul Horner

Elcho Island, NT, Paul Horner

Northern Shovel-nosed Snake *Brachyurophis roperi* (Kinghorn, 1931)

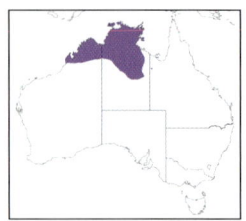

PRONUNCIATION *Brah-key-u-roe-fiss roe-per-e.*
ETYMOLOGY Roper short-tailed snake, pertaining to the type locality.
TYPE LOCALITY Roper River, N Australia.
APPEARANCE Small, robust snake with head indistinct from body. Dorsal colouration dark brown to charcoal. Thin, irregular cross-bands that are variable, from 1–5 scales wide; bands cream, yellow, orange or pinkish. Pupil round; iris dark. Tongue pale; pink inside mouth. Ventral colouration creamish-white. Both sexes reach 37cm TL. **Scalation** MB 15–17 rows, 150–180 VENT, SUB 15–24 and anal scale divided. Scales smooth. **Similar species** *B. approximans* (p. 930), (*B. campbelli* (p. 97).

RANGE Encountered from the Kimberley, WA, across to Borroloola, NT, south to the QLD border near Camooweal.

COMMENTS Nocturnal. Terrestrial. Lives on heavy clay to rocky soils. Shelters beneath timber and rocks in open tropical woodland and on rocky hillsides. **Diet** Reptile eggs. **Reproduction** Oviparous. Up to five per clutch. Neonates approximately 12cm long and hatch in December–January. **Disposition** Generally inoffensive unless threatened.

BITE/VENOM HARMFUL

IUCN LISTING Least Concern.

Barkly, NT, Ryan Francis

Lake Argyle, WA, Rob Valentic

Adelaide River, NT, Jules Farquhar

Juvenile, Robinson River, NT, Anders Zimny

SOUTHERN SHOVEL-NOSED SNAKE *Brachyurophis semifasciatus* Günther, 1863

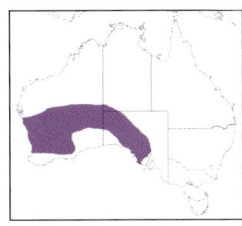

PRONUNCIATION *Brah-key-u-roe-fiss sem-e-fas-cee-ah-tus.*
ETYMOLOGY Half-banded short-tailed snake.
TYPE LOCALITY WA.
APPEARANCE Small, robust snake with head indistinct from body. Dorsal colouration reddish-brown to charcoal, with regular cross-bands variable, at 2–3 scales wide. Bands cream, yellow, orange or red. Pupil round; iris dark. Tongue pale; pink inside mouth. Ventral colouration creamish-white. Both sexes reach 37cm TL. **Scalation** MB 17 rows, 147–188 VENT, SUB 15–26 and anal scale divided. Scales smooth.

RANGE Encountered in southern WA, between Cockburn and Wiluna, across to SW NT, and south to the Eyre Peninsula of SA.

COMMENTS Nocturnal. Terrestrial. Lives on heavy clay to sandy soils. Shelters beneath timber and rocks in open woodland, mallee, sand ridge deserts and heaths. **Diet** Reptile eggs. **Reproduction** Oviparous. 1–7 per clutch. Neonates approximately 11cm long and hatch in February–April. **Disposition** Generally inoffensive unless threatened.

BITE/VENOM HARMFUL
IUCN LISTING Least Concern.

Gingin, WA, Danny Melville

Ilkurlka, WA, Brian Bush

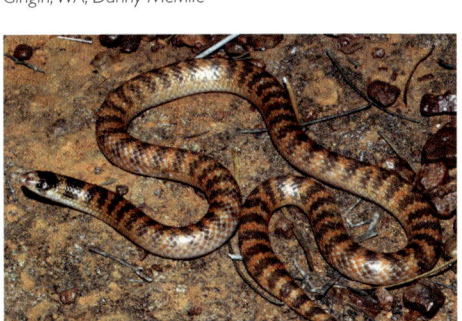

Ilkurlka, WA, Brian Bush

Kalbarri, WA, Adam Elliott

Crowned Snakes, genus *Cacophis* Günther, 1863

The genus *Cacophis* currently contains four species, all of which are endemic to Australia. They are usually encountered crossing roads at night or in and around houses. They specialize in eating lizards. and are egg-layers. **VENOM** Toxicity unknown, and there is no evidence of envenomations. **Species-level identification difficulty** − 3.

TYPE SPECIES *Cacophis krefftii.*
ETYMOLOGY Bad-snake.

Key to *Cacophis*

1. Ventral colouration grey..2
   Ventral colouration yellow or orange-red...3

2. Ventral scales 175 or fewer; pale collar 1–2 scales wide........................*C. churchilli* (opposite)
   Ventral scales 170 or more; pale collar 4 scales wide.............................*C. harriettae* (p. 106)

3. Ventral colouration orange-red;
   head marking does not enclose the nape..*C. squamulosus* (p. 108)
   Ventral colouration yellow; nape collar yellow, 1–2 scales wide................*C. krefftii* (p. 107)

Cacophis harriettae, Archerfield, QLD, Scott Eipper

NORTHERN CROWNED SNAKE *Cacophis churchilli* Wells & Wellington, 1985
(Northern Dwarf-crowned Snake)

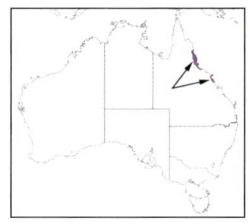

PRONUNCIATION *Kak-co-fiss chur-chill-e.*
ETYMOLOGY Churchill's bad-snake, pertaining to W. Churchill.
TYPE LOCALITY Black Mountain road, near Kuranda, QLD.
APPEARANCE Small, robust snake with head slightly distinct from body. Dorsal colouration charcoal-grey to blackish. Some individuals have yellow longitudinal stripes on first third of body. Top of head dark, while sides and rear are stippled with grey, brown and white. Conspicuous yellow or white band across nape. Pupils round; iris reddish-orange. Tongue dark pink; pink inside mouth. Ventral colouration grey. Adult females larger than males, reaching 50cm TL. **Scalation** MB 15 rows, 160–175 VENT, SUB 20–30 and anal scale divided. Scales smooth and glossy. **Similar species** *C. harriettae* (p. 106), *C. krefftii* (p. 107).

RANGE Restricted to NE QLD, from Mossman south to Paluma. Also records from the Whitsundays, south to Sarina.

COMMENTS Nocturnal. Terrestrial. Lives in rainforests and woodland. Shelters beneath rotten logs, rocks and man-made debris. **Diet** Exclusively lizards. **Reproduction** Oviparous. 7–9 per clutch. Neonate size unknown. **Disposition** Bluff strikes readily, but generally inoffensive.

BITE/VENOM HARMFUL
IUCN LISTING Least Concern.

Milaa Milaa, QLD, Matt Summerville

Tully Falls, QLD, Scott Eipper

Tully Falls, QLD, Scott Eipper

Tully Falls, QLD, Scott Eipper

WHITE-CROWNED SNAKE *Cacophis harriettae* Krefft, 1869

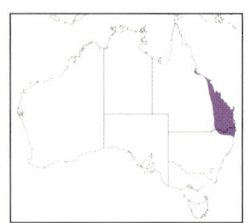

PRONUNCIATION *Kak-co-fiss ha-reet-ee.*

ETYMOLOGY Harriett's bad-snake. Pertains to Harriett Scott, the scientific illustrator.

TYPE LOCALITY Warro, Port Curtis, QLD.

APPEARANCE Small, robust snake with head slightly distinct from body. Dorsal colouration charcoal-grey to blackish. Top of head dark, while sides and rear are stippled with grey, brown and white. Conspicuous wide, pale yellow or white band across nape. Pupils round; iris reddish-orange. Tongue dark pink; pink inside mouth. Ventral colouration grey. Adult females larger than males, reaching 55cm TL. **Scalation** MB 15 rows, 170–200 VENT, SUB 25–45 and anal scale divided. Scales smooth and glossy in appearance. **Similar species** *C. churchilli* (p. 105) *C. krefftii* (opposite).

RANGE Encountered in QLD, from Kirrima south along coast south of Grafton. Also extends west as far as Glenmorgan.

COMMENTS Nocturnal. Terrestrial. Lives in rainforests, woodland and gardens. Shelters beneath rotten logs, rocks and man-made debris. **Diet** Exclusively lizards. **Reproduction** Oviparous. 2–10 per clutch. Neonates approximately 14.5cm TL hatching in October–January. **Disposition** Bluff strikes readily, but generally inoffensive.

BITE/VENOM HARMFUL

IUCN LISTING Least Concern.

Albany Creek, QLD, Scott Eipper

Archerfield, QLD, Scott Eipper

Albany Creek, QLD, Tie Eipper

Gatton, QLD, Scott Eipper

Dwarf Crowned Snake *Cacophis krefftii* Günther, 1863

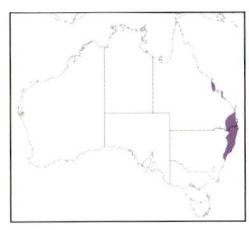

PRONUNCIATION *Kak-co-fiss kref-tee*.

ETYMOLOGY Krefft's bad-snake. Pertains to J. L. G. Krefft, past curator of the Australian Museum and zoologist.

TYPE LOCALITY North of Clarence River, NSW.

APPEARANCE Small, robust snake with head slightly distinct from body. Dorsal colouration charcoal-grey to blackish. Top of head dark, while sides and rear are stippled with grey, brown and white. Conspicuous thin yellow, orange or white band across nape. Pupils round; iris red. Tongue dark pink; pink inside mouth. Ventral colouration yellow. Adult females larger than males, reaching 32cm TL. **Scalation** MB 15 rows, 140–160 VENT, SUB 20–30 and anal scale divided. Scales glossy in appearance and smooth. **Similar species** *C. churchilli* (p. 105), *C. harriettae* (opposite).

RANGE Encountered in QLD, from the Whitsundays south to Gosford, NSW.

COMMENTS Nocturnal. Terrestrial. Lives in rainforests, woodland and gardens. Shelters beneath rotten logs, rocks and man-made debris. **Diet** Exclusively lizards. **Reproduction** Oviparous. 2–5 per clutch. Neonates approximately 10cm TL and hatch in October–February. **Disposition** Bluff strikes readily, but generally inoffensive.

BITE/VENOM HARMFUL

IUCN LISTING Least Concern.

Mt Tamborine, QLD, Scott Eipper

Mt Cougal, QLD, Scott Eipper

Mt Tamborine, QLD, Scott Eipper

Byron Bay, NSW, Scott Eipper

Golden Crowned Snake *Cacophis squamulosus*
(Duméril, Bibron & Duméril, 1854)

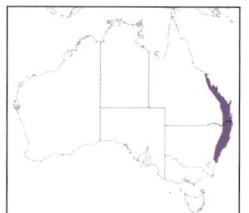

PRONUNCIATION *Kak-co-fiss squam-u-low-suss.*
ETYMOLOGY Scaled bad-snake.
TYPE LOCALITY Tasmania (in error).
APPEARANCE Small to medium-sized, robust snake with head slightly distinct from body. Dorsal colouration tan to brownish-black. Top of head dark, while sides and rear are stippled with grey, brown and white. Conspicuous orange to yellowish marking that often extends down from around head along forebody. Pupils vertically elliptic; iris reddish-orange. Tongue dark pink; pink inside mouth. Ventral colouration orange. Adult females larger than males, reaching 75cm TL. **Scalation** MB 15 rows, 165–185 VENT, SUB 25–40 and anal scale divided. Scales smooth and glossy.
RANGE Encountered in coastal QLD, from Eungella south to Wollongong, NSW.
COMMENTS Nocturnal. Terrestrial. Lives in rainforests, woodland and gardens. Shelters under cover such as rotting logs and rocks, or indoors when brought into homes by cats. **Diet** Exclusively lizards. **Reproduction** Oviparous. 2–15 per clutch. Neonates approximately 15.5cm TL and hatch in October–January. **Disposition** Bluff strikes readily, but generally inoffensive.
BITE/VENOM HARMFUL
IUCN LISTING Least Concern.

Mt Glorious, QLD, Scott Eipper

Goomburra, QLD, Scott Eipper

Mt Glorious, QLD, Tie Eipper

Mt Cougal, QLD, Scott Eipper

Small-eyed Snakes, genus *Cryptophis* Worrell, 1961

The genus *Cryptophis* currently contains five species, all found in Australia. Further research is warranted to determine if each species currently recognized is a distinct species, or better regarded as a subspecies. These are nocturnal snakes, usually encountered crossing roads at night, or when discovered in and around houses. They specialize in eating lizards and are livebearers. Two species are endemic to Australia. **VENOM** The venom is myotoxic with procoagulants. Envenomations from this genus can lead to symptoms such as headache, nausea, vomiting, swelling, pain and collapse. Bites have resulted in, or could cause, fatalities. Tiger Snake or Polyvalent antivenom is used to treat envenomations. **Species-level identification difficulty** (within Australia) – 3.

TYPE SPECIES *Hoplocephalus pallidiceps*.
ETYMOLOGY Hidden snake.

Key to *Cryptophis*

1. Nasal scale in contact with preocular...2
   Nasal scale separated from preocular...*C. boschmai* (p. 110)

2. Dorsal colouration pink to reddish with or without a dark vertebral stripe 3
   Dorsal colouration dark brown, grey or black 4

3. Pink to reddish with a dark head and vertebral stripe.................*C. nigrostriatus* (p. 113)
   Uniformly pink above...*C. incredibilis* (p. 111)

4. Head uniform in colour; found in eastern Australia......................*C. nigrescens* (p. 112)
   Head is usually lighter laterally;
   often a pale lower lateral stripe along the body...........................*C. pallidiceps* (p. 114)

Cryptophis incredibilis, Prince of Wales Island, QLD Justin Wright

Cryptophis nigrescens, Bellthorpe, QLD, Scott Eipper

CARPENTARIA SNAKE *Cryptophis boschmai* (Brongersma & Knaap-van Meeuwen, 1961)
(Carpentaria Whip Snake)

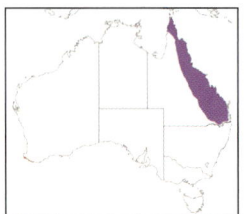

PRONUNCIATION *Crypt-toe-fis bosh-may.*

ETYMOLOGY Boschma's hidden snake. Pertains to H. Boschma, Dutch zoologist.

TYPE LOCALITY Merauke, Papua Province, Indonesia.

APPEARANCE Small, robust snake with head slightly distinct from body. Dorsal colouration blackish-brown to tan, usually lighter on flanks than on midline. This can form a longitudinal stripe running along lower flanks. Pupil vertically elliptic; iris black. Tongue and mouth pink. Ventral colouration white. Adult females larger than males, reaching 45cm TL. **Scalation** MB 15 rows, 145–190 VENT, SUB 20–35 and anal scale single. Scales glossy in appearance and smooth. **Similar species** *C. nigrescens* (p. 112).

RANGE Encountered in QLD, from western Cape York Peninsula to outskirts of Brisbane. Also found in PNG.

COMMENTS Nocturnal. Terrestrial. Lives in woodland, brigalow and tropical savannah. Shelters beneath rocks, fallen timber and man-made debris. **Diet** Lizards. **Reproduction** Viviparous. 5–11 per litter. Neonates approximately 12cm TL and born in December–March. **Disposition** Generally inoffensive, but readily bites if harassed.

BITE/VENOM VENOMOUS

IUCN LISTING Least Concern.

Dotswood, QLD, Scott Eipper

Cracow, QLD, Scott Eipper

Greenvale, QLD, Scott Eipper

Purga, QLD, Scott Eipper

Pink Snake *Cryptophis incredibilis* (Wells & Wellington, 1985)

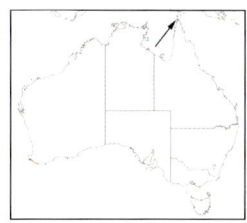

PRONUNCIATION *Crypt-toe-fis in-cred-bill-iss*.
ETYMOLOGY Pertaining to the spectacular appearance of the species.
TYPE LOCALITY Prince of Wales Island, Torres Strait.
APPEARANCE Small, moderately slender snake with head slightly distinct from body. Dorsal colouration bright pink (as the name suggests). Pupil vertically elliptic; iris black. Tongue and mouth pink. Ventral colouration whitish-cream. Both sexes reach 45cm TL. **Scalation** MB 15 rows, 180–185 VENT, SUB 50–65 and anal scale single. Scales glossy in appearance and smooth. **Similar species** *C. nigrostriatus* (p. 113).
RANGE Only encountered on Prince of Wales Island in the Torres Strait.
COMMENTS Nocturnal. Terrestrial. Most individuals have been found beneath flotsam on the shoreline that neighbours open woodland on sandy soil, or while active hunting at night. Probably just a colour variation of *C. nigrostriatus*. **Diet** Lizards. **Reproduction** Likely to give birth to live young. **Disposition** Generally inoffensive but will bite if threatened.
BITE/VENOM HARMFUL
IUCN LISTING Least Concern.

Prince of Wales Island, QLD, Justin Wright

Prince of Wales Island, QLD, Hal Cogger

Prince of Wales Island, QLD, Hal Cogger

Prince of Wales Island, QLD, Hal Cogger

SMALL-EYED SNAKE *Cryptophis nigrescens* (Günther, 1862)
(Eastern Small-eyed Snake)

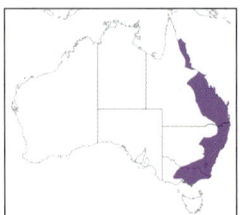

PRONUNCIATION *Crypt-toe-fis nigh-gress-sens.*
ETYMOLOGY Blackish hidden snake.
TYPE LOCALITY Environs of Sydney, NSW.
APPEARANCE Medium-sized, robust snake with head slightly distinct from body. Dorsal colouration grey to jet-black, with a black head. Pupil vertically elliptic; iris black. Tongue and mouth pink. Older individuals can develop a condition called macrocephaly, where the head enlarges past its normal size. Ventral colouration pink to orange-red, sometimes with grey flecks. Adult males larger than females, reaching 80cm TL. Exceptionally large animals can reach above a metre. **Scalation** MB 15 rows, 165–210 VENT, SUB 30–47 and anal single. Scales glossy in appearance and smooth. **Similar species** *C. boschmai* (p. 110), *Pseudechis porphyriacus* (p. 185) and *Stegonotus australis* (p. 71).
RANGE Encountered from Mossman, QLD, along the GDR to Melbourne, Vic.
COMMENTS Nocturnal. Terrestrial. Lives in forests, woodland and rocky heaths. Shelters beneath rocks, fallen timber and man-made debris. **Diet** Lizards, frogs and occasionally other snakes.
Reproduction Viviparous. 2–8 per litter. Neonates approximately 15cm TL and born in December–March. **Disposition** Generally inoffensive, but readily bites if harassed.
BITE/VENOM DANGEROUSLY VENOMOUS
IUCN LISTING Least Concern.

Carol Park, QLD, Scott Eipper

Crystal Cascades, QLD, Scott Eipper

Lamington NP, QLD, Scott Elpper

Tenterfield, NSW, Scott Eipper

Black Striped Snake Cryptophis nigrostriatus (Krefft, 1864)

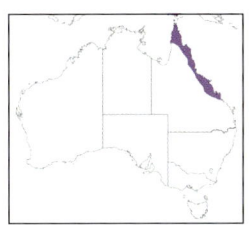

PRONUNCIATION *Crypt-toe-fis nigh-gro-stree-ah-tuss.*

ETYMOLOGY Black-striped hidden snake.

TYPE LOCALITY Neighbourhood of Rockhampton, QLD.

APPEARANCE Small, moderately slender snake with head slightly distinct from body. Dorsal colouration charcoal-grey to blackish, with a pair of broad pink stripes on flanks. Width of stripes variable. Top of head dark. Pupil vertically oval; iris dark reddish-brown. Tongue and mouth pink. Ventral colouration whitish-cream. Both sexes reach 50cm TL. **Scalation** MB 15 rows, 160–190 VENT, SUB 45–75 and anal scale single. Scales glossy in appearance and smooth. **Similar species** *C. incredibilis* (p. 111).

RANGE Encountered in eastern QLD, from Awoonga Dam, north to tip of Cape York Peninsula. Also encountered in PNG.

COMMENTS Nocturnal. Terrestrial. Lives in rainforests and woodland. Shelters beneath rocks, fallen timber and man-made debris. **Diet** Lizards. **Reproduction** Viviparous. 4–9 per litter. Neonates approximately 15cm TL and born in January–March. **Disposition** Generally inoffensive, but readily bites if harassed.

BITE/VENOM HARMFUL

IUCN LISTING Least Concern.

Weipa, QLD, Scott Eipper

Davies Creek, QLD, Scott Eipper

Herveys Range, QLD, Scott Eipper

Spring Creek Station, QLD, Scott Eipper

Northern Small-Eyed Snake *Cryptophis pallidiceps* (Günther, 1858)
(Secretive Snake, Western Carpentaria Snake)

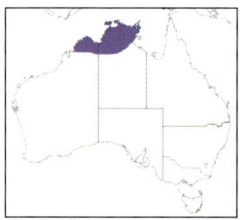

PRONUNCIATION *Crypt-toe-fis pal-lid-e-ceps*.

ETYMOLOGY Pale-headed hidden snake.

TYPE LOCALITY Port Essington, NT (in error); neotype east of Derby, WA.

APPEARANCE Small to medium-sized, robust snake with head slightly distinct from body. Dorsal colouration blackish-brown. Longitudinal pinkish-orange to yellow stripe runs along lower flanks. Pupil vertically elliptic; iris dark brown. Tongue and mouth pink. Ventral colouration white. Adult females larger than males, reaching 50cm TL. **Scalation** MB 15 rows, 160–180 VENT, SUB 35–55 and anal scale single. Scales glossy in appearance and smooth.

RANGE Encountered in WA, from the Kimberley region, across the northern section of the NT, to Groote Eylandt. Also found in Indonesia.

COMMENTS Nocturnal. Terrestrial. Lives in rocky woodland and tropical savannah. Shelters beneath rocks, fallen timber and man-made debris. **Diet** Lizards. **Reproduction** Viviparous. 2–5 per litter. Neonates approximately 11cm TL and born in December–March. **Disposition** Generally inoffensive, but readily bites if harassed.

BITE/VENOM VENOMOUS

IUCN LISTING Least Concern.

Mataranka, NT, Rob Valentic

Alligator River, NT, Ryan Francis

Darwin, NT, Steve Swanson

Arnhemland, NT, Anders Zimny

ELAPIDS | 115

Whip Snakes, genus *Demansia* Günther, 1858

This genus currently contains 15 species, all found in Australia. Fourteen are endemic to Australia. These fast-moving diurnal hunters are found across most of Australia. **VENOM** Envenomations from this genus can lead to symptoms such as headache, nausea, vomiting, pain, swelling, dizziness and collapse. Polyvalent antivenom has been used to treat severe envenomations, with mixed success. **Species-level identification difficulty** (within Australia) – 5.

TYPE SPECIES *Lycodon reticulatus*.

ETYMOLOGY Pertaining to A. van Diemen, the Dutch navigator.

Key to *Demansia*

1. Broad nape band edged yellow or white; sometimes with a secondary nuchal band (which may be bordered with white or yellow) 2
   No nape band, or if present not usually prominently pale edged 6

2. Anterior ventral scales lack broad dark spots forming a pair of stripes 3
   Anterior ventral scales have broad dark spots forming a pair of stripes ..... *D. rimacola* (p. 125)

3. First pale nuchal band the width of one scale 4
   First pale nuchal band 3–4 scales wide .... *D. flagellatio* (p. 119)

4. Found outside of Queensland 5
   Found in Queensland .... *D. torquata* (p. 129)

5. Dark nuchal band edged with yellow .... *D. shinei* (p. 127)
   Dark nuchal band edged with white .... *D. calodera* (p. 117)

6. No indication of a nape band 7
   Usually, an indication of a dark nape band, without a light margin .... *D. quaesitor* (p. 123)

7. More than 160 ventral scales 8
   Less than 155 ventral scales .... *D. simplex* (p. 128)

8. No dark markings centrally aligned on the anterior ventral scales 9
   Dark markings centrally aligned on the anterior ventral scales .... *D. olivacea* (p. 120)

9. No pale-edged dark bar running from the snout to nostrils 10
   Dark, usually pale-edged bar running from the snout to nostrils 11

10. Ventral scales 160–197 .... *D. vestigiata* (p. 130)
    Ventral scales 198–225 .... *D. papuensis* (p. 121)

11. Dark tear-drop marking extends on to sixth supralabial 12
    Dark tear-drop marking extends on to fifth supralabial 13

12. Greenish dorsum; found at Eighty Mile Beach and east;
    higher average subcaudal count (males 91, females 81) .... *D. angusticeps* (p. 116)
    Reddish dorsum; found at Marble Bar and west; lower average
    subcaudal count (males 77, females 72) .... *D. rufescens* (p. 126)

13. Scale margins without prominent dark edges (except in mid-east QLD);
    comma marking does not usually reach lower temporal scale 14
    Found in WA, scale margins are prominently dark-edged,
    comma marking extending to lower temporal scale .... *D. reticulata* (p. 124)

14. Scale margins without prominent dark edges; no reddish paravertebral
    stripes or reddish forebody; forebody grey to bluish grey grading to tan,
    or yellowish brown on posterior rear third, head and nape usually the
    same colour as posterior .... *D. cyanochasma* (p. 118)
    Scale margins with or without prominent dark edges;
    usually reddish paravertebral stripes or reddish forebody;
    some populations with a yellowish head and tail, while
    others are essentially concolorous; not found in WA .... *D. psammophis* (p. 122)

NARROW-HEADED WHIP SNAKE *Demansia angusticeps* (Macleay, 1888)

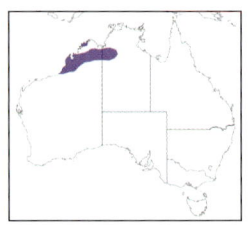

PRONUNCIATION *Dee-man-see-ah an-gust-e-ceps*.
ETYMOLOGY Narrow head Van-Diemen's snake.
TYPE LOCALITY Vicinity of King's Sound, NW Australia.
APPEARANCE Medium-sized, slender snake with head slightly distinct from body. Dorsal colouration brownish-olive to grey. Anterior edge of each midbody scale can be marked with black. Eye has dark brown, comma-shaped mark around it. Pale-edged brown line runs between nostrils, bisecting rostral scale. Lower flanks can have a yellow flush. Pupil round; iris orange. Tongue dark. Ventral colouration white, cream or yellow, often with grey flecks anteriorly. Adult males larger than females, reaching 90cm TL. **Scalation** MB 15 rows, 180–200 VENT, SUB 70–100 and anal scale divided. Scales matt in appearance and smooth. **Similar species** *D. olivacea* (p. 120).
RANGE Encountered in WA, from the Kimberley region across the far northwestern NT.
COMMENTS Diurnal. Terrestrial. Lives in rocky woodland and tropical savannah. Shelters beneath rocks, fallen timber and man-made debris. **Diet** Lizards. **Reproduction** Poorly known; likely to lay eggs. **Disposition** Quick to flee, generally inoffensive, but may bite if threatened.
BITE/VENOM VENOMOUS
IUCN LISTING Least Concern.

Fitzroy Crossing, WA, Jules Farquhar

160km west of Halls Creek, WA, Brian Bush

Broome, WA, Brad Maryan

Black-necked Whip Snake *Demansia calodera* Storr, 1978

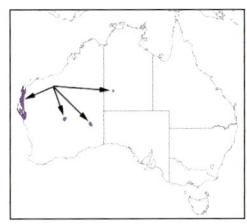

PRONUNCIATION *Dee-man-see-ah cal-o-der-rah.*
ETYMOLOGY Beautiful-necked Van-Diemen's snake.
TYPE LOCALITY Tamala, WA.
APPEARANCE Medium-sized, slender snake with head slightly distinct from body. Dorsal colouration grey to reddish-brown, and usually a conspicuous black band edged with cream or white on nape, which may fade to grey in old individuals. Anterior edge of each midbody scale can be marked with black. Eye has dark brown, comma-shaped mark around it. Pale-edged brown line runs between nostrils, bisecting rostral scale. Lower flanks can have yellow flush. There are reports of some individuals having dark posterior edges to ventral scales. Pupil round; iris orange. Tongue dark. Ventral colouration white, cream or pale yellow. Adult males larger than females, reaching 70cm TL. **Scalation** MB 15 rows, 170–195 VENT, SUB 65–90 and anal scale divided. Scales matt in appearance and smooth.
RANGE Encountered in WA, from Kalbarri to the Northwest Cape; also found in the Gibson and Little Sandy Deserts. Also found in the south-west NT.
COMMENTS Diurnal. Terrestrial. Lives in rocky woodland and tropical savannah. Shelters beneath rocks, fallen timber and man-made debris. **Diet** Lizards. **Reproduction** Probably lays eggs. **Disposition** Quick to flee, generally inoffensive, but may bite if threatened.
BITE/VENOM VENOMOUS
IUCN LISTING Least Concern.

Hamelin Station, WA, Brad Maryan

Three Bays, Tamala Station, WA, Brad Maryan

Juvenile, Quobba, WA, Brian Bush

Near Carnarvon, WA, Brian Bush

DESERT WHIP SNAKE *Demansia cyanochasma*
Nankivell, Maryan, Bush & Hutchinson, 2023

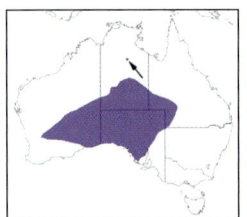

PRONUNCIATION *Dee-man-see-ah sigh-an-oh-caz-mah.*

ETYMOLOGY Blue gap Van-Diemen's snake; pertaining to forebody colouration, separating head and posterior third.

TYPE LOCALITY Moonaree Station, Gawler Ranges, SA.

APPEARANCE Medium-sized, slender snake with head slightly distinct from body. Head yellowish-brown to tan. Scale margins are without prominent dark edges; forebody grey to bluish-grey, grading to tan or yellowish-brown on posterior rear third. Around each eye, a white diagonal streak bordered posteriorly with black. Pupil round; iris orange to yellow. Tongue dark; mouth lining pinkish. Ventral colouration white to yellow. Adult males larger than females, reaching 89cm TL. **Scalation** MB 15 rows, 173–203 VENT, SUB 64–85 and anal scale divided. Scales matt in appearance and smooth. **Similar species** *D. psammophis* (p. 122), *D. reticulata* (p. 124).

RANGE Encountered continuously from the Eastern Goldfields of WA, through the southern NT, and across SA to Port Augusta, through to the Flinders Ranges to SW QLD.

COMMENTS Diurnal. Terrestrial. Lives in sand ridge deserts, mallee and mulga woodland. Shelters beneath rocks, fallen timber and man-made debris. **Diet** Lizards. **Reproduction** Oviparous. 5–8 per clutch. Neonate information unknown. **Disposition** Quick to flee, generally inoffensive, but may bite if threatened.

BITE/VENOM VENOMOUS

IUCN LISTING Least Concern.

Ellery Big Hole, NT, Scott Eipper

Great Victoria Desert, WA, Glen Gaikhorst

Whyalla, SA, Adam Elliott

Curtain Springs, NT, Scott Eipper

LONG-TAILED WHIP SNAKE *Demansia flagellatio* Wells & Wellington, 1985
(Ornate Whip Snake)

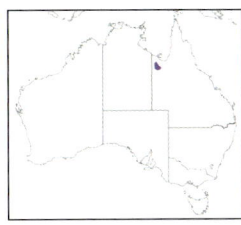

PRONUNCIATION *Dee-man-see-ah   fla-gell-ah-tee-o*.

ETYMOLOGY Whipping Van-Diemen's snake.

TYPE LOCALITY Mount Isa District, QLD.

APPEARANCE Medium-sized, slender snake with head slightly distinct from body. Dorsal colouration reddish-brown to bluish-grey. Head and nape dark grey to charcoal, with pale sides to head and two broad, pale orange to yellow bands. Eye has dark grey, comma-shaped mark around it. Dark line runs between nostrils, bisecting the rostral scale. Rear of body gradually becomes bright yellow. Pupil round; iris orange to yellow. Tongue dark; mouth lining pinkish. Ventral colouration white, cream or yellow. Adult males larger than females, reaching 71cm TL. **Scalation** MB 15 rows, 195–215 VENT, SUB 100–115 and anal scale divided. Scales matt in appearance and smooth. **Similar species** *D. shinei* (p. 127).

RANGE Encountered in the northwestern region of QLD along the Selwyn Range.

COMMENTS Diurnal. Terrestrial. Lives in tropical savannah and spinifex-dominated rocky woodland. Shelters beneath rocks, fallen timber and man-made debris. **Diet** Lizards. **Reproduction** Poorly known; likely to lay eggs. **Disposition** Quick to flee, generally inoffensive, but may bite if threatened.

BITE/VENOM VENOMOUS

IUCN LISTING Least Concern.

Lawn Hill, QLD, Gary Stephenson

Lawn Hill, QLD, Gary Stephenson

OLIVE WHIP SNAKE *Demansia olivacea* (Gray, 1842)
(Marbled-headed Whip Snake)

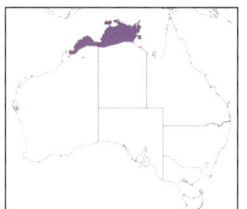

PRONUNCIATION *Dee-man-see-ah ol-e-vee-cee-ah*.

ETYMOLOGY Olive Van-Diemen's snake.

TYPE LOCALITY Port Essington, Australia.

APPEARANCE Medium-sized, slender snake with head slightly distinct from body. Dorsal colouration brownish-olive to grey. Anterior edge of each midbody scale can be marked with black. Head finely stippled with grey or brown. Dark-coloured streak from eye back along body towards corner of mouth. Pupil round; iris orange to yellow. Tongue dark; mouth lining pinkish. Ventral colouration white, cream or yellow, often with grey flecks anteriorly. Adult males larger than females, reaching 80cm TL. **Scalation** MB 15 rows, 160–210 VENT, SUB 65–110 and anal scale divided. Scales matt in appearance and smooth. **Similar species** *D. angusticeps* (p. 116).

RANGE Encountered in WA, from the Kimberley region across the NW NT, to the edge of the Gulf of Carpentaria.

COMMENTS Diurnal. Terrestrial. Lives in rocky woodland and tropical savannah. Shelters beneath rocks, fallen timber and man-made debris. **Diet** Lizards. **Reproduction** Oviparous. 3–5 per clutch. Neonate information unknown. **Disposition** Quick to flee, generally inoffensive, but may bite if threatened.

BITE/VENOM VENOMOUS

IUCN LISTING Least Concern.

North Kimberley, WA, Anders Zimny

Kununurra, WA, Phill Mangion

Coonawarra, NT, Paul Horner

GREATER BLACK WHIP SNAKE *Demansia papuensis* (Macleay, 1877)
(Papuan Whip Snake)

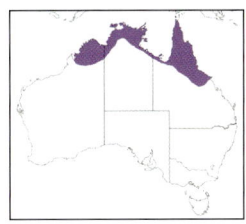

PRONUNCIATION *Dee-man-see-ah pap-u-en-sis.*

ETYMOLOGY Papuan Van-Diemens's snake.

TYPE LOCALITY New Guinea – in error.

APPEARANCE Large, slender snake with head slightly distinct from body. Dorsal colouration grey to black, but some individuals are tan to reddish-brown. Pupil round; iris orange to yellow. Tongue dark; mouth lining pinkish. Ventral colouration grey. Adult males larger than females, reaching 160cm TL. **Scalation** MB 15 rows, 198–225 VENT, SUB 75–110 and anal scale divided. Scales matt in appearance and smooth. **Similar species** *D. vestigiata* (p. 130).

RANGE Encountered from Mackay, QLD, up the east coast and across the north coast of Australia, to WA. The type specimen probably originated in QLD; incorrectly thought to be from southern PNG.

COMMENTS Diurnal but can be nocturnal in hot weather. Terrestrial. Found in forests, open woodland and grassland. Shelters beneath logs, rocks and man-made debris. **Diet** Predominantly lizards and frogs; occasionally other snakes. **Reproduction** Oviparous. 5–13 per clutch. Neonates approximately 30cm TL and hatch in December–March. **Disposition** Quick to flee, generally inoffensive, but may bite if threatened.

BITE/VENOM DANGEROUSLY VENOMOUS

IUCN LISTING Least Concern.

Charnley River Station, WA, Ian 'Bushrat' Bool

Kununurra, WA, Hal Cogger

Coen, QLD, Matt Summerville

Iron Range, QLD, Jesse Campbell

Yellow-Faced Whip Snake *Demansia psammophis* (Schlegel, 1837)

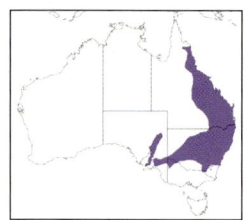

PRONUNCIATION *Dee-man-see-ah sam-o-fiss.*
ETYMOLOGY Sand Van-Diemen's snake.
TYPE LOCALITY Nouvelle Hollande (Australia).
APPEARANCE Medium-sized, slender snake with head slightly distinct from body. Dorsal colouration varies from olive, grey, brown and greenish, to reddish-orange and grey. Head can be yellow or same colour as body. Some populations have black-edged midbody scales. Northeastern individuals often have pair of reddish stripes on first third, blending into grey towards rear of body. Characteristic black, comma-shaped mark over eye, and pale-edged brown line that runs between nostrils bisecting rostral scale. Pupil round; iris orange to yellow. Tongue dark; mouth lining pinkish. Ventral colouration white to greenish-yellow; usually bright yellow under tail. Adult males larger than females, reaching 80cm TL. **Scalation** MB 15 rows, 165–230 VENT, SUB 60–105 and anal scale divided. Scales matt in appearance and smooth. **Similar species** *D. cyanochasma* (p. 118), *D. reticulata* (p. 124), *D. vestigiata* (p. 130).

RANGE Encountered from southeastern SA, through NW Vic, to the NSW coast, and up the eastern half of QLD to Mossman.

COMMENTS Diurnal. Terrestrial. Lives in rocky woodland and tropical savannah. **Diet** Lizards. **Reproduction** Oviparous. 2–17 per clutch. Neonates approximately 20cm TL and hatch in August–March. **Disposition** Quick to flee, generally inoffensive, but may bite if threatened.

BITE/VENOM VENOMOUS

IUCN LISTING Least Concern.

Mt Molloy, QLD, Shane Black

Alpha, QLD, Scott Eipper

Hattah, VIC, Scott Eipper

Hinze Dam, QLD, Scott Eipper

Sombre Whip Snake *Demansia quaesitor* Shea, 2007

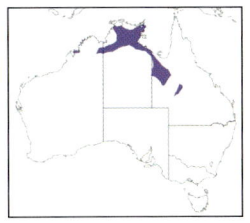

PRONUNCIATION *Dee-man-see-ah kway-sit-tore*.

ETYMOLOGY Judge (seeker) Van-Diemen's snake; pertaining to the sombre appearance of a judge and to the active seeker (foraging) habits of the genus.

TYPE LOCALITY 'Hodgson Downs', Mt Langdon, NT.

APPEARANCE Medium-sized, slender snake with head slightly distinct from body. Dorsal colouration brown to grey, with black band across nape. Eastern population rusty-red to grey, with or without reddish head, but lacks black nape band. Dark streak leading from eye towards rear of mouth. Pupil round; iris orange. Tongue dark; mouth lining pinkish. Ventral colouration white. Adult males larger than females, reaching 75cm TL. **Scalation** MB 15 rows, 180–200 VENT, SUB 60–100 and anal scale divided. Scales matt in appearance and smooth.

RANGE Encountered in WA, from the eastern Kimberley region across the NT, through the south of the Gulf of Carpentaria and along the Selwyn Range, QLD, to Dajarra.

COMMENTS Diurnal. Terrestrial. Lives in rocky woodland and tropical savannah. Shelters beneath rocks, fallen timber and man-made debris. **Diet** Lizards. **Reproduction** Oviparous. Six per clutch. Neonate information unknown. **Disposition** Quick to flee, generally inoffensive, but may bite if threatened.

BITE/VENOM VENOMOUS

IUCN LISTING Least Concern.

McArthur River, NT, Anders Zimny

Litchfield NP, NT, Jules Farquhar

Mt Isa, QLD, Reid Newell

Mt Isa, QLD, Ryan Francis

Reticulated Whip Snake *Demansia reticulata* (Gray, 1842)

PRONUNCIATION *Dee-man-see-ah ree-tic-u-lah-tah.*

ETYMOLOGY Reticulated Van-Diemen's snake.

TYPE LOCALITY Australia.

APPEARANCE Medium-sized, slender snake with head slightly distinct from body. First half of body greenish-grey to yellow. Remaining body and tail usually brown or grey. Posterior edges of each midbody scale usually edged with black. Eye has white diagonal streak bordered posteriorly with black. Pupil round; iris orange to yellow. Tongue dark; mouth lining pinkish. Ventral colouration white to yellow. Adult males larger than females, reaching 100cm TL. **Scalation** MB 15 rows, 165–217 VENT, SUB 70–102 and anal scale divided. Scales matt in appearance and smooth. **Similar species** *D. cyanochasma* (p. 118), *D. psammophis* (p. 122).

RANGE Encountered in WA, from Shark Bay south through western WA to Myalup, and inland to Kellerberrin.

COMMENTS Diurnal. Terrestrial. Lives in heaths, mallee, mulga and open woodland. Shelters beneath rocks, fallen timber and man-made debris. **Diet** Lizards. **Reproduction** Oviparous. Up to six per clutch. Neonate information unknown. **Disposition** Quick to flee, generally inoffensive, but may bite if threatened.

BITE/VENOM VENOMOUS

IUCN LISTING Least Concern.

Wooramel River, WA, Gary Stephenson

Badgingarra NP, WA, Brad Maryan

Newman, WA, Anders Zimny

Lancelin, WA, Glen Gaikhorst

CRACK-DWELLING WHIP SNAKE *Demansia rimicola* Scanlon, 2007
(Blacksoil Whip Snake)

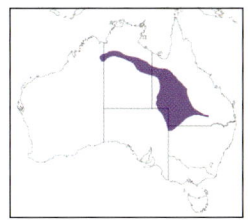

PRONUNCIATION *Dee-man-see-ah rim-nee-co-lah*.
ETYMOLOGY Crack-dwelling Van-Diemen's snake.
TYPE LOCALITY 61.4km N Muttaburra via Hughenden Highway, QLD.
APPEARANCE Medium-sized, slender snake with head slightly distinct from body. Dorsal colouration brownish-olive to brown. Can have reddish-brown stripes along lower flanks. Head and neck usually darker brown than body, particularly in young individuals. Usually two white bands both in front of and behind eyes, as well as two bands on nape. Head finely stippled with grey or black. Eye has dark-coloured streak from eye back along body towards corner of mouth. Pupil round; iris orange to yellow. Tongue dark; mouth lining pinkish. Ventral colouration orange to red, with two longitudinal dark brown spots forming parallel rows. Adult males larger than females, reaching 90cm TL. **Scalation** MB 15 rows, 175–205 VENT, SUB 65–110 and anal scale divided. Scales matt in appearance and smooth.

RANGE Encountered in WA, from the Kimberley region across the central and eastern NT, through into NE SA. Also occurs through western QLD to Roma, NW to Gregory.

COMMENTS Diurnal. Terrestrial. Occurs in mitchell grass-dominated, cracking claypans, grassland, black soil plains and tropical savannah. Shelters inside soil cracks, and beneath rocks, fallen timber and man-made debris. **Diet** Lizards. **Reproduction** Oviparous. Up to eight per clutch. Neonates approximately 17cm TL and hatch in March. **Disposition** Quick to flee, generally inoffensive, but may bite if threatened.

BITE/VENOM VENOMOUS

IUCN LISTING Least Concern.

Davenport Downs, QLD, Hal Cogger

Juvenile, 60km W Mica Creek, QLD, Scott Eipper

Kynuna, QLD, Scott Eipper

Boulia, QLD, Shane Black

RED WHIP SNAKE *Demansia rufescens* Storr, 1978
(Rufous Whip Snake)

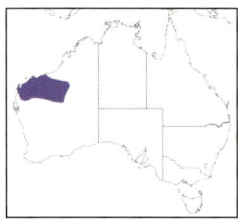

PRONUNCIATION *Dee-man-see-ah roo-fee-sens.*
ETYMOLOGY Reddish Van-Diemen's snake.
TYPE LOCALITY Marandoo mine site, near Mt Bruce, WA.
APPEARANCE Medium-sized, slender snake with head slightly distinct from body. Dorsal colouration reddish-brown, and head and neck grey. Eye has dark brown, comma-shaped mark around it. Pupil round; iris orange. Tongue dark; mouth lining pinkish. Ventral colouration white or cream. Adult males larger than females, reaching 68cm TL. **Scalation** MB 15 rows, 177–200 VENT, SUB 65–85 and anal scale divided. Scales matt in appearance and smooth.
RANGE Encountered in WA, from Port Hedland, through the Pilbara, to the Northwest Cape; also found on Barrow and Dolphin Islands.
COMMENTS Diurnal. Terrestrial. Lives in rocky woodland and gorges dominated by spinifex on rocky soils. Shelters beneath rocks, fallen timber and man-made debris. **Diet** Lizards. **Reproduction** Poorly known; likely to lay eggs. **Disposition** Quick to flee, generally inoffensive, but may bite if threatened.
BITE/VENOM VENOMOUS
IUCN LISTING Least Concern.

Mt Whaleback, WA, Brad Maryan

Karratha, WA, Brian Bush

Koodaideri, WA, Brian Bush

Marillana Station, WA, Brad Maryan

SHINE'S WHIP SNAKE *Demansia shinei* Shea, 2007

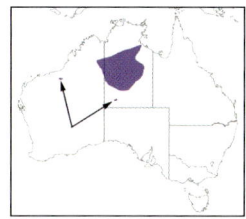

PRONUNCIATION *Dee-man-see-ah shy-nee*.

ETYMOLOGY Shine's Van-Diemen's snake; pertaining to R. Shine, Australian herpetologist.

Type locality Frewena, NT.

APPEARANCE Medium-sized, slender snake with head slightly distinct from body. Dorsal colouration brown to bluish-grey. Head and nape grey with pale sides to head and two broad, pale orange to yellow bands. Behind eye a broad yellow band over temporal region; band thinner in far western animals. Eye has dark grey, comma-shaped mark. Rear of body gradually becomes lighter. Pupil round; iris orange to yellow. Tongue dark; mouth lining pinkish. Ventral colouration pale yellow. Adult males larger than females, reaching 84cm TL. **Scalation** MB 15 rows, 177–207 VENT, SUB 69–99 and anal scale divided. Scales matt in appearance and smooth. **Similar species** *D. flagellatio* (p.119).

RANGE Encountered from Borroloola south to Yulara in the NT, NW to Halls Creek, WA. There is a presumably isolated population in the eastern Pilbara between the Little Sandy Desert and Nifty Mine.

COMMENTS Diurnal. Terrestrial. Lives in tropical savannah, spinifex-dominated woodland and sand ridge deserts. Shelters beneath rocks, fallen timber and man-made debris. **Diet** Lizards. **Reproduction** Poorly known; expected to lay eggs. **Disposition** Quick to flee, generally inoffensive, but may bite if threatened.

BITE/VENOM VENOMOUS

IUCN LISTING Least Concern.

Barkly Tablelands, NT, Reid Newell

Three Ways, NT, Rob Valentic

Barkly Tablelands, NT, Reid Newell

Grey Whip Snake *Demansia simplex* Storr, 1978

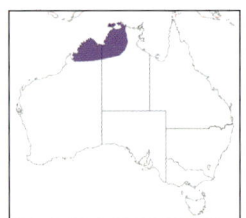

PRONUNCIATION *Dee-man-see-ah sim-plex*.
ETYMOLOGY Simple Van-Diemen's snake.
TYPE LOCALITY Kalumburu, WA.
APPEARANCE Medium-sized, moderately slender snake with head slightly distinct from body. Dorsal colouration plain grey. Vertebral zone slightly darker than flanks. Head grey to tan. Yellow stripe runs along each of the lower flanks. Eye has white-edged, dark grey, comma-shaped mark. Pupil round; iris orange to yellow. Tongue dark; mouth lining pinkish. Ventral colouration white to cream. Adult males larger than females, reaching 54cm TL. **Scalation** MB 15 rows, 140–150 VENT, SUB 55–65 and anal scale divided. Scales matt in appearance and smooth.
RANGE Encountered in WA, from the Kimberley region across the northern section of the NT to Jabiru.
COMMENTS Diurnal. Terrestrial. Lives in rocky woodland and tropical savannah. Shelters beneath rocks, fallen timber and man-made debris. **Diet** Lizards. **Reproduction** Poorly known; likely to lay eggs. **Disposition** Quick to flee, generally inoffensive, but may bite if threatened.
BITE/VENOM VENOMOUS
IUCN LISTING Least Concern.

Doongan Station, WA, Anders Zimny

Juvenile, Katherine, NT, Scott Eipper

Juvenile, Katherine, NT, Scott Eipper

Keep River, NT, Paul Horner

Collared Whip Snake *Demansia torquata* (Günther, 1862)

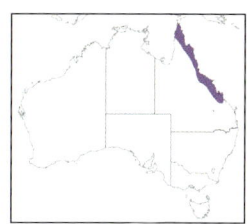

PRONUNCIATION Dee-man-see-ah tore-kwah-tah.
ETYMOLOGY Necklaced Van-Diemen's snake.
TYPE LOCALITY Percy Islands, QLD.
APPEARANCE Medium-sized, slender snake with head slightly distinct from body. Dorsal colouration brown to bluish-grey. Head and nape grey, with pale sides to head and two thin, pale yellow to orange bands. Eye has dark grey, comma-shaped mark. Rear of body gradually becomes lighter, particularly in young individuals. Pupil round; iris orange to yellow. Tongue dark; mouth lining pinkish. Juvenile and subadult specimens brighter in colour than adults. Ventral colouration reddish. Adult males larger than females, reaching 70cm TL. **Scalation** MB 15 rows, 185–220 VENT, SUB 70–90 and anal scale divided. Scales matt in appearance and smooth.
RANGE Encountered in eastern QLD from Gladstone and tip of Cape York Peninsula.
COMMENTS Diurnal. Terrestrial. Lives in tropical savannah and open woodland. Shelters beneath rocks, fallen timber and man-made debris. **Diet** Lizards. **Reproduction** Oviparous. 5–7 per clutch. Gravid in October–February. Neonates approximately 17cm TL hatching in December–March.
Disposition Quick to flee, generally inoffensive, but may bite if threatened.
BITE/VENOM VENOMOUS
IUCN LISTING Least Concern.

Mt Molloy, QLD, Shane Black

Conway, QLD, Scott Eipper

Juvenile, Almaden, QLD, Scott Eipper

Juvenile, Almaden, QLD, Scott Eipper

Lesser Black Whip Snake *Demansia vestigiata* (De Vis, 1884)

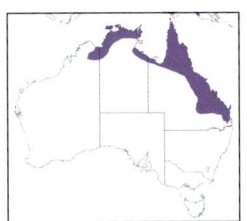

PRONUNCIATION *Dee-man-see-ah ves-tee-gee-ah-tah*.

ETYMOLOGY Marked Van-Diemen's snake; pertaining to pale markings on scales of holotype.

TYPE LOCALITY Unknown.

APPEARANCE Large, slender snake with head slightly distinct from body. Dorsal colouration grey to black, with each of the dorsal scales with a darker rear edge, giving the snake a variegated appearance. Some individuals have yellow forebody. Pupil round; iris orange to yellow. Tongue dark; mouth lining pinkish. Ventral colouration grey. Adult males larger than females, reaching 120cm TL. **Scalation** MB 15 rows, 165–197 VENT, SUB 70–95 and anal scale divided. Scales matt in appearance and smooth. **Similar species** *D. papuensis* (p. 121), *D. psammophis* (p. 122).

RANGE Encountered from Ipswich, QLD, up east coast and across north coast of Australia. Also found in southern PNG.

COMMENTS Diurnal. Terrestrial. Lives in dry forests, open woodland and grassland. Shelters beneath rocks, fallen timber and man-made debris. **Diet** Lizards, frogs, occasionally other snakes. **Reproduction** Oviparous. 5–15 per clutch. Neonates approximately 27cm TL and hatch in August–December. **Disposition** Quick to flee, generally inoffensive, but may bite if threatened.

BITE/VENOM DANGEROUSLY VENOMOUS

IUCN LISTING Least Concern.

Bohle, QLD, Scott Eipper

Mission River, QLD, Scott Eipper

Mission River, QLD, Scott Eipper

Juvenile, Robinson River, NT, Anders Zimny

Genus *Denisonia* Krefft, 1869

The genus *Denisonia* contains two species that are endemic to Australia. They are nocturnal, heavily built frog specialists, and are often found on cracking clay soils, which provide shelter for both the snakes and their prey, frogs and lizards. They are encountered crossing roads at night, particularly after rain, and are livebearers. **VENOM** Envenomations from this genus can lead to symptoms such as nausea, vomiting, swelling, pain, headache, dizziness and collapse. Polyvalent antivenom has been used to treat severe envenomations. **Species-level identification difficulty** – 2.

TYPE SPECIES *Denisonia ornata*.

ETYMOLOGY Pertains to W. Denison, Australian government official.

Key to *Denisonia*
1. Body patterned with bands or blotches .... *D. devisi* (p. 132)
   Body plain, without pattern .... *D. maculata* (p. 133)

Denisonia devisi, Tara, QLD, Scott Eipper

Denisonia devisi, Naree Station, NSW, Jesse Campbell

De Vis' Banded Snake *Denisonia devisi* Waite & Longman, 1920
(Mud Adder)

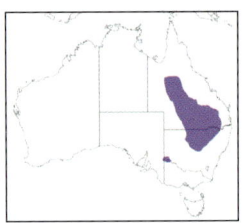

PRONUNCIATION *Den-iss-oh-nee-ah dee-vee-e.*

ETYMOLOGY Devis' Denison snake; pertains to Charles De Vis of the Queensland Museum.

TYPE LOCALITY Near Surat, QLD.

APPEARANCE Small to medium-sized, heavy, robust snake with head strongly distinct from body. Dorsal colouration brown to grey, with dark brown to black broad cross-bands. Some populations weakly banded or blotched. Head dark brown to grey with white and black-barred lips. Pupil vertically elliptic; iris light orange to tan. Tongue dark; mouth lining pinkish. Ventral colouration opaline-white. Adult females larger than males, reaching 70cm TL. **Scalation** MB 17 rows, 120–150 VENT, SUB 20–40 and anal scale single. Scales matt in appearance and smooth. **Similar species** Death adders *Acanthophis* spp. (p. 76) are frequently confused with this species. Distinguished easily by smooth tail-tip v a grub-like lure.

RANGE Encountered across inland QLD, south of Charters Towers, through the interior of NSW, to near Wagga Wagga. Separate population on the Murray River on the NSW/Vic border at Wallpolla Island, Vic. Also found in SE SA.

COMMENTS Nocturnal. Terrestrial. Commonly encountered after rain, crossing roads at night. Lives in open woodland, black soil plains, swamp and inland river margins. Shelters in soil cracks, and under logs, rocks and man-made debris. **Diet** Frogs, lizards, occasionally small rodents. **Reproduction** Viviparous. 3–12 per litter. Neonates approximately 15cm TL and born in March. **Disposition** Generally inoffensive, but readily bites if harassed.

BITE/VENOM VENOMOUS

IUCN LISTING Least Concern.

Dalby, QLD, Scott Eipper

Dalby, QLD, Tie Eipper

Juvenile, Glenmorgan, QLD, Scott Eipper

Muttaburra, QLD, Anders Zimny

Ornamental Snake *Denisonia maculata* (Steindachner, 1867)

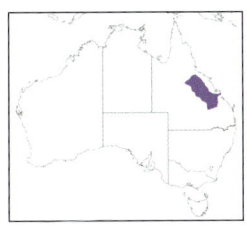

PRONUNCIATION *Den-iss-oh-nee-ah mac-u-la-tah.*
ETYMOLOGY Spotted Denison snake.
TYPE LOCALITY NSW (in error).
APPEARANCE Small, heavy, robust snake with head strongly distinct from body. Dorsal colouration brownish-grey with dark grey flecks. Head dark brown to grey, usually slightly darker than body. Labial scales mottled or barred with pale grey to black. Tail can be bright yellow in juveniles. Pupil vertically elliptic; iris greenish-grey. Tongue dark; mouth lining pinkish. Ventral colouration grey with dark flecking under head and forebody. Adult females larger than males, reaching 50cm TL. **Scalation** MB 17 rows, 120–150 VENT, SUB 20–40 and anal scale single. Scales matt in appearance and smooth.
RANGE Encountered in QLD, from Eidsvold through to Charters Towers.
COMMENTS Nocturnal. Terrestrial. Lives in open woodland, around swamps and inland river margins, usually with deep cracking soils. Shelters in soil cracks, and under logs, rocks and man-made debris. **Diet** Frogs and periodically lizards. **Reproduction** Viviparous. 3–11 per litter. Neonates approximately 16cm TL and born in February–May. **Disposition** Generally inoffensive, but readily bites if harassed.
BITE/VENOM VENOMOUS
IUCN LISTING Least Concern.

Dysart, QLD, Scott Eipper

Dysart, QLD, Scott Eipper

Moranbah, QLD, Alexander Davies

Genus *Drysdalia* Worrell, 1961

This genus contains three species, all endemic to Australia. They are among Australia's most cold-tolerant snakes. They are diurnal, usually found while crossing tracks, basking on vegetation or beneath cover, and can be quite common in some locations, with up to seven individuals using the same shelter site. They are lizard feeders, but will occasionally eat frogs, and are live bearers. This genus is adversely effected by fuel-reduction burning practices. This has led to the localised extinction of some, formerly common, populations. **VENOM** Envenomations from this genus can lead to symptoms such as swelling, pain, headache and collapse. Tiger Snake antivenom has been used to treat severe envenomations. **Species-level identification difficulty** – 2.

TYPE SPECIES *Hoplocephalus coronoides*.

ETYMOLOGY Drysdale's (snake), pertaining to G. R Drysdale, Australian painter.

Key to *Drysdalia*

1 Pale band across nape 2
  No pale band across nape .... *D. coronoides* (opposite)

2 White labial stripe edged with black; found west of Horsham, Vic .... *D. masterii* (p. 136)
  No white labial stripe, found in eastern NSW .... *D. rhodogaster* (p. 137)

Drysdalia coronoides, Ruined Castle, VIC, Scott Eipper

WHITE-LIPPED SNAKE *Drysdalia coronoides* (Günther, 1858)

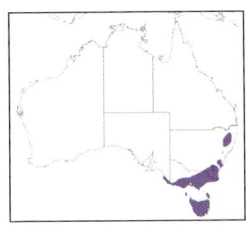

PRONUNCIATION *Drys-day-lee-ah co-ron-noy-dees.*
ETYMOLOGY Crowned Drysdale snake.
TYPE LOCALITY Tasmania.
APPEARANCE Small, moderately slender snake with head slightly distinct from body. Dorsal colouration highly variable, from tan to reddish, brown, grey or greenish. Prominent ragged white stripes along lips usually bordered with dark red, brown or black. Pupil round; iris reddish-copper. Tongue dark; inside of mouth pink. Ventral colouration yellow to orange-red. Adult females larger than males, reaching 45cm TL. **Scalation** MB 15 rows, 120–160 VENT, SUB 35–70 and anal scale single. Scales matt in appearance and smooth.
RANGE Encountered in far SE SA, through Tas, southern Vic, and highland areas in eastern NSW.
COMMENTS Diurnal. Terrestrial. Lives in grassland, heaths and forests. Often basks in grass tussocks and on vegetation. Shelters beneath rocks, fallen timber and man-made debris. Sometimes erroneously called 'whip snakes' in Tasmania. **Diet** Lizards and occasionally frogs. **Reproduction** Viviparous. 2–10 per litter. Neonates approximately 12cm TL and born in February–May.
Disposition Quick to flee, generally inoffensive, but may bite if threatened.
BITE/VENOM VENOMOUS
IUCN LISTING Least Concern.

Ballarat, VIC, Jules Farquhar

Budderoo, NSW, Jesse Campbell

Flinders Island, TAS, Shane Black

Pretty Valley, VIC, Scott Eipper

Masters' Snake *Drysdalia masterii* (Krefft, 1866)

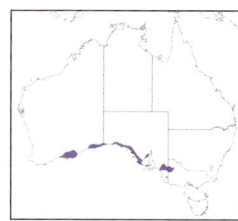

PRONUNCIATION *Drys-day-lee-ah mas-ter-e.*
ETYMOLOGY Masters' Drysdale snake; pertaining to G. Masters, English ornithologist.
TYPE LOCALITY Flinders Ranges, SA.
APPEARANCE Small, moderately slender snake with head slightly distinct from body. Dorsal colouration brownish to grey, often with fine dark flecking. Head dark brown to grey, usually slightly darker than body. Pale cream to orange or pale brown collar on nape. Labial scales mottled or barred with pale grey to black, often with a fine black stripe. Pupil round; iris yellow to brown. Tongue dark; pink inside mouth. Ventral colouration orange to yellow. Adult females larger than males, reaching 40cm TL. **Scalation** MB 15 rows, 140–160 VENT, SUB 40–55 and anal scale single. Scales matt in appearance and smooth.

RANGE Encountered in southern Australia in three separate populations. One extends from the Big Desert, Vic, to Tailem Bend, SA, the second on the southern Yorke Peninsula, and the third on the coastal regions of the Eyre Peninsula across to Esperance, WA.

COMMENTS Diurnal. Terrestrial. Lives in open woodland, mallee, heath and chenopod shrubland. Shelters beneath rocks, fallen timber and man-made debris. **Diet** Lizards. **Reproduction** Viviparous. 2–4 per litter. Gravid in October–May. Neonates approximately 9cm TL and born in February–May. **Disposition** Quick to flee, generally inoffensive, but may bite if threatened.

BITE/VENOM VENOMOUS
IUCN LISTING Least Concern.

Border Village area, SA, Jules Farquhar

Port Lincoln, SA, Shawn Scott

Port Lincoln, SA, Shawn Scott

Eucla, WA, Hal Cogger

MUSTARD-BELLIED SNAKE *Drysdalia rhodogaster* (Jan & Sordelli, 1873)
(Eastern Masters' Snake, Rose-bellied Snake)

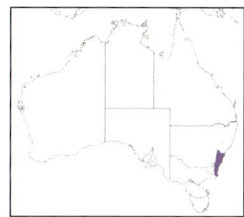

PRONUNCIATION *Drys-day-lee-ah roe-dough-gas-ter.*
ETYMOLOGY Rose-bellied Drysdale snake.
TYPE LOCALITY Australia.
APPEARANCE Small, moderately slender snake with head slightly distinct from body. Dorsal colouration brownish to grey, often with fine dark flecking. Head dark brown to grey, usually slightly darker than body. Pale cream to orange or pale brown collar on nape. Labial scales mottled or barred with pale grey to black, often with fine black stripe. Pupil round; iris reddish-orange. Tongue dark; pink mouth. Ventral colouration yellow to pink. Adult females larger than males, reaching 50cm TL. **Scalation** MB 15 rows, 140–160 VENT, SUB 40–55 and anal scale single. Scales matt in appearance and smooth.
RANGE Encountered in NSW, from the central coast region south to Vic border east of the GDR.
COMMENTS Diurnal. Terrestrial. Lives in woodland, grassland and heaths. Basks sitting in the centre of grass tussocks. Shelters beneath rocks, fallen timber and man-made debris. **Diet** Lizards. **Reproduction** Viviparous. 2–6 per litter. Neonates approximately 13cm TL and born in January–April. **Disposition** Quick to flee, generally inoffensive, but may bite if threatened.
BITE/VENOM VENOMOUS
IUCN LISTING Least Concern.

Dharwal, NSW, Jesse Campbell

Lawson, NSW, Scott Eipper

Lawson, NSW, Scott Eipper

Maddens Falls, NSW, Scott Eipper

Genus *Echiopsis* Fitzinger, 1843

Monotypic genus. The three disjunct populations have some minor morphological differences. They are crepuscular to nocturnal, usually seen crossing tracks or roads at night, and climb into low bushes and clumps of spinifex to bask. They are live bearers. **VENOM** Contains neurotoxins and procoagulants. Polyvalent antivenom has been used to treat severe envenomations. **Species-level identification difficulty** (within Australia) – 2.

TYPE SPECIES *Naja curta*.

ETYMOLOGY Viper-like.

Echiopsis curta, SA, Scott Eipper

Echiopsis curta, Esperance, WA, Scott Eipper

BARDICK *Echiopsis curta* (Schlegel, 1837)

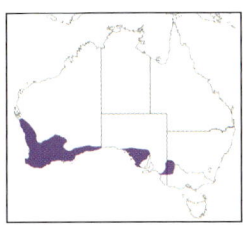

PRONUNCIATION *E-key-op-siss kerr-tah*.

ETYMOLOGY Short, viper-like snake.

TYPE LOCALITY Vicinity of King George's Sound, SW WA.

APPEARANCE Small to medium-sized, heavily robust snake with head that is strongly distinct from body. Dorsal colouration yellow to dark brown or grey. Some individuals reddish-orange, and some have barring on lips. Pupil vertically oval; iris brown. Tongue dark; mouth lining bluish-grey. Ventral colouration yellowish to cream. Adult females larger than males, reaching 70cm TL. **Scalation** MB 17–21 (usually 19) rows, 120–155 VENT, SUB 25–40 and anal scale single. Scales smooth and matt in appearance.

RANGE Three separate populations throughout Australia. One population occurs in SW WA, one on the Eyre Peninsula, SA, and another in western Vic, SW NSW and adjoining SA.

COMMENTS Mainly nocturnal. Terrestrial. Lives in open woodland, mallee, heath and coastal dune assemblages. Found sheltering under cover, for example beneath logs, in clumps of spinifex and leaf litter, and beneath man-made debris. **Diet** Lizards, frogs, small mammals, and periodically birds and insects. **Reproduction** Viviparous. 3–14 per litter. Neonates approximately 14cm TL and born in March–May. **Disposition** Generally inoffensive, but readily bites if harassed.

BITE/VENOM DANGEROUSLY VENOMOUS

IUCN LISTING Least Concern.

Cooljarloo, WA, Danny Melville

Lancelin, WA, Adam Elliott

Big Desert, VIC, Adam Elliott

Jurien, WA, Brian Bush

Genus *Elapognathus* Boulenger, 1896

This genus currently contains two species; both found in SW Australia. The species are diurnal, and usually found while crossing tracks, basking on vegetation or beneath cover. They mainly eat lizards and frogs, and are live bearers. **VENOM** Toxicity unknown, and there is no evidence of envenomations. **Species-level identification difficulty** –2.

TYPE SPECIES *Hoplocephalus minor*.

ETYMOLOGY Elap-jaw – pertaining to the genus of African garter-snakes *Elapsoidea* – the type genus for the family Elapidae.

Key to *Elapognathus*
1 Grey head with black nape collar..*E. coronatus* (opposite)
   Head the same colour as body; no nape collar.................................*E. minor* (p. 142)

Elapognathus minor, Two Peoples Beach, WA, Hal Cogger

Elapognathus coronatus, WA, Hal Cogger

CROWNED SNAKE *Elapognathus coronatus* (Schlegel, 1837)
(Werr)

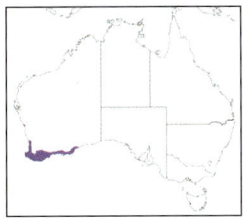

PRONUNCIATION *E-la-pog-nath-us cor-o-nah-tuss.*

ETYMOLOGY Crowned elap jaw.

TYPE LOCALITY King George's Sound, WA.

APPEARANCE Small to medium-sized, robust snake with head strongly distinct from body. Dorsal colouration tan, brown, olive to grey. Head silver to bluish-grey, bordered with black band across back of head. Labial scales bisected by black upper-edged, white stripe. Pupil round; iris reddish-orange. Tongue dark; mouth pink. Ventral colouration reddish-orange to yellow. Both sexes reach 70cm TL. **Scalation** MB 15 rows, 130–160 VENT, SUB 35–50 and anal scale single. Scales smooth and matt in appearance.

RANGE Encountered in SW WA, from Muchea south along the coast and east along the Great Australian Bight to Point Culver. Also found on the Archipelago of Recherche.

COMMENTS Diurnal. Terrestrial. Lives in open woodland, swamps and heaths. Shelters inside stick-ant nests, and beneath rocks, fallen timber and man-made debris. **Diet** Lizards, occasionally frogs and rodents. **Reproduction** Viviparous. 3–9 per litter. Gravid in October–May. Neonates approximately 14.5cm TL. **Disposition** Generally inoffensive, but readily bites if harassed.

BITE/VENOM VENOMOUS

IUCN LISTING Least Concern.

Forrestdale, WA, Robert Audcent

Banjup, WA David Robinson

Cape Le Grand, WA, Danny Melville

D'Entrecasteaux NP, WA, Steve Swanson

Short-nosed Snake *Elapognathus minor* (Günther, 1863)
(Little Brown Snake)

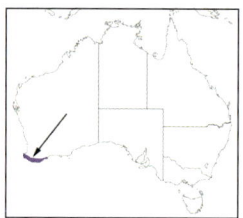

PRONUNCIATION E-la-pog-nath-us my-nor.
ETYMOLOGY Little elap jaw.
TYPE LOCALITY Swan River, SW Australia.
APPEARANCE Small, robust snake with head strongly distinct from body. As its name suggests, it has a short snout. Dorsal colouration brownish to grey. Skin between scales black, giving a reticulated appearance when inflated. Lower flanks can be reddish. Lips light grey with dark margins on rear of each supralabial. Dark band extending from nape through pale lateral margin behind head. Pupil round; iris burnt-gold. Tongue dark; mouth lining bluish-grey. Ventral colouration greenish-yellow, with orange-tinted stripes on lateral edges. Adult females larger than males, reaching 50cm TL. **Scalation** MB 15 rows, 115–130 VENT, SUB 40–55 and anal scale single. Scales smooth and matt in appearance.

RANGE Encountered in SW WA, from Busselton across to Two People's Bay.

COMMENTS Diurnal. Terrestrial. Lives in closed woodland, swamps and wet heaths. Shelters inside stick-ant nests, and beneath rocks, fallen timber and man-made debris. **Diet** Lizards. **Reproduction** Viviparous. 8–12 per litter. Neonates approximately 10cm TL. Gravid in November–February. **Disposition** Quick to flee, generally inoffensive, but may bite if threatened.

BITE/VENOM VENOMOUS

IUCN LISTING Least Concern.

Gull Rock NP, WA, Robert Audcent

Walpole, WA, Glen Gaikhorst

Gull Rock NP, WA, Robert Audcent

Two Peoples Beach, WA, Hal Cogger

Naped Snakes, genus *Furina* Duméril, 1853

This genus currently contains three species, all endemic to Australia. Distinct populations occur within *F. ornata*. However, the characteristics have considerable overlap with *F. barnardi* as currently understood, making the distinction between the two species unclear and subject to debate. Further research is needed to define whether they are distinct species, subspecies or just geographical variants. Some authors place the members of the genus *Glyphodon* in *Furina*.

These are small, nocturnal snakes that are found across much of Australia. Males are more brightly coloured than females. Occasionally they are mistaken for juvenile brown snakes *Pseudonaja* spp. They hunt predominantly lizards, which they will sometimes constrict while waiting for venom to take effect. Oviparous. **VENOM** Toxicity unknown; one bite from an adult *F. ornata* caused minor pain and swelling. **Species-level identification difficulty** – 5.
TYPE SPECIES: *Calamaria diadema*.
ETYMOLOGY Thief.

Key to *Furina*

1. Nasal scale undivided; usually 15 (rarely 17) midbody scale rows; prefrontal contacting the preocular............................2
   Nasal scale divided; 17 midbody scale rows; prefrontal not usually contacting the preocular................................*F. barnardi* (p. 144)

2. Orange-red nape band completely bisects black pigment (faded in old animals)..*F. ornata* (p. 146)
   Orange-red nape band mainly enclosed by black pigment.........................*F. diadema* (p. 145)

Furina barnardi, Middlemount, QLD, Tie Eipper

Yellow-naped Snake *Furina barnardi* (Kinghorn, 1939)

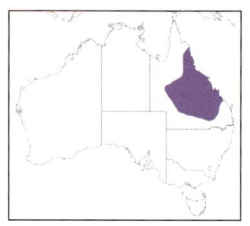

PRONUNCIATION *Few-ree-nah bar-narr-dee.*
ETYMOLOGY Barnard's thief; pertaining to H. Barnard, who collected the holotype.
TYPE LOCALITY 15m south of Duaringa, QLD.
APPEARANCE Small, slender snake with head slightly distinct from body. Dorsal colouration pale to dark brown, grey or almost black. Most individuals have thin, pale yellow to orange band across nape, which is absent in many older individuals. Head and nape chocolate-brown to black; supralabials usually pale. Pupils vertically oval; iris dark brown. Mouth lining and tongue pink. Ventral colouration white. Adult females larger than males, reaching 50cm TL. **Scalation** MB 17 rows, 170–200 VENT, SUB 35–50 and anal scale divided. Scales glossy in appearance and smooth.
Similar species *F. ornata* (p. 146).
RANGE Encountered in QLD, from Mareeba south to Gladstone and inland to around Barcaldine.
COMMENTS Nocturnal. Terrestrial. Lives in open woodland, brigalow and savannah. Shelters beneath rocks, fallen timber and man-made debris. **Diet** Lizards. **Reproduction** Oviparous. 7–10 per clutch. Neonates approximately 14cm TL hatching in December–March. **Disposition** Quick to flee, generally inoffensive, but may bite if threatened.
BITE/VENOM VENOMOUS
IUCN LISTING Least Concern.

Chullangun, QLD, Anders Zimny

Mareeba, QLD, Matt Summerville

Mareeba, QLD. Matt Summerville

Mitchell, QLD. Specimen courtesy of David Peica, Rob Valentic

Red-naped Snake *Furina diadema* (Schlegel, 1837)

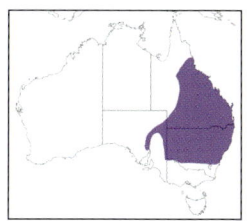

PRONUNCIATION *Few-ree-nah die-ah-dee-mah.*
ETYMOLOGY Diadem thief.
TYPE LOCALITY Australia.
APPEARANCE Small, slender snake with head slightly distinct from body. Dorsal colouration reddish to dark brown or brownish-grey, and edges of midbody scales darker, forming a reticulated appearance. Most individuals have orange to reddish mark enclosed within black nape. Head and nape black. Supralabials usually pale. Pupils vertically oval with dark brown irises. Mouth lining and tongue both pink. Ventral colouration white. Adult females larger than males, reaching 40cm TL. **Scalation** MB 15 rows, 160–210 VENT, SUB 35–70 and anal scale divided. Scales glossy in appearance and smooth. **Similar species** *F. ornata* (p. 146).

RANGE Encountered in QLD, from Rockhampton south and west, covering most of NSW, eastern SA, including the Flinders Ranges, to Port Augusta and far NW Vic.

COMMENTS Nocturnal. Terrestrial. Lives in open woodland, mulga, brigalow, sand ridge desert, rocky outcrops and mallee. Shelters beneath rocks, fallen timber and man-made debris. **Diet** Lizards. **Reproduction** Oviparous. 2–5 per clutch. Neonates approximately 12cm TL and hatch in November–April. **Disposition** Inoffensive, usually bluff striking towards antagonist, and rarely attempting to bite.

BITE/VENOM VENOMOUS
IUCN LISTING Least Concern.

Gatton, QLD, Adam Elliott

Dalby, QLD, Anders Zimny

Glenmorgan, QLD, Scott Eipper

White Rock, QLD, Scott Eipper

ORANGE-NAPED SNAKE *Furina ornata* (Gray, 1842)
(Moon Snake)

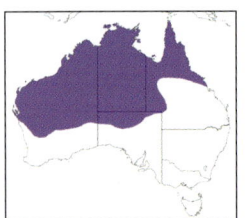

PRONUNCIATION *Few-ree-nah or-nah-tah.*

ETYMOLOGY Ornamented thief.

TYPE LOCALITY Australia (Swan River – in error).

APPEARANCE Small to medium-sized, slender snake with head slightly distinct from body. Dorsal colouration reddish to dark brown, grey or almost black, and edges of midbody scales darker, forming reticulated appearance. Most individuals have broad, orange to reddish band across nape, which is indistinct in some older females. Head and nape black. Supralabials usually pale. Pupils vertically oval; irises dark brown. Mouth lining and tongue pink. Ventral colouration white. Adult females larger than males, reaching 70cm TL. Northern individuals larger than southern ones. **Scalation** MB 15–17 rows, 160–240 VENT, SUB 35–70 and anal scale divided. Midbody scales in 15 rows in southern WA and parts of QLD. Scales glossy in appearance and smooth. **Similar species** *F. barnardi* (p. 144) and *F. diadema* (p. 145).

RANGE Encountered across northern and central Australia, from north of the goldfields, WA, across through NW SA, all of the NT, western and northern QLD, reaching the coast at Bowen. Also occurs on the Torres Strait Islands.

COMMENTS Nocturnal. Terrestrial. Lives in open woodland, mulga, deserts, black soil plains, gorges, brigalow and savannah. Shelters beneath rocks, fallen timber and man-made debris. **Diet** Lizards. **Reproduction** Oviparous. 3–6 per clutch. Neonates approximately 12cm TL and hatch in November–April. **Disposition** Inoffensive, usually bluff striking, and rarely attempting to bite.

BITE/VENOM VENOMOUS

IUCN LISTING Least Concern.

Weipa, QLD, Scott Eipper

Yandi, WA, Brian Bush

Mica Creek, QLD, Scott Eipper

Shay Gap, WA, Brian Bush

Genus *Glyphodon* Günther, 1858

This genus currently contains two species. *G. tristis* also occurs in NG. Interestingly, this population has white to cream v yellow nape bands seen in the Australian individuals. Further research is needed to define whether this is a distinct taxon or just a geographical variation. Some authors place the members of the genus *Glyphodon* in the genus *Furina*.

These are medium-sized, nocturnal snakes from eastern Australia. They are usually seen while crossing roads at night. They hunt predominantly lizards, but will also eat snakes, frogs and reptile eggs, and are oviparous. **VENOM** Envenomations from this genus can lead to symptoms such as headache, nausea, swelling, vomiting, abdominal cramps, vision disturbances, dizziness and collapse. Polyvalent antivenom has been used to treat severe envenomations. **Species-level identification difficulty** (within Australia) – 3.

TYPE SPECIES *Glyphodon tristis*.
ETYMOLOGY Notched tooth.

Key to *Glyphodon*

1. 17 midbody scale rows; usually a yellowish nape..*G. tristis* (p. 149)
   21 midbody scale rows; nape same colour as body...*G. dunmalli* (p. 148)

Glyphodon tristis, Cooktown, QLD, Shane Black

Glyphodon dunmalli eating *Gehyra*, Lake Broadwater, QLD, Scott Eipper

Dunmall's Snake *Glyphodon dunmalli* (Worrell, 1955)

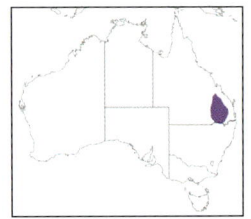

PRONUNCIATION *G-ly-fo-don dun-mall-ee*.

ETYMOLOGY Dunmall's notched tooth, pertaining to W. Dunmall, who collected the first specimen.

TYPE LOCALITY Glenmorgan, QLD.

APPEARANCE Medium-sized, heavy, robust snake with head that is strongly distinct from body. Dorsal colouration dark brown to greyish-black. Occasionally some yellowish flushing on supralabials and rear of head. Pupils vertically oval; irises black-brown. Mouth lining and tongue pink. Ventral colouration white to light grey. Adult males larger than females, reaching 75cm TL.

Scalation MB 21 rows, 175–190 VENT, SUB 35–50 and anal scale divided. Nasal scale divided. Scales glossy in appearance and smooth.

RANGE Encountered from the NSW/QLD border, north around Rockhampton, QLD.

COMMENTS Nocturnal. Terrestrial. Lives in dry forests, brigalow and scrubland, where it can be seen foraging on the ground or crossing roads. Shelters under cover, for example beneath logs and rocks. **Diet** Lizards, reptile eggs, frogs. **Reproduction** Oviparous. 5–9 per clutch. Neonates approximately 15cm TL hatching in February–April. **Disposition** Generally inoffensive, but readily bites if harassed.

BITE/VENOM VENOMOUS

IUCN LISTING Least Concern.

Westmar, QLD, Alexander Davies

Gayndah, QLD, Hal Cogger

Cecil Plains, QLD, Alexander Davies

Lake Broadwater, QLD, Scott Eipper

Brown-headed Snake *Glyphodon tristis* Günther, 1858

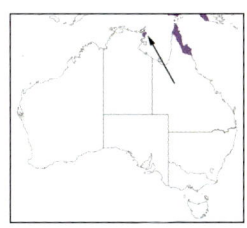

PRONUNCIATION *G-ly-fo-don tris-tiss.*
ETYMOLOGY Dull-coloured notched-tooth.
TYPE LOCALITY NE coast of Australia.
APPEARANCE Medium-sized, heavy, robust snake with head strongly distinct from body. Dorsal colouration brown to greyish-black. Prominent yellow band, usually 7–8 scales in width, across nape, which fades in older individuals. Head usually lighter brown than body. White skin between scales distinct, giving reticulated appearance. Pupils vertically oval; irises black-brown. Mouth lining and tongue pink. Ventral colouration white to light grey. Adult males larger than females, reaching 100cm TL. **Scalation** MB 19 rows, 160–190 VENT, SUB 30–60 and anal scale divided. Nasal scale divided. Scales glossy in appearance and smooth.

RANGE Found in QLD north of Edmonton. Also occurs in PNG. Suggested by some publications as occurring in Arnhem Land, NT, but no known specimens exist.

COMMENTS Nocturnal; usually only seen on the surface at night, foraging on the forest floor or crossing roads. Terrestrial. Lives in forests, vine thickets, savannah and scrubland. Shelters beneath rocks, fallen timber and man-made debris. **Diet** Lizards, reptile eggs, frogs. **Reproduction** Oviparous. 6–8 per clutch. Neonates approximately 16.5cm TL hatching in February–April. **Disposition** Generally inoffensive, but readily bites if harassed.

BITE/VENOM VENOMOUS
IUCN LISTING Least Concern.

Iron Range, QLD, Anders Zimny

Iron Range, QLD, Angus McNab

Weipa, QLD, Scott Eipper

Iron Range, QLD, Scott Eipper

Genus *Hemiaspis* Fitzinger, 1860

This genus currently contains two species endemic to Australia. Diurnal to crepuscular, they are usually found while crossing tracks, basking on clumps of vegetation or beneath cover. They are lizard feeders, but will occasionally eat frogs, and are live bearers. **VENOM** The venom contains procoagulants and neurotoxins. Envenomations from this genus can lead to symptoms such as headache, ptosis, nausea, vomiting, pain, dizziness and collapse. **Species-level identification difficulty** – 2.

TYPE SPECIES *Alecto signata*.
ETYMOLOGY Half-shield.

Key to *Hemiaspis*
1. Black nape collar, no pale streak behind eye..*H. daemelii* (opposite)
   No nape collar, pale streak behind eye..*H. signata* (p. 152)

Hemiaspis damelii, Quambone, NSW, Hal Cogger

Hemiaspis signata, Springbrook NP, QLD, Scott Eipper

Hemiaspis signata, Maddens Plains, NSW, Scott Eipper

Grey Snake *Hemiaspis damelii* (Günther, 1876)

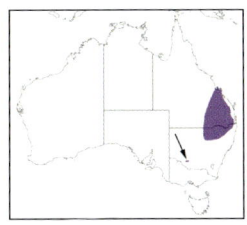

PRONUNCIATION Heem-e-as-pis day-mel-e.
ETYMOLOGY Marked half-shield.
TYPE LOCALITY Rockhampton and Peak Downs, QLD.
APPEARANCE Small, moderately robust snake with head slightly distinct from body. Dorsal colouration ranges from tan, to brown or grey. Nape has distinct black band that can extend on to top of head. Pupils round; irises dark. Tongue dark. Ventral colouration cream to light grey. Adult females larger than males, reaching 55cm TL. **Scalation** MB 17 rows, 140–170 VENT, SUB 35–50 single, and anal scale divided. Scales glossy in appearance and smooth.
RANGE Encountered in QLD, from Rockhampton through the southern brigalow belt to the Macquarie Marshes. Another isolated population in south-central NSW, near Waugorah and Balranald.
COMMENTS Crepuscular to nocturnal. Terrestrial. Lives in grassland, brigalow and open woodland, mainly around river systems. Shelters beneath rocks, fallen timber and man-made debris. Also found in soil cracks. **Diet** Frogs, periodically lizards. **Reproduction** Viviparous. 4–16 per litter. Gravid in December–April. Black head the young are born with fades with age. **Disposition** Relatively quick to flee, generally inoffensive, but may bite if threatened.
BITE/VENOM VENOMOUS
IUCN LISTING Least Concern.

Narrabri, NSW, Wes Read

Glenmorgan, QLD, Rob Valentic

Dalby, QLD, Scott Eipper

Dalby, QLD, Scott Eipper

Marsh Snake *Hemiaspis signata* (Jan, 1859)
(Black-bellied Swamp Snake, Swamp Snake)

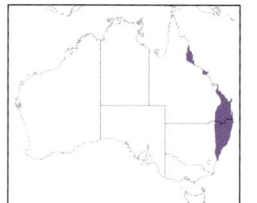

PRONUNCIATION *Heem-e-as-pis sig-nah-tah.*

ETYMOLOGY Marked half-shield.

TYPE LOCALITY Australia. Rrestricted to Sydney *fide* Boettger (1898:117).

APPEARANCE Small to medium-sized, moderately robust snake with head slightly distinct from body. Dorsal colouration highly variable, ranging from all shades of brown, greenish and grey to black. Head can be olive, becoming yellowish lower down towards ventral surface. Prominent ragged white stripes along lips and extending behind eye on to neck. Pupils round with reddish to gold irises. Tongue dark. Ventral colouration black to grey, or orange in some individuals. Adult males larger than females, reaching 90cm TL. **Scalation** MB 17 rows, 150–170 VENT, SUB 40–60 single, and anal scale divided. Scales glossy in appearance and smooth.

RANGE Encountered in northern QLD in isolated populations centred around Cairns and Eungella. Further south, extends from Gladstone, QLD, along the GDR to Nowra, NSW.

COMMENTS Diurnal. Terrestrial. Lives in rainforests, grassland, heaths and swamps. Often seen basking in grass tussocks and on vegetation, Shelters beneath rocks, fallen timber and man-made debris. **Diet** Frogs, occasionally lizards. **Reproduction** Viviparous. 4–16 per litter. Neonates approximately 14cm TL. Gravid in September–May. **Disposition** Relatively quick to flee, generally inoffensive, but may bite if threatened.

BITE/VENOM VENOMOUS

IUCN LISTING Least Concern.

Helensburgh, NSW, Malcolm Campbell

Myall Lakes NP, NSW, Angus McNab

Watagans NP, NSW, Malcolm Campbell

Mt Lewis, QLD, Shane Black

Broad-headed Snakes, genus *Hoplocephalus* Wagler, 1803

This genus currently contains three species; phylogenetic assessment shows that *Hoplocephalus* is closely related to *Paroplocephalus* and possibly synonymous. The snakes are predominantly nocturnal, but will bask cryptically. They are usually found while crossing tracks, or basking on clumps of vegetation or beneath cover. Some members are threatened due to habitat destruction. They are primarily lizard feeders, but occasionally eat frogs and small mammals, and are viviparous. All three species are endemic to Australia. **VENOM** Contains procoagulants, neurotoxins and haemolysins. Several people have suffered severe effects that commenced very quickly after a bite. Some of these included loss of vision, ptosis, vomiting, severe headache, dizziness, loss of bowel function, loss of consciousness, severe pain and heavy sweating. Bites have resulted in, or could cause, fatalities in humans. Tiger Snake or Polyvalent antivenom is used to treat envenomations. **Species-level identification difficulty** (within Australia) – 3.

TYPE SPECIES *Naja bungaroides*.

ETYMOLOGY Armoured head.

Key to *Hoplocephalus*

1. Head and body covered in white to yellow flecks forming irregular bands ....*H. bungaroides* (p. 155)
   Body plain or with regular, distinct bands (rarely striped) 2

2. Body plain; ventral scales 190–225 ....*H. bitorquatus* (p. 154)
   Body with regular, distinct bands (rarely striped or without banding); ventral scales 225–250 ....*H. stephensii* (p. 156)

Hoplocephalus stephensii, Manorina, Mt Glorious

Hoplocephalus stephensii, Coffs Harbour, NSW, Scott Eipper

PALE-HEADED SNAKE *Hoplocephalus bitorquatus* (Jan, 1859)

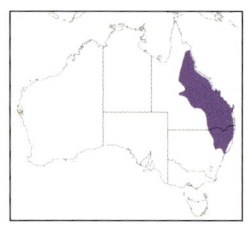

PRONUNCIATION *Hop-low-keff-al-us bi-torr-quart-tus.*
ETYMOLOGY Double-necklaced armoured-head.
TYPE LOCALITY Australia (Sydney – in error)
APPEARANCE Medium-sized, muscular snake with head strongly distinct from body. Dorsal colouration silver-grey to black. Conspicuous white band across nape of juvenile and young individuals that fades to pale saddle in old ones. Face marked with black spots. Pupil round; iris light tan. Tongue dark; pink inside mouth. Ventral colouration grey. Adult females larger than males, reaching 65cm TL. **Scalation** MB 19–21 rows, 190–225 VENT, SUB 40–65 and anal scale single. Scales smooth. **Similar species** *H. stephensii* (p. 156).

RANGE Disjunct distribution, with one population in north QLD, and another further south in southern QLD into NSW.

COMMENTS Nocturnal and arboreal. Lives in open woodland, brigalow and scrubland. Often seen sitting exposed on trunks of large trees or crossing roads. Shelters under cover, such as logs and rock slabs, in tree hollows and beneath bark of standing trees. **Diet** Lizards, frogs, small mammals. **Reproduction** Viviparous. 2–17 per litter. Neonates approximately 26cm TL, born in October– March. **Disposition** Nervous, and inoffensive unless threatened. Will readily bite if harassed.

BITE/VENOM DANGEROUSLY VENOMOUS
IUCN LISTING Least Concern.

Almaden, QLD, Scott Eipper

Glenmorgan, QLD, Scott Eipper

Lake Broadwater, QLD, Scott Eipper

Pilliga, NSW, Scott Eipper

Broad-headed Snake *Hoplocephalus bungaroides* (Schlegel, 1837)

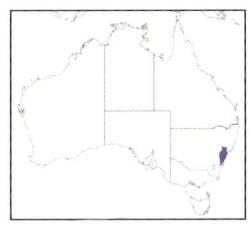

PRONUNCIATION *Hop-low-keff-al-us bun-gar-roid-dees.*

ETYMOLOGY 'Bungarus-like' armour head, *Bungarus* is the genus for the kraits of Asia.

TYPE LOCALITY Port Jackson, NSW.

APPEARANCE Medium-sized, muscular snake with head strongly distinct from body. Dorsal colouration black, with yellow to white flecks that form irregular thin cross-bands. Pupil round; iris gold. Tongue dark; pink inside mouth. Ventral colouration grey. Adult females larger than males, reaching 90cm TL. **Scalation** MB 21 rows, 200–230 VENT, SUB 40–65 and anal scale single. Scales smooth.

RANGE Encountered in Sydney and surrounding areas of NSW.

COMMENTS Nocturnal. Predominantly terrestrial. Lives in dry forest and heaths on sandstone escarpments with rock exfoliations. Occasionally encountered partially exposed by day, basking, or while moving on rock faces at night. Shelters under cover such as rock slabs, in tree hollows and rock crevices, and beneath bark of standing trees. **Diet** Lizards, small mammals. **Reproduction** Viviparous. 2–12 per litter. Neonates approximately 26cm TL born in October–March. **Disposition** Nervous, and inoffensive unless threatened. Will readily bite if harassed.

BITE/VENOM DANGEROUSLY VENOMOUS

IUCN LISTING Least Concern.

Sydney Basin, NSW, Jules Farquhar

Tianjara Falls, NSW, Scott Eipper

Tianjara Falls, NSW, Scott Eipper

Waterfall, NSW, Scott Eipper

STEPHENS' BANDED SNAKE *Hoplocephalus stephensii* Krefft, 1869

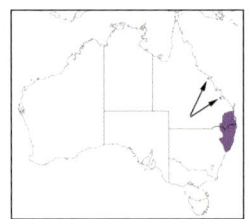

PRONUNCIATION *Hop-low-keff-al-us stee-fen-zee-e.*

ETYMOLOGY Stephens' armoured-head, pertaining to W. J Stephens, of the Australian Museum.

TYPE LOCALITY Port Macquarie, on the Hastings River, NSW.

APPEARANCE Medium to large, muscular snake with head strongly distinct from body. Dorsal colouration dark grey to black. Bands more distinct in young and juvenile individuals than in adults, and unbanded populations are known. Face blotched with black spots, and supralabials marked with black and white. Pupil round; iris brown. Tongue dark; pink inside mouth. Ventral colouration grey. Adult females larger than males, reaching 90cm TL. **Scalation** MB 21 rows, 225–250 VENT, SUB 50–70 and anal scale single. Scales smooth. **Similar species** *H. bitorquatus* (p. 154).

RANGE Encountered from Maryborough, QLD, to Ourimbah, NSW. Two isolated populations occur further north at Eungella and Kroombit Tops, QLD.

COMMENTS Nocturnal. Predominantly terrestrial. Lives in forests, granite outcrops and scrubland. Shelters under cover such as logs and rock slabs, and in tree hollows and beneath the bark of standing trees. **Diet** Lizards, frogs, small mammals. **Reproduction** Viviparous. 2–17 per litter. Neonates approximately 24cm TL born in October–March. **Disposition** Nervous, and inoffensive unless threatened. Will readily bite if harassed.

BITE/VENOM DANGEROUSLY VENOMOUS

IUCN LISTING Least Concern.

Manorina, Mt Glorious, QLD, Scott Eipper

Tenterfield, NSW, Scott Eipper

Kroombit Tops, QLD, Scott Eipper

Mt Tamborine, QLD, Scott Eipper

Genus *Narophis* Worrell, 1961

This monotypic genus, restricted to Australia, was recently split from *Neelaps* following phylogenetic evidence showing that both species are not closely related despite morphological similarities. These are nocturnal, fossorial snakes that are occasionally encountered crossing roads at night. They specialize in eating fossorial skinks, and are oviparous. **VENOM** Toxicity unknown, and there is no evidence of envenomations. **Species-level identification difficulty** – 2.

TYPE SPECIES *Furina bimaculata*.
ETYMOLOGY Narrow snake.

Narophis bimaculatus, Tamala Station, Shark Bay, WA, Hal Cogger

Narophis bimaculatus, Yanchep, WA, Scott Eipper

Narophis bimaculatus, Coolgardie, WA, Brian Bush

Black-naped Burrowing Snake *Narophis bimaculatus*
(Duméril, Bibron & Duméril, 1854)

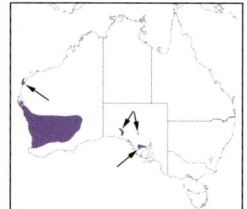

PRONUNCIATION *Nah-ro-fiss buy-mac-u-lah-tuss.*
ETYMOLOGY Twin-spotted narrow snake.
TYPE LOCALITY Australia.
APPEARANCE Small, slender snake with head indistinct from body. Dorsal colouration yellowish-orange to reddish-brown. Each midbody scale has darker edges forming a reticulated pattern. Black mark on snout, over eyes on head and over nape. Between snout and eye marking, usually creamy-yellow. Pupil vertically oval; iris black. Tongue pink; buccal cavity pale. Ventral colouration cream to white. Males and females reach a similar size of TL 45cm. **Scalation** MB 15 rows, 176–228 VENT, SUB 19–30 and anal scale divided. Scales shiny in appearance and smooth.
RANGE Encountered on the Eyre Peninsula, around Kingoonya, SA. Further west found throughout the goldfields of WA, to Bunbury, and north to Shark Bay. Isolated population on the Northwest Cape.
COMMENTS Nocturnal. Fossorial. Lives in mallee, coastal heaths and sand dunes, usually beneath the surface. Shelters in leaf litter and in the top few centimetres of soil beneath vegetation. **Diet** Small burrowing skinks. **Reproduction** Oviparous. 2–6 per clutch. Hatchlings approximately 16cm TL. Gravid in October–January. **Disposition** Inoffensive but will bite if threatened.
BITE/VENOM HARMFUL
IUCN LISTING Least Concern.

South of Coolgardie, WA, Brian Bush

Yanchep, WA Scott Eipper

Moore River, WA, Rob Valentic

Yanchep, WA, Scott Eipper

Genus *Neelaps* Günther 1863

Monotypic genus, restricted to Australia. Recently, *Narophis bimaculatus* was removed from the genus *Neelaps* following phylogenetic evidence showing that the species are not closely related, despite morphological similarities. These are nocturnal, fossorial snakes that are occasionally encountered crossing roads at night. They specialize in eating fossorial skinks, and are oviparous.
VENOM Toxicity unknown, and there is no evidence of envenomations. **Species-level identification difficulty** – 2.
TYPE SPECIES *Furina calonotus*.
ETYMOLOGY Not elaps – pertaining to the genus of African garter-snakes *Elapsoidea* – the type genus for the family Elapidae.

Neelaps calonotos, Yanchep, WA, Scott Eipper

Neelaps calonotos, Mayla Lake, WA, Scott Eipper

Black-striped Burrowing Snake *Neelaps calonotos*
(Duméril, Bibron & Duméril, 1854)

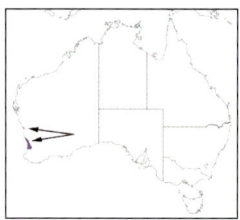

PRONUNCIATION *Nee-laps cal-on-not-oss.*
ETYMOLOGY Beautiful back not elaps.
TYPE LOCALITY Australia.
APPEARANCE Small, slender snake with head indistinct from body. Dorsal colouration yellowish-orange to reddish-brown. Most individuals have a continuous black stripe extending the length of the body, but in some it is reduced to a few black vertebral spots. Each midbody scale has white spot in centre. Black mark on snout, over eyes on head and over nape. Between the dark markings, usually creamy-yellow. Pupil vertically oval; iris black. Tongue pink; buccal cavity pale. Ventral colouration cream to white. Adult females larger than males, reaching 30cm TL. **Scalation** MB 15 rows, 126–143 VENT, SUB 23–35 and anal scale divided. Scales shiny in appearance and smooth.

RANGE Encountered in WA, from Mandurah to Lancelin, with isolated populations at Dongara and Eneabba.

COMMENTS Nocturnal. Fossorial. Lives in coastal heaths, woodland and sand dunes, usually beneath the surface. Shelters in leaf litter and in the top few centimetres of soil beneath vegetation. **Diet** Small burrowing skinks. **Reproduction** Oviparous. 2–5 per clutch. Hatchlings approximately 12cm TL. Gravid in October–January. **Disposition** Inoffensive but will bite if threatened.

BITE/VENOM HARMFUL
IUCN LISTING Least Concern.

Moore River, WA, Rob Valentic

Yanchep, WA, Scott Eipper

Whiteman Park, WA, Danny Melville

Lancelin, WA, Hal Cogger

Tiger Snakes, genus *Notechis* Boulenger, 1896

Monotypic genus, comprising a single species with a further five subspecies. The subspecies are distinguished on morphology and distribution alone. There is little evidence of genetic support, but no evidence of gene flow between the populations, hence the retention of subspecific recognition. Some authorities do not recognize these subspecies. Tiger snakes are active hunters with excellent vision. They are found in southern Australia, usually around water, in forests or grassland. They are primarily diurnal but occasionally active after dark, usually in hot weather, and can be found in large numbers close to people, including in the inner city. Male combat has been recorded in the genus, and the snakes are livebearers. **VENOM** Venom strongly neurotoxic, with powerful coagulants and weakly haemolytic and cytotoxic, myotoxic activity. Bites have resulted in, or could cause, fatalities in humans. Tiger Snake or Polyvalent antivenom is used to treat envenomations. **Species-level identification difficulty** – 2.

TYPE SPECIES *Naja (Hamadryas) scutata*.

ETYMOLOGY Southern snake.

Key to *Notechis*

1. Found outside of WA 2
   Found in WA .... *N. scutatus occidentalis* (p. 165)

2. Found outside of Tas and islands in the Bass Strait 4
   Occurs in Tas and associated islands in the Bass Strait 3

3. Only found on the Furneaux group of islands including Flinders & Chappell Islands .... *N. s. serventyi* (p. 168)
   Found in Tasmania, excluding the Furneaux group islands .... *N. s. humphreysi* (p. 163)

4. Usually black with or without white bands (except Kangaroo Island); usually 17 midbody scale rows; restricted to SA 5
   Not usually black; with or without bands; usually 19 midbody scale rows .... *N. s. scutatus* (p. 166)

5. 17 midbody scale rows; found in SA between Crystal Brook and Mt Remarkable .... *N. s. ater* (p. 162).
   Found in SA on islands in Spencer Gulf, western Eyre Peninsula, southern Yorke Peninsula and Kangaroo Island .... *N. s. niger* (p. 164)

Notechis scutatus scutatus, Pearl Beach, NSW, Hal Cogger

Krefft's Tiger Snake *Notechis scutatus ater* (Krefft, 1866)

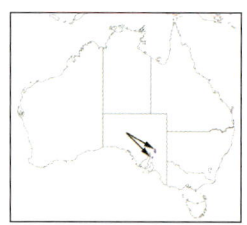

PRONUNCIATION *No-tech-is scu-tar-tus a-ter.*
ETYMOLOGY Black-shielded southern-snake.
TYPE LOCALITY Flinders Ranges, SA.
APPEARANCE Large, moderately robust snake with head distinct from body. Dorsal colouration black, with or without pale white, yellowish to grey cross-bands. Pupil round; iris reddish-brown, grey to yellow. Tongue dark; mouth lining pink. Ventral colouration blackish to grey. Adult males larger than females, reaching 90cm TL, and shiny in appearance.
Scalation MB 17 rows, 163–173 VENT, SUB 41–50 and anal scale single. Scales smooth.
RANGE Encountered in the southern Flinders Ranges between Crystal Brook and Mt Remarkable NP, SA.
COMMENTS Diurnal. Terrestrial. Lives along streams in dry open forests and scrubland. Shelters beneath rocks, fallen timber and man-made debris. Threatened due to habitat destruction and predation by feral species. **Diet** Lizards, frogs, tadpoles, birds, small mammals. **Reproduction** Viviparous. 6–15 per litter. Neonates born in November–May. **Disposition** Generally inoffensive, but bites if harassed.
BITE/VENOM DANGEROUSLY VENOMOUS
IUCN LISTING Least Concern.

Mt Remarkable, SA, Scott Eipper

Southern Flinders Ranges, SA, Scott Eipper

Southern Flinders Ranges, SA, Scott Eipper

Southern Flinders Ranges, SA, Scott Eipper

Tasmanian Tiger Snake *Notechis scutatus humphreysi* Worrell, 1963

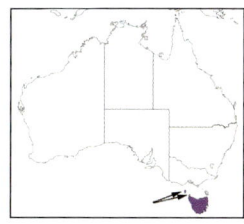

PRONUNCIATION *No-tech-iss scu-tar-tus hum-free-zee.*

ETYMOLOGY Humphrey's shielded southern snake, pertaining to R. Humphrys, a naturalist from Bundaberg, QLD.

TYPE LOCALITY New Year Island, Tas.

APPEARANCE Large to very large, moderately robust snake with head distinct from body. Very variable in both colour and pattern. Many different colour forms known, from whitish with grey head, to yellow, jet-black and all shades of brown. Some individuals banded or speckled, while others are immaculate. Pupil round; iris reddish-brown, grey to yellow. Tongue dark; mouth lining pink. Ventral colouration yellow to grey. Adult males larger than females, reaching 200cm TL. **Scalation** MB 15–19 rows, 161–174 VENT, SUB 48–53 and anal scale single. Scales smooth.

RANGE Encountered across Tas, and also on the King Island group in Bass Strait.

COMMENTS Diurnal. Terrestrial. Lives in urban environments, swamps, forests, grassland, rainforests, heaths and scrubland. Shelters inside animal burrows, and beneath rocks, fallen timber and man-made debris. **Diet** Lizards, frogs, birds, small mammals. **Reproduction** Viviparous. 8–64 per litter. Neonates approximately 22cm TL born in October–June. **Disposition** Generally inoffensive, but bites if harassed.

BITE/VENOM DANGEROUSLY VENOMOUS

IUCN LISTING Least Concern.

Juvenile, Meander, TAS, Ryan Francis

Little Pine Lake, TAS, Ryan Francis

Captive bred, Scott Eipper

Captive bred, Scott Eipper

Peninsula Tiger Snake *Notechis scutatus niger* Kinghorn, 1921

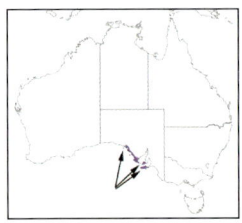

PRONUNCIATION *No-tech-iss scu-tar-tus nigh-ger*.
ETYMOLOGY Black-shielded southern snake.
TYPE LOCALITY Kingscote, Kangaroo Island, SA.
APPEARANCE Large, moderately robust snake with head distinct from body. Dorsal colouration coppery-brown to black, and head usually grey or black. Some populations are banded with thin white to cream bands, while others are immaculate. Pupil round; iris reddish-brown, grey to yellow. Tongue dark; mouth lining pink. Ventral colouration grey to black. Adult males larger than females, reaching 120cm TL, and shiny in appearance. **Scalation** MB 17–21 rows, 160–184 VENT, SUB 45–54 and anal scale single. Scales smooth.

RANGE Encountered in SA, from Kangaroo Island, the southern half of the Yorke Peninsula and along the western coast of the Eyre Peninsula. Also occurs on islands in the Spencer Gulf.

COMMENTS Diurnal. Terrestrial. Lives in swamps, forests, grassland, samphire heaths and scrubland. Shelters inside animal burrows, and beneath rocks, fallen timber and man-made debris. **Diet** Lizards, frogs, birds, small mammals. **Reproduction** Viviparous. 6–38 per litter. Neonates approximately 23cm TL born in September–April. **Disposition** Generally inoffensive, but bites if harassed.

BITE/VENOM DANGEROUSLY VENOMOUS
IUCN LISTING Least Concern.

Kangaroo Island, SA, Scott Eipper

Kangaroo Island, SA, Tie Eipper

Coffin Bay, SA, Shane Black

Kangaroo Island, SA, Scott Eipper

WESTERN TIGER SNAKE *Notechis scutatus occidentalis* Glauert, 1948

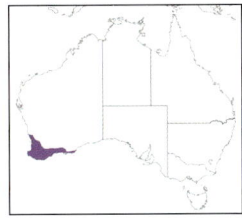

PRONUNCIATION *No-tech-iss scu-tar-tus occ-sid-dent-tah-liss.*
ETYMOLOGY Western shielded southern snake.
TYPE LOCALITY Bassendean, WA.
APPEARANCE Large, moderately robust snake with head distinct from body. Dorsal colouration black or brown, with lower flanks yellow, orange or even whitish; usually vibrant yellow cross-bands. Generally darker on rear third of body. Pupil round; iris reddish-brown, grey to yellow. Tongue dark; mouth lining pink. Ventral colouration yellow and orange with grey flecks. Adult males larger than females, reaching 160cm TL, and shiny in appearance. **Scalation** MB 17–19 rows, 140–165 VENT, SUB 36–51 and anal scale single. Scales smooth.
RANGE Encountered in SW WA, from Gingin to Israelite Bay, WA. Also occurs on Garden and Carnac Islands.
COMMENTS Diurnal. Terrestrial. Lives around water courses in swamps, forests and scrubland. Often found in urban areas. Shelters under rocks, fallen timber and man-made debris. **Diet** Lizards, frogs, birds, small mammals. **Reproduction** Viviparous. 15–35 per litter. Neonates approximately 20cm TL born in November–June. **Disposition** Generally inoffensive, but bites if harassed.
BITE/VENOM DANGEROUSLY VENOMOUS
IUCN LISTING Least Concern.

Herdsman Lake, WA, Scott Eipper

Juvenile, Perth, WA, Tie Eipper

Busselton, WA, Scott Eipper

Jandakot, WA, Adam Elliott

COMMON TIGER SNAKE *Notechis scutatus scutatus* (W. C. H Peters, 1861)

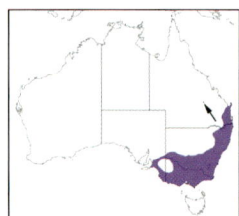

PRONUNCIATION *No-tech-iss scu-tar-tus scu-tar-tus*.
ETYMOLOGY Shielded southern-snake.
TYPE LOCALITY Java, Indonesia (in error). Neotype: 'The Brothers', 10km NE of Benambra, Vic.
APPEARANCE Large, moderately robust snake with head distinct from body. Very variable in both colour and pattern. Dorsal colouration can be any shade of brown, grey, black or yellow, with or without crossbands. Pupil round; iris reddish-brown, grey to yellow. Tongue dark; mouth lining pink. Ventral colouration yellow to grey. Adult males larger than females, reaching 170cm TL. **Scalation** MB 17–19 rows, 140–190 VENT, SUB 35–65 and anal scale single. Scales smooth and shiny in appearance.
RANGE Encountered from southern QLD, through eastern NSW, most of Vic and SE SA.
COMMENTS Usually diurnal but may become nocturnal in hot weather and during summer. Terrestrial. Lives in swamps, forests, grassland, rainforests, wallum, open woodland, heaths and scrubland. Common in urban areas, often in gardens and houses. Shelters beneath rocks, fallen timber and man-made debris. **Diet** Lizards, frogs, birds, small mammals. **Reproduction** Viviparous. 5–49 per litter. Neonates approximately 22.5cm TL born in November–May. **Disposition** Generally inoffensive, but bites if harassed.
BITE/VENOM DANGEROUSLY VENOMOUS
IUCN LISTING Least Concern.

Beerwah, QLD, Scott Eipper

Stoney Rises, VIC, Adam Elliott

Beerwah, QLD, Scott Eipper

Juvenile, Maddens Falls, NSW, Scott Eipper

Ottways, VIC, Scott Eipper

Lake George, NSW, Scott Eipper

Stony Rises, VIC, Scott Eipper

Mount Gambier, SA, Scott Eipper

Mt Gambier, SA, Tie Eipper

Rockbank, VIC, Scott Eipper

Chappell Island Tiger Snake *Notechis scutatus serventyi* Worrell, 1963

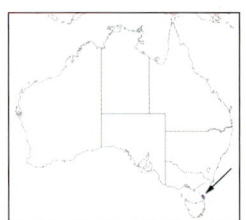

PRONUNCIATION *No-tech-iss scu-tar-tus ser-ven-tee-e.*

ETYMOLOGY Serventy's shielded southern snake, after D. L. Serventy, Australian naturalist.

TYPE LOCALITY Chappell Island, Tas.

APPEARANCE Large to very large, moderately robust snake with head distinct from body. Body yellowish, grey, dark brown to black, and head usually grey or black. Some individuals are banded with thin white to cream bands, while others are immaculate. Lower flanks can be yellow. Pupil round; iris reddish-brown, grey to yellow. Tongue dark; mouth lining pink. The largest of all tiger snakes. Ventral colouration grey to black. Adult males larger than females, reaching 210cm TL.

Scalation MB 17 rows, 160–171 VENT, SUB 47–52 and anal scale single. Scales smooth and shiny in appearance.

RANGE Encountered in the Furneaux group of islands that includes the Flinders and Chappell Islands.

COMMENTS Diurnal. Terrestrial. Lives in swamps, grassland, heaths and scrubland. Shelters inside animal burrows, and beneath rocks, fallen timber and man-made debris. **Diet** Muttonbird chicks as adults; juveniles consume small lizards and frogs. **Reproduction** Viviparous. 6– 38 per litter. Neonates approximately 27cm TL born in October–March. **Disposition** Generally inoffensive, but bites if harassed.

BITE/VENOM DANGEROUSLY VENOMOUS

IUCN LISTING Least Concern.

Chappell Island, TAS, Scott Eipper

Chappell Island, TAS, Scott Eipper

Chappell Island, TAS, Shane Black

Flinders Island, TAS, Adam Elliott

Taipans, genus *Oxyuranus* Kinghorn, 1923

This genus currently contains three species, two of which are endemic to Australia. The retention of the Papuan subspecies is tentative, as it is distinguished on colouration alone. There is no evidence of genetic support, however, the last paper reviewing the systematics of the genus did not synonymize the Papuan subspecies. This genus exhibits marked seasonal colour change, becoming darker in winter and lighter in summer – a trait that is most prominent in arid-dwelling species. The snakes are active hunters with excellent vision, found in deserts, grassland and open forests. They are primarily diurnal but occasionally active after dark, usually in hot weather, and are oviparous. Male combat has been recorded in the genus. Taipans are among the world's most dangerous snakes, with extremely toxic venom, long fangs and a large venom yield. **VENOM** Strongly neurotoxic and with prothrombin activity, as well as myotoxic and weakly haemolytic. Effects from bites can be very rapid, with humans being significantly affected within minutes of a bite. Bites have resulted in, and can cause, fatalities. Taipan or Polyvalent antivenom is used to treat envenomations.
Species-level identification difficulty (within Australia) – 2.

TYPE SPECIES *Oxyuranus maclennani*.

ETYMOLOGY Sharp tail.

Key to *Oxyuranus*

1. Scales on neck smooth 2
   Scales on neck keeled 3

2. 21 midbody scale rows; 1 temporal scale
   between parietal and fifth supralabial scale .... *O. temporalis* (p. 174)
   23 midbody scale rows; 2 temporal scales
   between parietal and fifth supralabial scale .... *O. microlepidotus* (p. 170)

3. Found in NG, Torres Strait Islands and the tip of
   Cape York Peninsula; usually with a reddish-orange
   vertebral stripe that broadens posteriorly .... *O. scutellatus canni* (p. 171)
   Found in Australia; sometimes with reddish-
   orange vertebral stripe .... *O. s. scutellatus* (p. 172)

Oxyuranus scutellatus canni, Tie Eipper

Oxyuranus temporalis, Wannan, WA, Vik Dunis

INLAND TAIPAN *Oxyuranus microlepidotus* (McCoy, 1879)
(Fierce Snake, Small-scaled Snake, Western Taipan)

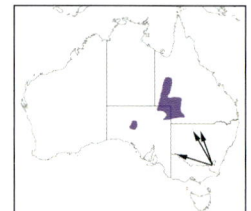

PRONUNCIATION *Oxy-u-ran-nus my-crow-lepi-dot-tus.*
ETYMOLOGY Small-scaled sharp-tail.
TYPE LOCALITY Junction of Murray and Darling Rivers, Vic.
APPEARANCE Large to very large, moderately robust snake with head slightly distinct from body. Variable; usually yellow to chocolate-brown with black reticulations. Head usually glossy black. Pupil round; iris black-brown. Mouth lining and tongue dark in colour. Ventral colouration bright yellow with dark flecks. Adult males larger than females, reaching 210cm TL. **Scalation** MB 23 rows, 220–251 VENT, divided SUB 55–70 and anal scale single. Scales smooth and glossy in appearance.
RANGE Encountered in far western QLD, NE SA, SW corner of the NT, and into western NSW. Historical records are from the junction of the Darling and Murray Rivers, but the species has not been seen there for many years.
COMMENTS Diurnal but can be crepuscular in hot weather. Inhabits black soil plains, gibber desert, grassland and savannah. Terrestrial. Found basking on edges of cracks in soil. Lives and hunts for small mammals within these cracks. This way of life has probably been the reason why it has evolved to be the world's most toxic snake to mice. **Diet** Small mammals. **Reproduction** Oviparous. 8–23 per clutch. Neonates approximately 47cm TL hatching in December–April.
Disposition Generally inoffensive, but readily bites if harassed.
BITE/VENOM DANGEROUSLY VENOMOUS
IUCN LISTING Least Concern.

Goyder Lagoon, SA, Scott Eipper

Morney Plain, QLD, Shane Black

Goyder Lagoon, SA, Scott Eipper

Goyder Lagoon, SA, Scott Eipper

PAPUAN TAIPAN *Oxyuranus scutellatus canni* (Slater, 1956)

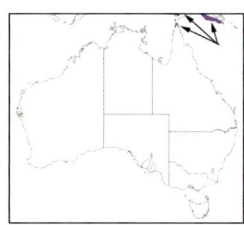

PRONUNCIATION *Oxy-u-ran-nus scoo-tell-ah-tuss can-ee.*
ETYMOLOGY Cann's small-shielded sharp-tail; for George Cann, Senior Curator of reptiles, Taronga Zoo.
TYPE LOCALITY Napa Napa, Port Moresby, PNG.
APPEARANCE Large to exceptionally large, moderately robust snake with head distinct from body. Very variable in colour. Top of body dark brown, grey or black, usually with reddish-orange stripe along spine that widens towards rear. Head usually lighter, often cream to white. Pupil round; iris red. Mouth lining pink; dark tongue. Ventral colouration yellow to orange, with or without red flecking. Adult males larger than females, reaching 240cm TL (exceptions up to 300cm).
Scalation MB 21–23 rows, 220–250 VENT, divided SUB 45–80 and anal scale single. Scales smooth and glossy in appearance. **Similar subspecies** *O. s. scutellatus* (p. 172).

RANGE Encountered on islands in the Torres Strait and PNG. Also reported from far north QLD, on the Cape York Peninsula. Possibly just a colour variation of the Coastal taipan (p. 172) as there are no morphological or genetic characteristics splitting the two 'subspecies.'

COMMENTS Usually diurnal but becomes nocturnal during warm weather. Terrestrial. Lives in dry forests, grassland, savannah and open woodland. Shelters inside animal burrows, and beneath rocks, fallen timber and man-made debris. **Diet** Small mammals, occasionally birds. **Reproduction** Oviparous. 9–17 per clutch. Neonates approximately 43cm TL hatching in November–May. **Disposition** Very alert, nervous and shy unless provoked. When threatened or cornered, will defend itself with vigour, unlike almost any other snake.

BITE/VENOM DANGEROUSLY VENOMOUS
IUCN LISTING Least Concern.

PNG, Scott Eipper

Captive bred, Scott Eipper

Captive bred, Scott Eipper

Captive bred, Scott Eipper

COASTAL TAIPAN *Oxyuranus scutellatus scutellatus* (Peters, 1867)

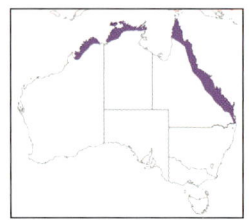

PRONUNCIATION *Oxy-u-ran-nus scoo-tell-ah-tuss scoo-tell-ah-tuss*.
ETYMOLOGY Small-shielded sharp-tail.
TYPE LOCALITY Rockhampton, QLD.
APPEARANCE Large to exceptionally large, moderately robust snake with head distinct from body. Very variable in colour. Top of body can be any shade of brown, grey, black or yellow; some individuals have reddish-orange stripe along spine that widens towards rear. Head usually lighter, often cream to white. Pupil round; iris red. Tongue dark; mouth lining pink. Ventral colouration yellow to orange, with or without red flecking. Adult males larger than females, reaching 240cm TL (exceptions up to 300cm). **Scalation** MB 21–23 rows, 220–250 VENT, divided SUB 45–80 and anal scale single. Scales smooth and glossy in appearance. **Similar subspecies** *O. s. canni* (p. 171).

RANGE Encountered from the QLD/NSW border region, across the northern coastline into the Kimberley region, WA.

COMMENTS Usually diurnal but becomes nocturnal during warm weather. Terrestrial. Lives in dry forests, grassland, savannah, open woodland, heaths and rainforest verges. Shelters inside animal burrows, and beneath rocks, fallen timber and man-made debris. Very alert and seemingly quite uncommon in southern part of its range. **Diet** Small mammals, rarely birds. **Reproduction** Oviparous. 5–22 per clutch. Neonates approximately 46cm TL hatching in November–May. **Disposition** Nervous and inoffensive unless threatened or cornered. Can defend itself with vigour, unlike almost any other snake.

BITE/VENOM DANGEROUSLY VENOMOUS
IUCN LISTING Least Concern.

Cairns, QLD, Shane Black

Cooktown, QLD, Scott Eipper

ELAPIDS 173

Coen, QLD, Shane Black

Innisfail, QLD, Shane Black

Tully, QLD, Scott Eipper

Airlie Beach, QLD, Scott Eipper

Cooktown, QLD, Scott Eipper

Cooktown, QLD, Scott Eipper

Weipa, QLD, Hal Cogger

Mt Molloy, QLD, Shane Black

WESTERN DESERT TAIPAN *Oxyuranus temporalis*
Doughty, Maryan, Donnellan & Hutchinson, 2007
(Central Ranges Taipan)

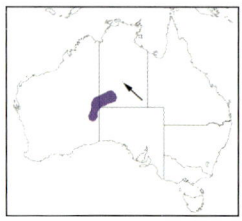

PRONUNCIATION *Oxy-u-ran-nus temp-pur-rah-liss.*

ETYMOLOGY Temporal sharp-tail, in reference to temporal scales.

TYPE LOCALITY East of the Walter James Range, WA.

APPEARANCE Large, moderately robust snake with head slightly distinct from body. Very variable in colour. From the few specimens known, yellow to dark brown. The nape often has white or black markings. Pupil round; iris black-brown. Tongue dark; mouth lining pink;. Ventral colouration yellow to cream, with or without orange flecking. Adult males larger than females, reaching 200cm TL. **Scalation** MB 21 rows, 240–252 VENT, divided SUB 56–61 and anal scale single. Scales smooth and glossy in appearance.

RANGE Encountered in far eastern WA, from Illurlka and the Walter James Range, across to near Kings Canyon, the George Gill Range and as far North as Ti Tree and the southern Tanami region, NT.

COMMENTS Diurnal. Terrestrial. Lives in deserts and mallee regions with sand ridges, gravel beds, spinifex and shrubs. Shelter sites used are unknown, but thought to be similar to those of other taipans. **Diet** Small mammals. **Reproduction** Oviparous. No published data on clutch size. **Disposition** Generally inoffensive, but readily bites if harassed.

BITE/VENOM DANGEROUSLY VENOMOUS

IUCN LISTING Least Concern.

Mt Liebig, NT, David Berger

30km west of Ilkurlka, WA, Brian Bush

Warburton, WA, Scott Kickham

30km W Ilkurlka, WA, Brian Bush

Genus *Paroplocephalus* Keogh, 2000

Monotypic, endemic genus. Phylogenetic assessment shows that *Paroplocephalus* is closely related to *Hoplocephalus* and possibly synonymous. Predominantly nocturnal, but will bask cryptically. Usually found beneath rocks, or while hunting on trees. Threatened due to habitat destruction. These snakes are lizard feeders that will occasionally eat frogs and small mammals, and are viviparous. **VENOM** Contains procoagulants. Within 30 minutes, an envenomation caused severe headache, excessive sweating and vomiting. Polyvalent antivenom has been used to treat severe envenomations.
Species-level identification difficulty (within Australia) – 2.
TYPE SPECIES *Brachyaspis atriceps*.
ETYMOLOGY Pertains to the similarity with *Hoplocephalus*.

Paroplocephalus atriceps, East of Forrestania, WA, Brian Bush

Paroplocephalus atriceps, Lake Cronin, WA, Scott Eipper

Lake Cronin Snake *Paroplocephalus atriceps* (Storr, 1980)
(Black-headed Bardick)

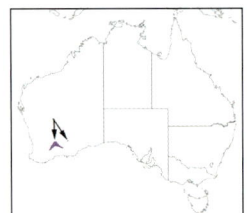

PRONUNCIATION *Pa-ro-plo-keff-a-luss at-ree-ceps.*
ETYMOLOGY Black-headed.
TYPE LOCALITY Lake Cronin, WA.
APPEARANCE Medium-sized, muscular snake with head strongly distinct from body. Dorsal colouration silver-grey to brown, and head jet-black. Lips barred with white. Pupil round; iris reddish-orange. Tongue dark; mouth lining pink. Ventral colouration grey. Adult females larger than males, reaching 75cm TL. **Scalation** MB 17–19 rows, 175–185 VENT, SUB 45–50 and anal scale single. Scales smooth and glossy.
RANGE Encountered around Lake Cronin, WA.
COMMENTS Nocturnal. Terrestrial to arboreal. Lives in dry forests, granite outcrops and adjoining habitats. Found sitting exposed on trunks of large salmon gums or crossing roads. Shelters under cover such as rocks and loose bark, and inside tree hollows. **Diet** Lizards, frogs, small mammals. **Reproduction** Viviparous; litter size expected to be similar to *Hoplocephalus*. **Disposition** Generally inoffensive, but readily bites if harassed.
BITE/VENOM DANGEROUSLY VENOMOUS
IUCN LISTING Least Concern.

Lake Cronin, WA, Brian Bush

Lake Cronin, WA, Scott Eipper

Black & Mulga Snakes, genus *Pseudechis* Wagler, 1830

This genus currently contains nine accepted species, with eight found in Australia, and seven endemic to Australia. Phylogenetic relationships within the genus suggest that some unresolved species complexes require further research to determine their identity. Some species are primarily diurnal but occasionally active after dark, usually in hot weather, while others are nocturnal. Found across almost all terrestrial habitats, with at least one species present across almost all of mainland Australia. Some species are very well known, occurring in urban environments. Male combat has been recorded in the genus. All species are oviparous, except one. When threatened, they flare the neck, looking larger and more dangerous. **VENOM** Contains haemolysins and cytotoxins, with mild neurotoxicity and myotoxicity. Envenomations cause significant pain, swelling, nausea and vomiting, and can lead to loss of consciousness. Localized effects including necrosis can occur, including the loss of fingers and in severe cases limbs. Bites have resulted in, or could cause fatalities. Black snake or Polyvalent antivenom is used to treat envenomations. **Species-level identification difficulty** (within Australia) – 5.

TYPE SPECIES *Coluber porphyricus*.
ETYMOLOGY False adder.

Key to *Pseudechis*

1. Midbody scales in 19–21 rows 2
   Midbody scales in 17 rows 4

2. Ventral scales 205–240, if 205 restricted to
   PNG and Torres Strait islands 3
   Ventral scales 175–205, not found in PNG or
   Torres Strait Islands .... *P. guttatus* (p. 182)

3. Midbody scales in 19 rows; with irregular cross-bands .... *P. colletti* (p. 181)
   Midbody scales in 19 (rarely 21) rows; no irregular
   cross-bands; restricted to PNG and Torres Strait Islands .... *P. papuanus* (p. 184)

4. Body completely black above with lowest lateral
   midbody scales with white, pink or crimson; ventral
   scales have a black posterior edge .... *P. porphyriacus* (p. 185)
   Any shade of yellow-brown to black above; lowest
   lateral midbody scales are not white, pink or crimson 5

5. Dark grey to black above with irregular patterns
   formed by pale spots within each patterned scale .... *P. butleri* (p. 180)
   Any shade of yellow-brown to black without an irregular
   pattern formed by pale spots within each patterned scale
   or if a pattern is present, it is formed by dark pigment
   on posterior edge of the scale 6

6. Rostral scale pointed at the centre point between internasal
   scales; subcaudal scales undivided, anal scale entire; usually
   marked with purple-brown streaks on nape 7
   Rostral scale rounded at the centre point between
   internasal scales; subcaudal scales undivided but usually
   with at least some divided; anal scale divided; usually
   lacks purple-brown streaks on nape .... *P. australis* (p. 178)

9. Purple-brown streaks restricted to nape and head scalation;
   remaining body unpatterned; found east of Borroloola .... *P. pailsei* (p. 183)
   Purple-brown markings usually prominent on head
   and nape, sometimes extending on first anterior
   quarter of body; some individuals partially reticulated;
   found west of Borroloola .... *P. weigeli* (p. 186)

Mulga Snake *Pseudechis australis* (Gray, 1842)
(King Brown Snake)

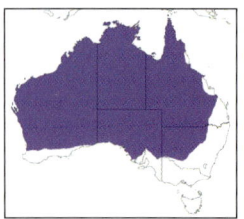

PRONUNCIATION *Sue-deck-iss os-trah-liss*.

ETYMOLOGY Southern false adder.

TYPE LOCALITY Port Essington, NT.

APPEARANCE Large to very large, robust elapid with head slightly distinct from body. Colouration very variable, from black to pale yellow, to reddish-purple above. Southern populations darker than northern ones. Some individuals have a variegated appearance, with anterior edge of scale lighter than rear. Pupil round; iris red to coppery-brown. Tongue dark; mouth lining pink. Ventral colouration yellow to cream, without orange flecking. In some individuals, underside of tail is pale orange. Adult males larger than females, reaching 305cm TL. **Scalation** MB 17 rows, 185–225 VENT, SUB 50–80 usually single with last few divided and anal scale divided. Scales smooth and glossy in appearance. **Similar species** *P. butleri* (p. 180), *P. pailsei* (p. 183), *P. weigeli* (p. 186).

RANGE Encountered all over Australia's drier regions throughout almost all of WA, all of the NT, most of SA, western NSW and QLD, west of the Great Dividing Range.

COMMENTS Diurnal to nocturnal; temperature dependent. Terrestrial. Encountered in most habitats in Australia, except closed forests and rainforests. Shelters under rocks, logs and ground debris, including man-made rubbish, rock crevices and deep cracks in the soil. **Diet** Geckos, legless lizards, dragons, skinks, small monitors, elapids, colubrids, pythons, frogs, birds, mammals, roadkill. **Reproduction** Oviparous. 4–23 per clutch. Neonates approximately 29cm TL hatching in December–March. **Disposition** Generally inoffensive, but bites if harassed.

BITE/VENOM DANGEROUSLY VENOMOUS

IUCN LISTING Least Concern.

St George, QLD, Jesse Campbell

Lockhart River, QLD, Shane Black

Cloncurry, QLD, Scott Eipper

Ilukurlka, WA, Scott Kickham

Kununurra, WA, Tie Eipper

Kununurra, WA, Scott Eipper

20km west of St George, QLD, Scott Eipper

Exmouth, WA, Adam Elliott

Alice Springs, NT, Scott Eipper

Renmark, SA, Scott Eipper

Spotted Mulga Snake *Pseudechis butleri* L. A. Smith, 1982
(Butler's Snake)

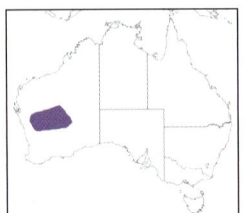

PRONUNCIATION *Sue-deck-iss butt-ler-e.*

ETYMOLOGY Butler's false adder; pertains to W. H. Butler, Australian naturalist.

TYPE LOCALITY 19km SE of Yalgoo, WA.

APPEARANCE Large, moderately robust snake with head slightly distinct from body. Dorsal colouration black to dark brown, with cream to yellow spots. Juveniles grey with shiny black head. Pupil round; iris grey to dark brown. Tongue dark; mouth lining pink. Ventral colouration yellow, and ventral scales can have black spots and black edges. Adult males larger than females, reaching 175cm TL. **Scalation** MB 17 rows, 200–225 VENT, SUB 50–70 usually single with last few divided and anal scale divided. Scales smooth and glossy in appearance. **Similar species** *P. australis* (p. 178).

RANGE Encountered in the goldfields region of southern inland WA.

COMMENTS Diurnal to nocturnal, depending on the temperature. Terrestrial. Found in mulga and acacia woodland, mallee and agricultural land. Shelters in logs, and under rocks and man-made debris. **Diet** Lizards, other snakes, small mammals, frogs, birds, bird and reptile eggs. **Reproduction** Oviparous. 7–17 per clutch. Neonates approximately 27.5cm TL hatching in January–March. **Disposition** Generally inoffensive, but bites if harassed.

BITE/VENOM DANGEROUSLY VENOMOUS

IUCN LISTING Least Concern.

Juvenile, captive bred, WA, Scott Eipper

Yalgoo, WA, Scott Eipper

Paynes Find, WA, Jake Meney

Juvenile, captive bred, WA, Scott Eipper

Collett's Snake *Pseudechis colletti* Boulenger, 1902
(Downs' Tiger Snake)

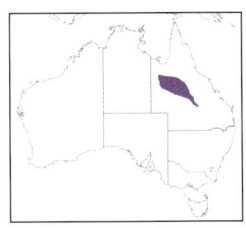

PRONUNCIATION *Sue-deck-iss col-let-e.*

ETYMOLOGY Collett's false adder; pertaining to R. Collett, a director of the Oslo Museum.

TYPE LOCALITY QLD.

APPEARANCE Large to very large, robust elapid with head slightly distinct from body. Dorsal colouration pale brown to black, with yellow, orange and pink cross-bands and spotting. Juveniles much brighter than adults, with orange and jet-black cross-bands. Pupil round; iris reddish-brown. Tongue dark; mouth lining pink. Ventral colouration orange, with or without dark brown to black flecking. Adult males larger than females, reaching 200cm TL. **Scalation** MB 19 rows, 215–235 VENT, SUB 45–65 usually single, last few divided and anal scale divided. Scales smooth and glossy in appearance.

RANGE Encountered in central western QLD.

COMMENTS Diurnal to nocturnal, depending on the temperature. Terrestrial. Lives on black soil plains and grassland. Often shelters in burrows, beneath man-made debris, under logs and rocks, and in crevices formed in cracking soils. Possibly synonymous with *P. guttatus*. **Diet** Lizards, frogs, snakes, small mammals. **Reproduction** Oviparous. 7–18 per clutch. Neonates approximately 37cm TL hatching in November–April. **Disposition** Generally inoffensive, but bites if harassed.

BITE/VENOM DANGEROUSLY VENOMOUS

IUCN LISTING Least Concern.

Captive bred, Tie Eipper

Juvenile, captive bred, Adam Elliott

Longreach, QLD, Scott Eipper

50km E of winton, QLD, Scott Eipper

Spotted Black Snake *Pseudechis guttatus* De Vis, 1905
(Blue-bellied Black Snake)

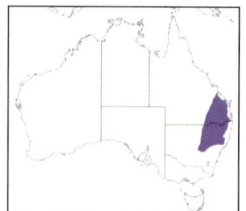

PRONUNCIATION *Sue-deck-iss goo-tah-tus.*
ETYMOLOGY Spotted false adder.
TYPE LOCALITY Cecil Plains, QLD.
APPEARANCE Large, robust snake with head slightly distinct from body. Colouration very variable. Can range from jet-black to grey, and sometimes light-toned brown. Several specimens are flecked with yellow, red, orange or cream. Pupil round; iris blackish-brown. Tongue dark; mouth lining pink. Ventral colouration grey to yellowish, with darker flecking. Adult males larger than females, reaching 180cm TL. **Scalation** MB 19 rows, 175–205 VENT, SUB 45–65 usually single, with last few divided and anal scale divided. Scales smooth and glossy in appearance.

RANGE Encountered in NSW and QLD, almost exclusively west of the GDR. In places such as west Brisbane, it reaches over the foothills into the black soil valleys, making it through the range.

COMMENTS Diurnal to nocturnal; temperature dependent. Terrestrial. Found in dry woodland, savannah, grassland, brigalow and agricultural land. Shelters in burrows, beneath man-made debris, under logs and rocks, and in cracks in the soil. Possibly synonymous with *P. colletti*. **Diet** Small mammals, frogs, lizards, other snakes. **Reproduction** Oviparous. 5–16 per clutch. Neonates approximately 28cm TL hatching in November–April. **Disposition** Generally inoffensive, but bites if harassed.

BITE/VENOM DANGEROUSLY VENOMOUS
IUCN LISTING Least Concern.

Glenmorgan, QLD, Scott Eipper

Gundy, NSW, Scott Eipper

Gundy, NSW, Scott Eipper

Tara, QLD, Scott Eipper

Eastern Pygmy Mulga Snake *Pseudechis pailsei* (Hoser, 1998)

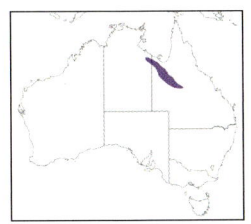

PRONUNCIATION *Sue-deck-iss pail-see-e.*

ETYMOLOGY Pails' false adder; pertaining to R. Pails, who first noticed the difference from other similar snakes.

TYPE LOCALITY East Leichhardt Dam, near Mt Isa, QLD.

APPEARANCE Large, moderately robust snake with head slightly distinct from body. Dorsal colouration pale yellow to golden brown, usually with magenta markings on nape. Narrow head thought to be adapted for squeezing into crevices. Pupil round; iris reddish-brown. Tongue dark; mouth lining pink. Ventral colouration yellow to cream. Adult males larger than females, reaching 120cm TL. **Scalation** MB 17 rows, 210–235 VENT, SUB 50–80 and anal scale single. Scales smooth and glossy in appearance. **Similar species** *P. australis* (p. 178), *P. weigeli* (p. 186).

RANGE Encountered along the Selwyn range in western QLD, from Winton to Hell's Gate. Probably occurs in the eastern NT, although there are no individuals currently known, despite habitat being continuous through from Hell's Gate to Nicholson.

COMMENTS Nocturnal but has been found crossing roads in the morning. Terrestrial. Lives in dry woodland, rocky gorges and hillsides. Shelters in rock crevices, tree hollows, beneath debris, or under logs and rocks. **Diet** Lizards, frogs, occasionally small mammals. **Reproduction** Oviparous. 5–11 per clutch. Neonates seen in March–April. **Disposition** Generally inoffensive, but readily bites if harassed.

BITE/VENOM DANGEROUSLY VENOMOUS

IUCN LISTING Least Concern.

Adel Grove, QLD, Scott Eipper

Adel Grove, QLD, Scott Eipper

Mt Isa, QLD, Ryan Francis

Juvenile, Mt Isa, QLD, Ryan Francis

Papuan Black Snake *Pseudechis papuanus* W. C. H. Peters & Doria, 1878

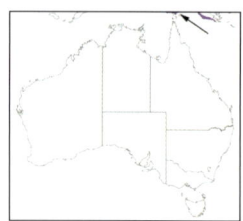

PRONUNCIATION *Sue-deck-iss pap-u-an-nus.*
ETYMOLOGY Papuan false adder.
TYPE LOCALITY Between Epa and Yule Island, PNG.
APPEARANCE Large to very large, robust snake with head slightly distinct from body. Dorsal colouration black to dark grey, occasionally flushed with dark red on some individuals. Pupil round; iris grey to dark brown. Tongue dark; mouth lining pink. Ventral colouration dark grey. Adult males larger than females, reaching 210cm TL. **Scalation** MB 19 (rarely 21) rows, 205–230 VENT, SUB 49–63 usually single with last few divided and anal scale divided. Scales smooth and glossy in appearance.
RANGE Encountered on Saibai and Boigu islands in the northern Torres Strait. Also occurs throughout southern PNG.
COMMENTS Diurnal, and occasionally nocturnal in hot weather. Terrestrial. Lives in swamps, forest edges and moist grassland. Shelters in burrows, and beneath debris, logs and rocks. Declining over much of its range due to habitat destruction and introduction of Cane toad. **Diet** Frogs, lizards, small mammals, other snakes. **Reproduction** Oviparous. 7–18 per clutch. Neonate information unknown. **Disposition** Generally inoffensive, but readily bites if harassed.
BITE/VENOM DANGEROUSLY VENOMOUS
IUCN LISTING Least Concern.

Saibai Island, QLD, David Williams

Central Province, PNG, Mark O'Shea

Saibai Island, QLD, David Williams

Western Province, PNG, Mark O'Shea

Red-bellied Black Snake *Pseudechis porphyriacus* (Shaw, 1794)
(Common Black Snake, Black Snake)

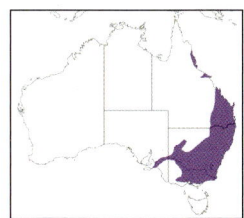

PRONUNCIATION *Sue-deck-iss poor-fee-ree-ah-cus*.

ETYMOLOGY Purple false adder.

Type locality Unknown, but presumably Botany Bay, NSW.

APPEARANCE Large to very large, robust snake with head slightly distinct from body. Dorsal colouration jet-black, with orange, red or maroon markings along lower flanks. Some individuals marked with white. Head, in particular around snout, can be brown in some locations. Pupil round; iris black-brown. Tongue dark; mouth lining pink. Ventral colouration duller compared to top, red to pinkish and marked with black bands. Tail black underneath. Adult males larger than females, reaching 200cm TL. **Scalation** MB 17 rows, 175–215 VENT, SUB 40–65 usually single, with last few divided and anal scale divided. Scales smooth and glossy in appearance. **Similar species** *Cryptophis nigrescens* (p. 112).

RANGE Encountered from the Adelaide Hills, SA, into southern QLD. Predominantly found along the GDR and south into Gippsland, Vic. However, the species can penetrate quite far inland along river systems. Three separate populations exist in QLD, one in the SE, a second further north around Mackay to Proserpine, and a third from Townsville to Cooktown.

COMMENTS Diurnal; occasionally nocturnal during hot weather. Terrestrial. Inhabits swamps, creeks, forests, rainforests, urban environments and grassland, particularly around waterbodies. Shelters in burrows, or under logs, debris and rocks. Gravid females sometimes bask communally. **Diet** Frogs, tadpoles, fish, dragons, skinks, legless lizards, geckos, mammals, other elapids, colubrids, blind snakes. **Reproduction** Viviparous. 5–23 per litter. Neonates approximately 25cm TL born in January–April. **Disposition** Generally inoffensive, but bites if harassed.

BITE/VENOM DANGEROUSLY VENOMOUS

IUCN LISTING Least Concern.

Ryde, NSW, Tie Eipper

Dalby, QLD, Scott Eipper

Maddens Plains, NSW, Scott Eipper

Paluma, QLD, Scott Eipper

Western Pygmy Mulga Snake *Pseudechis weigeli* (Wells & Wellington, 1987)

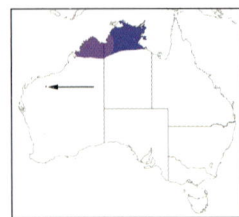

PRONUNCIATION *Sue-deck-iss why-gell-e.*

ETYMOLOGY Weigel's false adder; pertaining to J. Weigel, collector of the holotype.

TYPE LOCALITY Mitchell River, WA.

APPEARANCE Large, moderately built snake with head slightly distinct from body. Dorsal colouration pale yellow to golden-brown or grey, with magenta markings on nape. These darker reticulations can extend all the way down the body. They are usually more prominent in western population. Pupil round; iris reddish-brown. Tongue dark; mouth lining pink. Ventral colouration yellow to cream. Adult males larger than females, reaching 120cm TL. **Scalation** MB 17 rows, 210–230 VENT, SUB 50–80 single and anal scale single. Scales glossy in appearance and smooth. **Similar species** *P. australis* (p. 178) and *P. pailsei* (p. 183).

RANGE Encountered in the Kimberley region of WA and the top end of the NT. Disjunct population in the Pilbara region of WA is almost certainly a new, as yet undescribed species. Distribution map reflects both taxa individually with *P. weigeli* in purple and *P.* cf. *weigeli* in blue

COMMENTS Nocturnal. Terrestrial. Lives in dry woodland, grassland, rocky gorges and hillsides. It is probable that three species are currently included within the Western pygmy mulga snake. Two are genetically distinct but require further research to determine the validity of the taxa. **Diet** Lizards, frogs, periodically small mammals. **Reproduction** Oviparous. 7–12 per clutch. Neonates approximately 21cm TL hatching in February–April. **Disposition** Generally inoffensive, but readily bites if harassed.

BITE/VENOM DANGEROUSLY VENOMOUS

IUCN LISTING Least Concern.

Kununurra, WA, Scott Eipper

Kununurra, WA, Scott Eipper

P. cf. weigeli, Katherine, NT, Scott Eipper

P. cf. weigeli, Hayes Creek, NT, Rob Valentic

Brown Snakes, genus *Pseudonaja* Günther, 1858

This genus currently contains eight species, all found in Australia. Seven are endemic to Australia. Phylogenetic relationships within the genus suggest that some unresolved species complexes require further research to determine their identity. The insular subspecies of *P. affinis* may be nothing more than localized variants, and may prove to be synonymous with the nominate species. The snakes are primarily diurnal but occasionally active after dark, usually in hot weather. They are found across most dry terrestrial habitats, with at least one species present across almost all of mainland Australia. Some species are very well known, occurring in urban environments. Male combat has been recorded in the genus. All species are oviparous. When threatened they raise the forebody, showing the bright colouration beneath. **VENOM** Brown snakes have killed more people than any other Australian snake genus. Venom contains strong procoagulants, and neurotoxins; weakly haemolytic and possibly nephrotoxic. Envenomations from both adults and juveniles can lead to rapid collapse and death. Brown snake or Polyvalent antivenom is used to treat envenomations. **Species-level identification difficulty** (within Australia) – 4.
TYPE SPECIES: *Pseudonaja nuchalis*.
ETYMOLOGY False *Naja*, after the Cobra genus *Naja*.

Key to Pseudonaja

1. More than 178 ventrals 2
   Less than 176 ventrals .... *P. modesta* (p. 198)

2. 7 infralabial scales 3
   6 infralabial scales 4

3. Midbody scales in 17 rows .... *P. ingrami* (p. 195)
   Midbody scales in 19–21 rows .... *P. guttata* (p. 193)

4. Lower edge of the postocular scale is lower
   than the level of the eye 7
   Lower edge of the postocular scale is
   higher than or equal to the level of the eye 5

5. Midbody scales in 19 rows; found on Rottnest,
   Boxer and Figure Eight Islands, WA 6
   Midbody scales in 17–19 (rarely 21) rows; restricted to mainland .... *P. affinis affinis* (p. 188)

6. No suture between the parietal and lower postocular;
   found on Rottnest Island, WA .... *P. a. exilis* (p. 190)
   Suture between the parietal and lower postocular;
   found on Boxer and Figure Eight Islands, WA .... *P. a. tanneri* (p. 191)

7. Buccal cavity (mouth lining) pink 8
   Buccal cavity (mouth lining) dark 9

8. Ventral surface grey without flecks .... *P. inframacula* (p. 194)
   Ventral surface usually pale cream or yellowish with
   yellow, orange or brown flecks or dark posterior margins on
   each ventral scale, rarely dark grey-brown with pale flecks .... *P. textilis* (p. 200)

9. Snout chisel-shaped when viewed from above 10
   Snout rounded when viewed from above .... *P. mengdeni* (p. 196)

10. Tail length equal to or greater than 15%
    of body length; head somewhat rounded in profile .... *P. nuchalis* (p. 199)
    Tail length less than 15% of body length;
    head somewhat angular in profile .... *P. aspidorhyncha* (p. 192)

DUGITE *Pseudonaja affinis affinis* Günther, 1872
(Spotted Brown Snake)

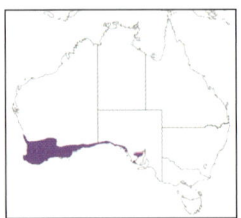

PRONUNCIATION *Sue-doh-neigh-ah af-fin-iss af-fin-iss.*
ETYMOLOGY Related false cobra.
TYPE LOCALITY Australia.
APPEARANCE Large to very large, moderately slender snake with head indistinct from body. Colouration very variable, from pale cream to almost black. Usually a shade of brown with random dark spots that sometimes coalesce into blotches. Head can be darker or lighter than body. Juvenile and immature individuals have a black head and black band on nape. Pupil round; iris copper. Tongue dark; mouth lining pink. Ventral colouration yellow, to cream, to light brown, with orange-red spots and blotches. Adult males larger than females, reaching 210cm TL. **Scalation** MB 19–(rarely) 21 rows, 190–230 VENT, SUB 50–70 and anal scale both divided. Eastern animals have 17 MB rows. Scales slightly glossy in appearance and smooth. **Similar species** *P. inframacula* (p. 194), *P. mengdeni* (p. 196).

RANGE Encountered in southern WA, across the southern coast to the western edge of the Eyre Peninsula, SA.

COMMENTS Usually diurnal but active at night during hot weather. Terrestrial. Lives in open woodland, coastal dune associations, heathland, grassland and urban areas. Shelters among vegetation, beneath debris, or under logs and rocks. **Diet** Small mammals and reptiles as adults, but juveniles are almost exclusive reptile predators. **Reproduction** Oviparous. 3–31 per clutch. Neonates approximately 25cm TL hatching in February–March. **Disposition** Nervous, and inoffensive unless threatened. Will readily bite if harassed.

BITE/VENOM DANGEROUSLY VENOMOUS
IUCN LISTING Least Concern.

Perth region, WA, Scott Eipper

Stoneville, WA, Brian Bush

Juvenile, Perth, WA, Scott Eipper

Lancelin, WA, Scott Eipper

Kallaroo, WA, Matt Summerville

Kings Park, WA, Scott Eipper

Perth region, WA, Scott Eipper

ROTTNEST ISLAND DUGITE *Pseudonaja affinis exilis* Storr, 1989

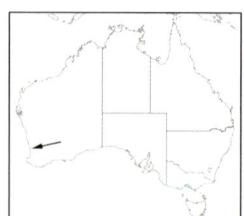

PRONUNCIATION *Sue-doh-neigh-ah af-fin-iss x-i-liss.*
ETYMOLOGY Thin-related false cobra.
TYPE LOCALITY Rottnest Island, WA.
APPEARANCE Large, moderately slender snake with head indistinct from body. Colouration dark brown to black both above and below. Juvenile and immature individuals have a black head and black band on nape. Pupil round; iris copper. Tongue dark; mouth lining pink. Adult males larger than females, reaching 130cm TL. Possibly better regarded as a geographic variant than a subspecies. **Scalation** MB 19 rows, 207–219 VENT, SUB 48–57 and anal scale both divided. Scales slightly glossy in appearance and smooth.
RANGE Encountered only in Rottnest Island, WA.
COMMENTS Diurnal. Terrestrial. Lives in dry woodland, coastal dune associations and heathland. Shelters among vegetation, beneath debris, or under logs and rocks. **Diet** Small mammals and reptiles as adults, but juveniles are almost exclusive reptile predators. **Reproduction** Thought to be egg layers. **Disposition** Generally inoffensive, biting if harassed.
BITE/VENOM DANGEROUSLY VENOMOUS
IUCN LISTING Least Concern.

Rottnest Island, WA, Gary Stephenson

Rottnest Island, WA, Gary Stephenson

Tanner's Brown Snake *Pseudonaja affinis tanneri* (Worrell, 1961)

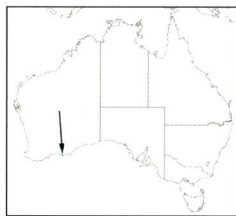

PRONUNCIATION *Sue-doh-neigh-ah af-fin-iss tan-er-ee.*
ETYMOLOGY Tanner's related false cobra; pertaining to C. Tanner, Australian herpetologist.
TYPE LOCALITY Boxer Island, WA.
APPEARANCE Large, moderately slender snake with head indistinct from body. Dorsal colouration chestnut-brown to dark brown. Juvenile and immature individuals have a black head and black band on nape. Pupil round; iris copper. Tongue dark; mouth lining pink. Ventral colouration lighter, with dark brown spotting. Adult males larger than females, reaching 120cm TL. Possibly better regarded as a geographic variant than a subspecies. **Scalation** MB 19 rows, 208–212 VENT, SUB 56–60 and anal scale both divided. Scales slightly glossy in appearance and smooth.

RANGE Encountered only in Boxer and Figure of Eight Islands in the Archipelago of Recherche off southern WA.

COMMENTS Diurnal. Terrestrial. Lives in dry woodland and heathland. Shelters among vegetation or beneath rocks. **Diet** Small mammals and reptiles as adults, but juveniles are almost exclusive reptile predators. **Reproduction** Oviparous. 12–15 per clutch. Neonates approximately 19cm TL hatching in February. **Disposition** Generally inoffensive, biting if harassed.

BITE/VENOM DANGEROUSLY VENOMOUS

IUCN LISTING Least Concern.

Boxer Island, WA, Rob Valentic

Figure Eight Island, WA, Brad Maryan

Figure Eight Island, WA, Brian Bush

STRAP-SNOUTED BROWN SNAKE *Pseudonaja aspidorhyncha* McCoy, 1879

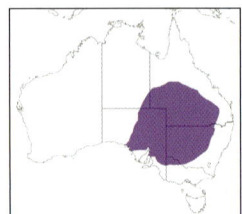

PRONUNCIATION *Sue-doh-neigh-ah as-pid-door-rink-car*.
ETYMOLOGY Shield-snouted false cobra.
TYPE LOCALITY Junction of Murray River and Darling River, Vic.
APPEARANCE Large, moderately slender snake with head indistinct from body. Colouration very variable, from pale cream to dark brown. Usually a shade of brown with or without dark bands. Almost always a black speck on nape. Juvenile and immature individuals have a black head and black band on nape. Pupil round; iris copper. Tongue dark; mouth lining blackish. Ventral colouration yellow with orange-red spots and blotches. Adult males larger than females, reaching 175cm TL. **Scalation** MB 17–19 rows, 200–230 VENT, SUB 45–70 and anal scale both divided. Scales slightly glossy in appearance and smooth. **Similar species** *P. inframacula* (p. 194), *P. mengdeni* (p. 196), *P. modesta* (p. 198), *P. nuchalis* (p. 199), *P. textilis* (p. 200).

RANGE Encountered in inland eastern Australia, including eastern SA, NW Vic, western NSW and southern west QLD.

COMMENTS Diurnal to nocturnal, depending on the temperature. Terrestrial. Lives in open woodland, brigalow, mallee and desert, and around agricultural areas. Shelters among vegetation, beneath debris, or under logs and rocks. **Diet** Small mammals and reptiles as adults; juveniles are almost exclusive reptile predators. **Reproduction** Oviparous. 9–14 per clutch. Neonates approximately 28cm TL hatching in December–March. **Disposition** Nervous, and inoffensive unless threatened. Will readily bite if harassed.

BITE/VENOM DANGEROUSLY VENOMOUS
IUCN LISTING Least Concern.

Marree, SA, Scott Kickham

Marree, SA, Shane Black

Gundabooka NP, NSW, Michael Payne

Bollon, QLD, Scott Eipper

Speckled Brown Snake *Pseudonaja guttata* (Parker, 1926)

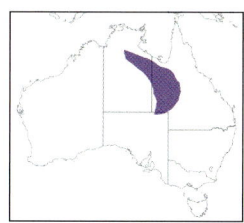

PRONUNCIATION *Sue-doh-neigh-ah goo-ta-tah.*
ETYMOLOGY Speckled false cobra.
TYPE LOCALITY Winton, QLD.
APPEARANCE Large, moderately slender snake with head indistinct from body. Colouration very variable – usually a shade of brown, yellow-cream or orange, with or without dark bands; many individuals flecked with black. Juveniles and immature individuals have black head and black band on nape. Pupil round; iris reddish-brown to yellow. Both tongue and mouth are dark. Ventral colouration yellow with orange-red spots and blotches. Adult males larger than females, reaching 120cm TL. **Scalation** MB 19–21 rows, 190–220 VENT, SUB 45–70 and anal scale both divided. Scales slightly glossy in appearance and smooth. **Similar species** *P. ingrami* (p. 195), *P. nuchalis* (p. 199), *P. textilis* (p. 200).
RANGE Encountered in inland northern QLD, NE SA and the NT.
COMMENTS Diurnal, and occasionally crepuscular in hot weather. Terrestrial. Lives in grassland and black soil plains. Shelters among vegetation, beneath debris and in soil cracks. **Diet** Frogs, lizards, occasionally small mammals; juveniles are almost exclusive reptile and frog predators. **Reproduction** Oviparous. 3–17 per clutch. Neonates 25cm TL hatching in February–April. **Disposition** Nervous, and inoffensive unless threatened. Will readily bite if harassed.

BITE/VENOM **DANGEROUSLY VENOMOUS**
IUCN LISTING Least Concern.

Kynuna, QLD, Anders Zimny

60km W Mica Creek, QLD, Scott Eipper

Barkly Tableland, NT, Shane Black

Cammoweal, QLD, Scott Eipper

Peninsula Brown Snake *Pseudonaja inframacula* (Waite, 1925)

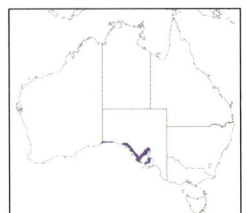

PRONUNCIATION *Sue-doh-neigh-ah in-fra-mack-u-la.*
ETYMOLOGY Below-spot false cobra.
TYPE LOCALITY Northern end of Coffin Bay, SA.
APPEARANCE Large, moderately slender snake with head indistinct from body. Colouration very variable, ranging from yellow-brown to purplish-black. Many specimens flecked with black, sometimes coalescing into blotches. Juveniles and immatures have black head and black band on nape. Pupil round; iris mainly blackish-brown with thin white rim edge. Tongue dark; inside of mouth pinkish. Ventral colouration grey. Adult males larger than females, reaching 160cm TL. **Scalation** MB 17–19 rows, 190–230 VENT, SUB 50–65 and anal scale both divided. Isolated population on the Nullarbor Plain of WA has 19 midbody rows. Scales slightly glossy in appearance and smooth. **Similar species** *P. affinis* (p. 188), *P. aspidorhyncha* (p. 192), *P. textilis* (p. 200).
RANGE Encountered in southern SA, on the Yorke and Eyre Peninsulas.
COMMENTS Diurnal. Terrestrial. Lives in open woodland, heaths, grassland, coastal dunes and around agricultural areas. Shelters among vegetation, beneath debris, in burrows, or under logs and rocks. **Diet** Small mammals, frogs and reptiles as adults; juveniles almost exclusively eat reptiles. **Reproduction** Oviparous. Up to 12 per clutch. Neonates 25cm TL hatching in January–March. **Disposition** Nervous, and inoffensive unless threatened. Will readily bite if harassed.
BITE/VENOM DANGEROUSLY VENOMOUS
IUCN LISTING Least Concern.

Head of Bight area, SA, Jules Farquhar

Juvenile, Yorke Peninsula, SA, Ryan Francis

Nullarbor, SA, Shawn Scott

Marion Bay, SA, Matt Summerville

INGRAM'S BROWN SNAKE *Pseudonaja ingrami* (Boulenger, 1908)

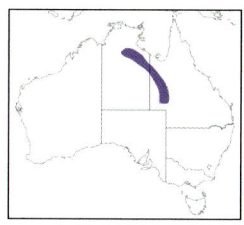

PRONUNCIATION *Sue-doh-neigh-ah in-gram-ee.*

ETYMOLOGY Ingram's false cobra; pertains to C. Ingram, English ornithologist.

TYPE LOCALITY Alexandria Station, NT.

APPEARANCE Large to very large, moderately slender snake with head indistinct from body. Colouration very variable, from pale yellow, orange, reddish-brown or dark brown, to black. Pupil round; iris dark orange-brown. Tongue dark; inside of mouth mainly black. Ventral colouration yellow with orange-red spots typically arranged in straight lines down forebody. Adult males larger than females, reaching 200cm TL. **Scalation** MB 17 rows, 190–220 VENT, SUB 55–70 and anal scale both divided. Scales slightly glossy in appearance and smooth. **Similar species** *P. guttata* (p. 193), *P. nuchalis* (p. 199), *P. textilis* (p. 200).

RANGE Encountered in inland northern QLD and the NT.

COMMENTS Diurnal. Terrestrial. Lives in grasslands and blacksoil plains. Seeks shelter in soil cracks, among vegetation, and beneath logs and rocks. **Diet** Small mammals and reptiles as adults; juveniles are almost exclusive reptile predators. **Reproduction** Oviparous. 5–18 per clutch. Neonates approximately 28cm TL hatching in January. **Disposition** Nervous, and inoffensive unless threatened. Will readily bite if harassed.

BITE/VENOM DANGEROUSLY VENOMOUS

IUCN LISTING Least Concern.

60km W Mica Creek, QLD, Scott Eipper

60km W Mica Creek, QLD, Scott Eipper

Boulia, QLD, Reid Newell

Juvenile, Boulia, QLD, Scott Eipper

Western Brown Snake *Pseudonaja mengdeni* Wells & Wellington, 1985

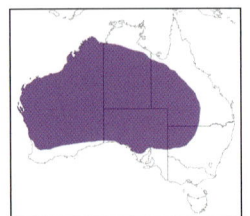

PRONUNCIATION *Sue-doh-neigh-ah meng-den-ee.*

ETYMOLOGY Mengden's false cobra; pertains to G. Mengden, American herpetologist.

TYPE LOCALITY 2km east of Maryvale, NT.

APPEARANCE Large, moderately slender snake with head indistinct from body. Colouration very variable, from pale cream to dark brown. Many colour morphs, including banded individuals, those marked with herringbone pattern, and bright orange ones with jet-black head and neck. Juveniles and immature individuals have black head and black band on nape; others are completely banded. Pupil round; iris copper. Tongue dark; mouth lining blackish. Ventral colouration yellow to creamish, with orange-red spots and blotches. Adult males larger than females, reaching 140cm TL. **Scalation** MB 17 rows, 190–220 VENT, SUB 55–70 and anal scale both divided. Scales slightly glossy in appearance and smooth. **Similar species** *P. affinis* (p. 188), *P. aspidorhyncha* (p. 192), *P. ingrami* (p. 195), *P. nuchalis* (p. 199), *P. textilis* (p. 200).

RANGE Encountered over much of arid and west Australia, including SA, western NSW, SW QLD, arid regions of the NT and most of WA, including the Kimberley region.

COMMENTS Diurnal but can also be nocturnal in warm conditions. Terrestrial. Lives in open woodland, mallee, grassland, desert and around agricultural areas. Shelters among vegetation, beneath man-made debris, or under logs and rocks. **Diet** Small mammals and reptiles as adults; juveniles are almost exclusive reptile predators. **Reproduction** Oviparous. 7–22 per clutch. Neonates approximately 23cm TL hatching in January–April. **Disposition** Nervous, and inoffensive unless threatened. Will readily bite if harassed.

BITE/VENOM DANGEROUSLY VENOMOUS

IUCN LISTING Least Concern.

Windarra, WA, Brian Bush

Alice Springs, NT, Scott Eipper

Port Hedland, WA, Brian Bush

Kumarina, WA, Brian Bush

Shay Gap, WA, Brian Bush

Kalgoolie, WA, Scott Eipper

Port Hedland, WA, Scott Eipper

Kings Canyon, WA, Jesse Campbell

Juvenile, Port Hedland, WA, Scott Eipper

Hamelin Station, WA, Adam Elliott

Ringed Brown Snake *Pseudonaja modesta* (Günther, 1872)
(Five-ringed Snake)

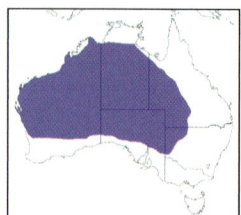

PRONUNCIATION *Sue-doh-neigh-ah mo-dess-tah.*

ETYMOLOGY Calm false cobra.

TYPE LOCALITY Perth, WA.

APPEARANCE Medium-sized, moderately slender snake with head indistinct from body. Dorsal colouration grey to reddish-brown, with 4–11 black cross-bands. Juveniles have more prominent bands than adults. Pupil round; iris copper. Tongue dark; mouth lining blackish. Ventral colouration yellow with orange-red spots and blotches. Adult males and females reach a similar size of 60cm TL. **Scalation** MB 17 rows, 145–175 VENT, SUB 35–50 and anal scale both divided. Scales slightly glossy in appearance and smooth. **Similar species** *P. aspidorhyncha* (p. 192).

RANGE Encountered in arid central Australia, to the western WA coastline.

COMMENTS Cathermal. Terrestrial. Lives in open woodland, mallee, grassland, deserts and around agricultural areas. Shelters among vegetation, beneath man-made debris, in abandoned burrows, or under logs and rocks. Probably a species complex. **Diet** Lizards, rarely other snakes. **Reproduction** Oviparous. 7–20 per clutch. Neonates approximately 18cm TL hatching in January–March.
Disposition Generally inoffensive, but readily bites if harassed.

BITE/VENOM VENOMOUS

IUCN LISTING Least Concern.

Eridunda Roadhouse, NT, Rob Valentic

Ora Banda, WA Scott Eipper

Danggali, SA, Shane Black

Winton, QLD, Scott Eipper

Northern Brown Snake *Pseudonaja nuchalis* Günther, 1858

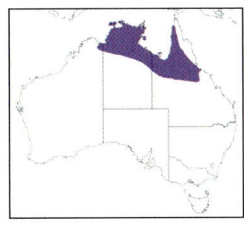

PRONUNCIATION *Sue-doh-neigh-ah new-car-liss.*
ETYMOLOGY Neck-marked false cobra.
TYPE LOCALITY Port Essington, NT.
APPEARANCE Large, moderately slender snake with head indistinct from body. Colouration very variable, from pale brown to gold or dark brown. Some individuals have dark bands, while others have dark nape band. Usually black speck or mark on nape. Juveniles and immature individuals have black head and black band on nape, and are sometimes completely banded. Pupil round; iris reddish. Tongue dark; mouth lining blackish. Ventral colouration yellow with orange-red spots and blotches. Adult males larger than females, reaching 140cm TL.
Scalation MB 17 (rarely 19) rows, 180–230 VENT, SUB 50–70 and anal scale both divided. Scales slightly glossy in appearance and smooth. **Similar species** *P. aspidorhyncha* (p. 192), *P. guttata* (p. 193), *P. ingrami* (p. 195), *P. mengdeni* (p. 196), *P. textilis* (p. 200).
RANGE Encountered across northern Australia, west of the Great Dividing Range, into the Kimberleys of WA.
COMMENTS Diurnal to nocturnal depending on the temperature. Terrestrial. Lives in tropical savannah, grassland, rocky outcrops and desert. Shelters among vegetation, beneath man-made debris, in burrows, or under logs and rocks. **Diet** Small mammals and reptiles as adults; juveniles are almost exclusive reptile predators. **Reproduction** Oviparous. 8–16 per clutch. Neonates approximately 23cm TL hatching in January–March. **Disposition** Nervous, and inoffensive unless threatened. Will readily bite if harassed.
BITE/VENOM DANGEROUSLY VENOMOUS
IUCN LISTING Least Concern.

Coen, QLD, Anders Zimny

Laura, QLD, Matt Summerville

Katherine, NT, Scott Eipper

Mt Isa, QLD, Scott Eipper

Eastern Brown Snake *Pseudonaja textilis*
(A. M. C. Duméril, Bibron & Duméril, 1854)
(Common Brown Snake)

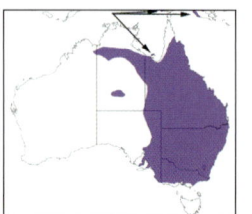

PRONUNCIATION *Sue-doh-neigh-ah tex-till-iss.*
ETYMOLOGY Woven false cobra.
TYPE LOCALITY Australia.
APPEARANCE Large to very large, moderately slender snake with head indistinct from body. Colouration very variable, from pale cream to black. Usually a shade of brown, with or without dark bands. Juveniles and immature individuals have black head and black band on nape, and some are completely banded. Pupil round; iris gold to brown. Tongue dark; mouth lining pink. Ventral colouration yellow to cream, with orange-red spots and blotches, sometimes marked with grey. Adult males larger than females, reaching 200cm TL. Alice Springs animals genetically similar to the southern Papuan snakes and may prove to be distinct. **Scalation** MB 17 rows, 180–235 VENT, SUB 45–75 and anal scale both divided. Scales slightly glossy in appearance and smooth. **Similar species** *P. aspidorhyncha* (p. 192), *P. guttata* (p. 193), *P. inframacula* (p. 194), *P. ingrami* (p. 195), *P. mengdeni* (p. 196), *P. nuchalis* (p. 199).

RANGE Encountered in eastern Australia over all of NSW, most of Vic and QLD, and south-eastern SA. Isolated populations in the NT around Alice Springs and the Victoria River district, WA. Also found in southern PNG and Indonesia.

COMMENTS Diurnal to nocturnal depending on the temperature. Terrestrial. Inhabits woodland, brigalow, mallee, grassland, deserts and agricultural areas. Shelters among vegetation, beneath man-made debris, or under logs and rocks. **Diet** Small mammals and reptiles as adults; juveniles are almost exclusive reptile predators. **Reproduction** Oviparous. 6–28 per clutch. Neonates approximately 32cm TL hatching in December–April. **Disposition** Nervous, and inoffensive unless threatened. Will readily bite if harassed.

BITE/VENOM DANGEROUSLY VENOMOUS
IUCN LISTING Least Concern.

Mt Carbine, QLD, Shane Black

Jimboomba, QLD, Tie Eipper

Warandyte, VIC, Scott Eipper

Kenmore, QLD, Scott Eipper

Royal NP, NSW, Scott Eipper

Juveniles, Southern QLD, Shane Black

Mt Stuart, QLD, Scott Eipper

West Macdonnell Ranges, NT, Jules Farquhar

Genus *Rhinoplocephalus* Müller, 1885

Monotypic genus, restricted to south-west Australia. Nocturnal, and usually found while crossing tracks, basking on vegetation or beneath cover. These snakes mainly eat lizards and frogs. Viviparous.
VENOM Toxicity unknown, and there is no evidence of envenomations. **Species-level identification difficulty** (within Australia) – 5.
TYPE SPECIES: *Rhinoplocephalus bicolor*.
ETYMOLOGY Snout – *Hoplocephalus* – pertains to a genus of Australian elapids.

Rhinoplocephalus bicolor, Coomalbidgup, WA, Brian Bush

Rhinoplocephalus bicolor, Albany, WA, Hal Cogger

Square-nosed Snake *Rhinoplocephalus bicolor* Müller, 1885
(Muller's Snake)

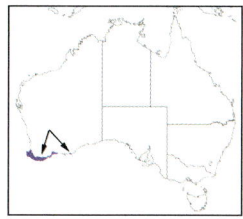

PRONUNCIATION *Rye-no-plo-keff-ah-luss by-co-lore.*

ETYMOLOGY Two-coloured snout – *Hoplocephalus*; pertains to the stark contrast between dorsal and ventral colouration.

TYPE LOCALITY Southern Australia.

APPEARANCE Small, robust snake with head slightly distinct from body. Dorsal colouration purplish-grey to blackish-brown. Lower flanks pink to pale orange. Pupil vertically oval; iris dark reddish-brown. Tongue and mouth lining pinkish. Ventral colouration usually creamy-white. Adult males larger than females, reaching 40cm TL. **Scalation** MB 15 rows, 135–165 VENT, SUB 20–45 and anal scale both single. Scales glossy in appearance and smooth.

RANGE Encountered in SW WA, from Esperance to Busselton.

COMMENTS Predominantly nocturnal but will bask on vegetation during mornings if conditions are suitable. Terrestrial. Lives in woodland and scrubland. Shelters beneath fallen grass trees, rocks and man-made debris, and inside stick-ant nests. **Diet** Lizards, frogs. **Reproduction** Viviparous. 1–5 per litter. Neonates approximately 14cm TL. Gravid in October–January. **Disposition** Inoffensive but will bite if harassed.

BITE/VENOM HARMFUL

IUCN LISTING Least Concern.

Cheynes Beach, WA, Angus McNab

Cheynes Beach, WA, Angus McNab

Mt Romance, WA, Robert Audcent

Banded Snakes, genus *Simoselaps* Jan, 1859

This genus currently contains four species, all of which are endemic. Further work is required to determine species relationships. These nocturnal, fossorial snakes are usually encountered crossing roads and paths at night. They specialize in eating lizards, which they constrict while the venom subdues the prey. These snakes are oviparous. **VENOM** Toxicity unknown, as there is no evidence of envenomations. **Species-level identification difficulty** (within Australia) – 3.

TYPE SPECIES *Elaps bertholdi*.

ETYMOLOGY Snub-nosed *Elaps* – pertaining to the genus of African garter-snakes *Elapsoidea* – the type genus for the family Elapidae.

Key to *Simoselaps*

1. Body with bands 2
   Body without bands .... *S. minimus* (p. 208)

2. 6 supralabial scales .... *S. anomalus* (opposite)
   5 supralabial scales 3

3. Dark nape collar 5 or more scales wide;
   occipital band edged with black or brown .... *S. bertholdi* (p. 206)
   Dark nape collar 3 or fewer scales wide;
   occipital band not edged with black or brown .... *S. littoralis* (p. 207)

Simoselaps bertholdi, Warnbro, WA, Scott Eipper

Simoselaps bertholdi, Warnbro, WA, Scott Eipper

DESERT BANDED SNAKE *Simoselaps anomalus* (Sternfield, 1919)

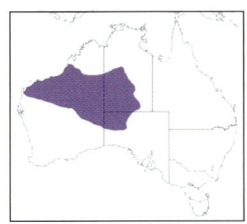

PRONUNCIATION *Si-mo-see-laps an-o-ma-luss.*

ETYMOLOGY Irregular snub-nosed *Elaps*.

TYPE LOCALITY Hermannsburg Mission, upper Finke River south of Macibbekk Distributions, NT.

APPEARANCE Small, moderately robust snake with head indistinct from body. Dorsal colouration yellow with regular thin, dark brown to black, straight-edged cross-bands. Head and neck background colouration white. Pupil vertically oval; iris black. Tongue and mouth lining pink. Ventral colouration creamish-white. Adult females larger than males, reaching 31cm TL. **Scalation** MB 15 rows, 115–130 VENT, SUB 15–30 and anal scale both divided. Scales glossy in appearance and smooth. **Similar species** *S. bertholdi* (p. 206), *S. littoralis* (p. 207).

RANGE Encountered in central Australia, including NW SA and southwestern half of the NT, through inland WA, reaching the coast between Exmouth and Derby.

COMMENTS Nocturnal. Terrestrial. Lives on sandy soils. Shelters beneath cover such as fallen timber and leaf litter, in mulga woodland and sand ridge deserts. **Diet** Ground-dwelling skinks. **Reproduction** Oviparous. 2–3 per clutch. Hatchlings approximately 9.5cm TL. **Disposition** Inoffensive but will bite if harassed.

BITE/VENOM HARMFUL

IUCN LISTING Least Concern.

Near Port Hedland, WA, Brad Maryan

De Grey River, WA, Brian Bush

Sandfire, WA, Jules Farquhar

Kings Canyon, NT, Shane Black

Jan's Banded Snake *Simoselaps bertholdi* (Jan, 1859)
(Southern Desert Banded Snake)

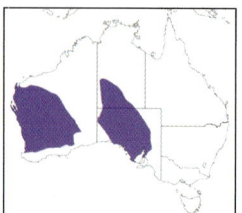

PRONUNCIATION *Si-mo-see-laps ber-thold-ee.*

ETYMOLOGY Berthold's snub-nosed *Elaps* pertains to A. Berthold, German zoologist.

TYPE LOCALITY Adelaide, SA.

APPEARANCE Small, moderately robust snake with head indistinct from body. Dorsal colouration orange with regular thin, dark brown to black, straight-edged cross-bands, sometimes coalescing forming a broken vertebral stripe. Head and neck background colouration whitish to cream. Pupil vertically oval; iris red-orange. Tongue and mouth lining pink. Ventral colouration creamish-white. Adult females larger than males, reaching 32cm TL. **Scalation** MB 15 rows, 115–135 VENT, SUB 15–30 and anal scale both divided. Scales glossy in appearance and smooth. **Similar species** *S. anomalus* (p. 205), *S. littoralis* (opposite).

RANGE Encountered in SA, from the western Eyre Peninsula, across South Australia, to WA's western coast, south of Yannarie.

COMMENTS Nocturnal. Terrestrial. Lives on sandy soils. Shelters beneath cover such as fallen timber and leaf litter, in mallee, coastal dune assemblages, mulga woodland and sand ridge deserts. **Diet** Ground-dwelling skinks, pygopods. **Reproduction** Oviparous. 1–8 per clutch. Hatchlings approximately 8.5cm TL. **Disposition** Inoffensive but will bite if harassed.

BITE/VENOM HARMFUL

IUCN LISTING Least Concern.

Bold Park, WA, Danny Melville

Port Augusta, SA, Shawn Scott

40km W of Ceduna, SA, Rob Valentic

WEST COAST BURROWING SNAKE *Simoselaps littoralis* (Storr, 1968)

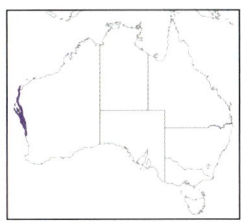

PRONUNCIATION *Si-mo-see-laps litt-or-rah-liss*.
ETYMOLOGY Shore snub-nosed *Elaps*.
TYPE LOCALITY Coastal dunes 11km south of Geraldton, WA.
APPEARANCE Small, moderately robust snake with head indistinct from body. Dorsal colouration pale yellow, with regular thin, dark brown to black, straight-edged cross-bands. Head and neck background colouration whitish to cream. Pupil vertically oval; iris silver-grey. Tongue and mouth lining pink. Ventral colouration creamish-white. Adult females larger than males, reaching 31cm TL. **Scalation** MB 15 rows, 100–125 VENT, SUB 15–25 and anal scale both divided. Scales glossy in appearance and smooth. **Similar species** *S. anomalus* (p. 205), *S. bertholdi* (opposite).
RANGE Encountered in coastal WA, from Cervantes north to Onslow.
COMMENTS Nocturnal. Terrestrial. Lives on sandy soils. Shelters beneath cover such as fallen timber and leaf litter, in coastal dune assemblages, heaths and sand ridge deserts. **Diet** Ground-dwelling skinks. **Reproduction** Oviparous. 3–4 per clutch. Hatchlings approximately 7.5cm TL.
Disposition Inoffensive but will bite if harassed.
BITE/VENOM HARMFUL
IUCN LISTING Least Concern.

Jurien Bay, WA, Danny Melville

Shark Bay, WA, Shawn Scott

Shark Bay, WA, Shawn Scott

Jurien, WA, Brian Bush

Dampierland Burrowing Snake *Simoselaps minimus* (Worrell, 1960)

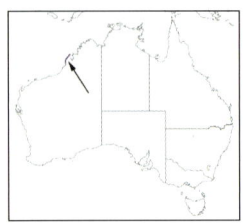

PRONUNCIATION *Si-mo-see-laps min-e-mus.*
ETYMOLOGY Least snub-nosed *Elaps*.
TYPE LOCALITY Broome, WA.
APPEARANCE Small, moderately robust snake with head indistinct from body. Dorsal colouration pale yellowish-brown. Edges of midbody scales darker, forming reticulated pattern. Head has black band over nostrils and eye, as well as nuchal band. Head and neck background colouration whitish to cream. Pupil vertically oval; iris black. Tongue and mouth lining pink. Ventral colouration creamish-white. Adult females larger than males, reaching 41cm TL. **Scalation** MB 15 rows, 125–135 VENT, SUB 19–25 and anal scale both divided. Scales glossy in appearance and smooth.
RANGE Encountered in northern WA on the Dampier Peninsula north of Broome.
COMMENTS Nocturnal. Terrestrial. Lives on sandy soils. Shelters beneath cover such as fallen timber and leaf litter, in coastal dune assemblages, heaths and sand ridge deserts. Suggested to possibly be a patternless variation of *S. anomalus*. **Diet** Ground-dwelling skinks. **Reproduction** Likely to lay eggs. **Disposition** Inoffensive but will bite if harassed.
BITE/VENOM HARMFUL
IUCN LISTING Least Concern.

Cape Leveque, Broome, WA, Brian Bush

Broome, WA, Brad Maryan

Curl & Hooded Snakes, genus *Suta* Worrell, 1961

This genus currently contains 10 species, all endemic to Australia. Phylogenetic relationships within the genus suggest that there are several species complexes that require further research to define whether they are distinct species, subspecies or just geographical variants. A recent revision placed the genus *Parasuta* into the synonymy, shifting its members into *Suta*. Nocturnal and robust, these snakes can be found close to people. They generally live in drier habitats such as woodland, grassland and deserts. They are seen active, crossing roads and paths at night, particularly after rain, and are livebearers. **VENOM** Contains neurotoxins, myotoxins and procoagulants. Envenomations from this genus can lead to symptoms such as headache, nausea, vomiting, abdominal pain, diarrhoea, dizziness, collapse and convulsions. Toxicity is unknown for many members of the genus. Polyvalent antivenom has been used to treat severe envenomations. A bite from one species resulted in the death of a human – probably due to anaphylaxis. **Species-level identification difficulty** (within Australia) – 5.

TYPE SPECIES *Hoplocephalus sutus*.
ETYMOLOGY Stitched.

Key to *Suta*

1. 19 midbody scale rows or more 2
   17 midbody scale rows or less 3

2. Dark streak from nostril through eye to temporal region .... *S. suta* (p. 220)
   No dark streak from the nostril through eye to temporal region .... *S. ordensis* (p. 217)

3. 17 midbody scale rows 4
   15 midbody scale rows or less 6

4. Body plain, without irregular cross-bands 5
   Body plain, with irregular cross-bands .... *S. fasciata* (p. 211)

5. Black hood from frontal scale on to nape
   with black nasal scales and rostral scale .... *S. flagellum* (p. 212)
   Head, nape and forebody with dark streaks
   and spots (fading with age) .... *S. punctata* (p. 218)

6. Black hood usually divided between nostrils
   and eyes (sometimes connecting in the centre) 7
   Complete black hood from snout on to nape 8

7. Anterior edges of scales brown to black,
   darker than the centre of the scale, forming a reticulum .... *S. spectabilis* (p. 219)
   Scales usually with darker centre than edges .... *S. gouldii* (p. 214)

8. Dark hood strongly contrasting with the body
   without signs of a dark vertebral stripe 9
   Dark hood tapers into a dark vertebral stripe .... *S. nigriceps* (p. 216)

9. Reddish-brown, usually with a single posterior temporal
   scale (except in Pilbara and Hammersley regions of WA) 10
   Brown to greyish, usually with a pair of posterior
   temporal scales; not found in WA .... *S. dwyeri* (p. 210)

10. Hood extends 1–4 (usually 3) scales on to nape,
    usually single posterior temporal scale, ... *S. monachus* (p. 215)
    Hood extends 4–6 (usually 5) scales on to nape,
    usually two posterior temporal scales, ... *S. gaikhorstorum* (p. 213)

Dwyer's Snake *Suta dwyeri* (Worrell, 1956)

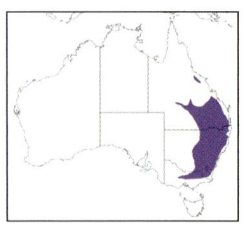

PRONUNCIATION *Sue-tah dwhy-err-e.*
ETYMOLOGY Dwyer's stitched; pertains to J. Dwyer, Australian naturalist.
TYPE LOCALITY Glenmorgan area, QLD.
APPEARANCE Small, robust snake with head distinct from body. Dorsal colouration brown to greyish. Head black on top, without black stripe running along spine. Pupil vertically oval; iris dark reddish-brown. Tongue pale; inside of mouth pink. Northern animals lighter bodied than their southern counterparts. Ventral colouration white to cream. Adult males larger than females, reaching 40cm TL. **Scalation** MB 15 rows, 135–170 VENT, SUB 20–40 and anal scale both single. Scales glossy in appearance and smooth. **Similar species** *S. flagellum* (p. 212), *S. nigriceps* (p. 216).
RANGE Encountered in southern QLD, from Longreach, across to Rockhampton, south through inland NSW west of the GDR, into northern Vic.
COMMENTS Nocturnal. Terrestrial. Lives in brigalow, open woodland, grassland and scrubland. Shelters beneath rocks and logs, and in crevices and man-made debris. Eastern population of *S. nigriceps* may be synonymous with this species, but further research is required. **Diet** Lizards, frogs. **Reproduction** Viviparous. 2–7 per litter. Neonates approximately 13cm TL born in December–April. **Disposition** Inoffensive, but will bite if threatened.
BITE/VENOM VENOMOUS
IUCN LISTING Least Concern.

Narromine, NSW, Scott Eipper

Roma, QLD, Scott Eipper

Kogan, QLD, Scott Eipper

Ryan's Lookout, VIC, Scott Eipper

Rosen's Snake *Suta fasciata* (Rosén, 1905)

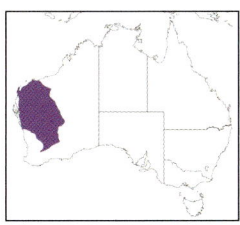

PRONUNCIATION Sue-tah fas-c-ah-tah.
ETYMOLOGY Banded stitched.
TYPE LOCALITY West Australia.
APPEARANCE Small, robust snake with head distinct from body. Dorsal colouration reddish-brown to orange or cream, with maroon, brown or grey irregular markings and spots. Pupil vertically oval; iris reddish-orange. Tongue and mouth pink. Ventral colouration white to cream. Both sexes reach 45cm TL. **Scalation** MB 17–19 rows, 140–165 VENT, SUB 20–40 and anal scale both single. Scales glossy in appearance and smooth. **Similar species** *Antaresia childreni stimsoni* (p. 31) differs by lacking labial pits.
RANGE Encountered in western and central WA, from Karratha to Kalgoorlie.
COMMENTS Nocturnal. Terrestrial. Lives on heavy clay and sandy soils in mulga, rocky gorges, open woodland and deserts. Shelters beneath rocks and logs, and in crevices and man-made debris. **Diet** Lizards, small mammals, frogs. **Reproduction** Viviparous. 1–7 per litter. Neonates approximately 19cm TL born in February–May. **Disposition** Generally inoffensive, but readily bites if harassed.
BITE/VENOM VENOMOUS
IUCN LISTING Least Concern.

Murrin Murrin, WA, Angus McNab

Wooramel, WA, Brian Bush

Laverton, WA, Scott Eipper

Juvenile, Laverton, WA, Scott Eipper

LITTLE WHIP SNAKE *Suta flagellum* (F. McCoy, 1878)

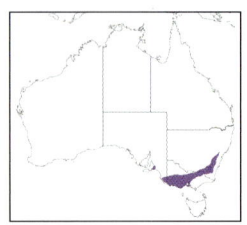

PRONUNCIATION *Sue-tah fla-gel-umm*.
ETYMOLOGY Whip-snout stitched.
TYPE LOCALITY Boroondara, Vic.
APPEARANCE Small, robust snake with head distinct from body. Dorsal colouration brown to greyish. Head black above, with pale bar between nostril and eyes. Pupil vertically oval; iris dark. Tongue and mouth pink. Ventral colouration white to cream. Both sexes reach 40cm TL. **Scalation** MB 17 rows, 125–150 VENT, SUB 20–40 and anal scale both single. Scales glossy in appearance and smooth. **Similar species** *S. dwyeri* (p. 210), *S. nigriceps* (p. 216), *S. spectabilis* (p. 219).

RANGE Encountered from eastern SA, throughout Vic, into southern NSW.

COMMENTS Nocturnal. Terrestrial. Lives in dry forests, open woodland and grassland. Shelters beneath rocks and logs, and in crevices and man-made debris. **Diet** Lizards, periodically frogs. **Reproduction** Viviparous. 2–11 per litter. Neonates approximately 13cm TL born in December–April. **Disposition** Inoffensive, but will bite if threatened.

BITE/VENOM VENOMOUS A bite from this species caused the death of a human in 2007, probably due to an anaphylactic reaction.

IUCN LISTING Least Concern.

Goulburn, NSW, Malcolm Campbell

Juvenile, Cape Clear, VIC, Adam Elliott

Daylsesford, VIC, Angus McNab

One Tree Hill, VIC, Scott Eipper

Pilbara Hooded Snake *Suta gaikhorstorum*
Maryan, Brennan, Hutchinson & Geidans, 2020

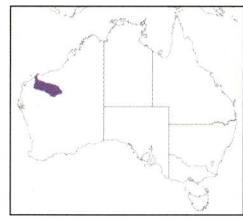

PRONUNCIATION *Sue-tah gay-cor-stor-rum.*

ETYMOLOGY Gaikhorst's stitched; pertains to K. & M. Gaikhorst, Australian naturalists.

TYPE LOCALITY 5km south of Tom Price Mine, WA.

APPEARANCE Small, robust snake with head distinct from body. Dorsal colouration orange-brown to orange or brick-red. In some individuals, posterior edge to midbody scales is tipped with black. Head black or dark brown. Lower flanks, lips and underside pearly-white. Pupil vertically oval; iris dark. Tongue and mouth pink. Both sexes reach 46cm TL. **Scalation** MB 15 rows, 160–168 VENT, SUB 23–34 and anal scale both single. Scales glossy in appearance and smooth. **Similar species** *S. monachus* (p. 215).

RANGE Encountered in the Pilbara region of WA, between Mt Whaleback, Cape Preston and Kooline Homestead.

COMMENTS Nocturnal. Terrestrial. Lives on heavy stony soils with open woodland or mulga, both with an understorey of *Triodia*. Shelters beneath rocks and man-made debris. **Diet** Lizards. **Reproduction** Probably gives birth to live young. Neonates size unknown. **Disposition** Inoffensive, but will bite if threatened.

BITE/VENOM VENOMOUS

IUCN LISTING Least Concern.

Mt Stuart Station, Duck Creek, WA, Brad Maryan

Newman, WA, Jules Farquhar

Newman, WA, Brian Bush

Yandagee Gorge, WA, Brian Bush

GOULD'S HOODED SNAKE *Suta gouldii* (Gray, 1841)
(Gould's Snake)

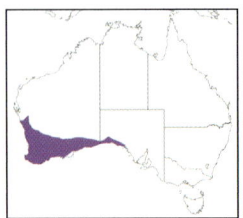

PRONUNCIATION *Sue-tah ghoul-dee.*
ETYMOLOGY Gould's stitched; pertains to J. Gould, British zoologist.
TYPE LOCALITY WA.
APPEARANCE Small, robust snake with head distinct from body. Dorsal colouration orange-brown to greyish. Head black or dark brown on top, narrowing between nostrils and eyes. Pupil vertically oval; iris dark. Tongue and mouth pink. Ventral colouration white to cream. Adult males larger than females, reaching 50cm TL. **Scalation** MB 15 rows, 136–174 VENT, SUB 25–41 and anal scale both single. Scales glossy in appearance and smooth.
Similar species *S. nigriceps* (p. 216), *S. spectabilis* (p. 219).
RANGE Encountered in WA, from Kalbarri across southern WA and into SA, where it occurs along the coast of the Great Australian Bight, to the western edge of the Eyre Peninsula at Denial Bay.
COMMENTS Nocturnal. Terrestrial. Lives in heaths, open woodland, grassland and mallee. Shelters under rocks, logs and other debris. **Diet** Lizards, small snakes, frogs. **Reproduction** Viviparous. 1–6 per litter. Gravid in August–March. **Disposition** Inoffensive, but will bite if threatened.
BITE/VENOM VENOMOUS
IUCN LISTING Least Concern.

Yanchep, WA, Danny Melville

Bickley, WA, Brian Bush

Casuarina, WA, Brad Maryan

Forrestania, WA, Brian Bush

INLAND HOODED SNAKE *Suta monachus* (Storr, 1964)
(Monk Snake)

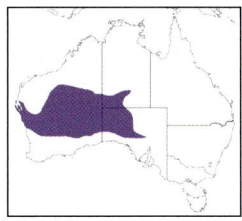

PRONUNCIATION *Sue-tah mo-nah-chuss*.
ETYMOLOGY Monk stitched.
TYPE LOCALITY Kalgoorlie, WA.
APPEARANCE Small, robust snake with head distinct from body. Dorsal colouration orange-brown to orange or brick-red. In some individuals, posterior edge to midbody scales tipped with black. Head black or dark brown. Lower flanks, lips and underside pearly-white. Pupil vertically oval; iris dark. Tongue and mouth pink. Both sexes reach 46cm TL. **Scalation** MB 15 rows, 150–174 VENT, SUB 23–32 and anal scale both single. Scales glossy in appearance and smooth. **Similar species** *S. gaikhorstorum* (p. 213).

RANGE Encountered from southwestern NT and central SA, through arid WA to Carnarvon. Does not occur in the Pilbara.

COMMENTS Nocturnal. Terrestrial. Lives in mallee, sand ridge deserts, open woodland and mulga. Shelters under logs and rocks, and in abandoned burrows of lizards and mammals. **Diet** Small lizards, small snakes. **Reproduction** Viviparous. 1–5 per litter. Neonates approximately 15cm TL born in December–April. **Disposition** Inoffensive, but will bite if threatened.

BITE/VENOM VENOMOUS
IUCN LISTING Least Concern.

Yulara, NT, Shawn Scott

Yulara, NT, Shawn Scott

Kalbarri, WA, Robert Audcent

Laverton, WA, Scott Eipper

MITCHELL'S SHORT-TAILED SNAKE Suta nigriceps (Günther, 1863)
(Black-backed Snake)

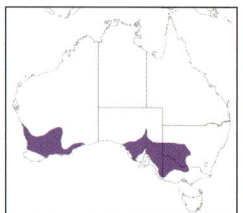

PRONUNCIATION *Sue-tah nig-ree-ceps*.
ETYMOLOGY Black-headed stitched.
TYPE LOCALITY Probably from Australia.
APPEARANCE Small, robust snake with head distinct from body. Dorsal colouration brown to greyish. Head black on top, with black stripe running along spine. Stripe sometimes coalesces into lighter flanks. Pupil vertically oval; iris dark. Tongue and mouth pink. Ventral colouration white to cream. Adult males larger than females, reaching 60cm TL. **Scalation** MB 15 rows, 145–175 VENT, SUB 18–38 and anal scale both single. Scales glossy in appearance and smooth. **Similar species** *S. dwyeri* (p. 210), *S. flagellum* (p. 212), *S. gouldii* (p. 214), *S. spectabilis* (p. 219).

RANGE Encountered in southern inland NSW and northern Vic, through the Flinders Ranges and northern Eyre Peninsula. Also occurs in WA, from the southern edge of the Nullarbor Plain, across to Cervantes and south to Albany.

COMMENTS Nocturnal. Terrestrial. Lives on rocky hillsides, open woodland, dry forest, mallee and scrubland. Shelters beneath rocks and logs, and in crevices and man-made debris. Animals from eastern half of range closer genetically to *S. dwyeri* than to the nominate population in WA. **Diet** Lizards, blind snakes, small elapids. **Reproduction** Viviparous. 1–7 per litter. Neonates approximately 16cm born in December–April. **Disposition** Inoffensive, but will bite if threatened.

BITE/VENOM HARMFUL
IUCN LISTING Least Concern.

Forrestania, WA, Brian Bush

Hattah-kulkyne NP, VIC, Angus McNab

Yanchep, WA, Danny Melville

Ngarkat, SA, Angus McNab

Ord River Curl Snake *Suta ordensis* (Storr, 1984)

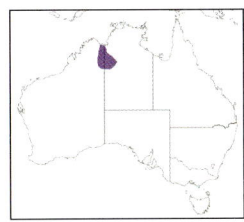

PRONUNCIATION *Sue-tah or-den-n-siss*.
ETYMOLOGY Ord (river) stitched.
TYPE LOCALITY Argyle Downs, WA.
APPEARANCE Small, robust snake with head distinct from body. Dorsal colouration yellow-brown to dark brown or grey. Some individuals have dark edges to posterior of scales, giving a reticulated appearance. Head usually darker than rest of body, but this fades with maturity. Some individuals have light barring on lips. Pupil vertically oval; iris dark. Tongue and mouth pink. Ventral colouration white to cream. Both sexes reach 35cm TL. **Scalation** MB 19 rows, 165–185 VENT, SUB 30–40 and anal scale both single. Scales glossy in appearance and smooth. **Similar species** *S. suta* (p. 220).
RANGE Encountered in WA in the Ord and Victoria River catchments.
COMMENTS Nocturnal. Terrestrial. Very poorly known. Lives in tropical woodland and black soil grassland. Ecology thought to be similar to that of *S. suta*. **Diet** Lizards. **Reproduction** Viviparous. **Disposition** Inoffensive, but will bite if threatened.
BITE/VENOM VENOMOUS
IUCN LISTING Least Concern.

Gordon Downs, WA, Paul Horner

Kalkarinji area, NT, Jules Farquhar

Kalkarinji area, NT, Jules Farquhar

Little Spotted Snake *Suta punctata* (Boulenger, 1896)

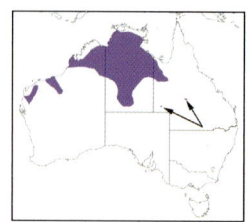

PRONUNCIATION *Sue-tah punk-ta-ta.*
ETYMOLOGY Spotted stitched.
TYPE LOCALITY Port Walcott, WA.
APPEARANCE Small, robust snake with head distinct from body. Dorsal colouration yellow-brown, orange to dark brown. Head, nape and forebody have a series of dark spots and blotches that usually fade with maturity. Usually a dark brown stripe running along lower flanks. Some individuals have dark edges to posterior of scales, giving a reticulated appearance. Pupil vertically oval; iris dark orange to brown. Tongue and mouth pink. Ventral colouration white to cream. Both sexes reach 60cm TL. **Scalation** MB 15 rows, 150–215 VENT, SUB 20–40 and anal scale both single. Scales glossy in appearance and smooth.
RANGE Encountered in WA north of Quobba, through most of the NT, to NW QLD.
COMMENTS Nocturnal. Terrestrial. Lives in open woodland, gorges and escarpments, mulga and deserts. Found actively hunting or crossing roads at night. Shelters beneath rocks, logs and man-made debris. **Diet** Lizards, frogs, small mammals. **Reproduction** Viviparous. 2–5 per litter. Neonates approximately 14cm TL born in December–April. **Disposition** Generally inoffensive, but readily bites if harassed.
BITE/VENOM VENOMOUS
IUCN LISTING Least Concern.

Mt Barnett Station, WA, Jake Meney

Newman, WA, Brian Bush

Alice Springs, NT, Scott Eipper

60km W Mica Creek, QLD, Scott Eipper

Port Lincoln Snake *Suta spectabilis* (Krefft, 1869)
(Spectacled Hooded Snake)

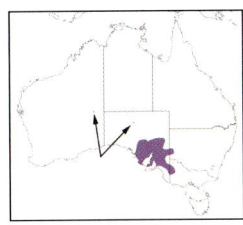

PRONUNCIATION *Sue-tah spec-ta-bil-liss*.
ETYMOLOGY Spectacled stitched.
TYPE LOCALITY Port Lincoln, SA.
APPEARANCE Small, robust snake with head distinct from body. Dorsal colouration brown to greyish. Head black on top, with pale bar between nostril and eyes. Black extends on to nape. Pupil vertically oval; iris dark. Tongue and mouth pink. Ventral colouration white, cream or yellow. Adult males larger than females, reaching 40cm TL. **Scalation** MB 15 rows, 138–168 VENT, SUB 21–33 and anal scale both single. Scales glossy in appearance and smooth. **Similar species** *S. flagellum* (p. 212), *S. gouldii* (p. 214), *S. nigriceps* (p. 216).
RANGE Encountered from western Vic to Fowler's Bay, SA. Outlying populations north of Coober Pedy, SA, and in the Great Victoria Desert, WA.
COMMENTS Nocturnal. Terrestrial. Lives in mallee, open woodland, grassland and heaths. Shelters under cover such as rocks and logs. Two individuals currently assigned to this taxa from the Great Victoria Desert, WA. appear to be a distinct taxa. **Diet** Lizards, occasionally frogs. **Reproduction** Viviparous. 2–5 per litter. Neonates approximately 14cm TL born in December–April. **Disposition** Inoffensive, but will bite if threatened.

BITE/VENOM HARMFUL
IUCN LISTING Least Concern.

Port Wakefield, SA, Shawn Scott

Port Wakefield, SA, Shawn Scott

Port Wakefield, SA, Shawn Scott

Port Germain, SA, Scott Eipper

Curl Snake *Suta suta* (Peters, 1863)
(Myall Snake)

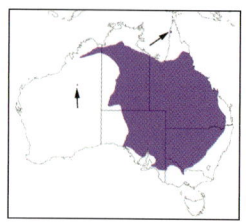

PRONUNCIATION *Sue-tah sue-tah*.

ETYMOLOGY Stitched.

TYPE LOCALITY Loos, 4.5km west of Gawler, SA.

APPEARANCE Small to medium-sized, robust snake with head distinct from body. Dorsal colouration yellow-brown to dark brown or grey. Some individuals have dark edges to posterior of scales, giving a reticulated appearance. Head usually darker than rest of body, but this fades with maturity. Some individuals have light barring on lips and broken yellow stripe along side of head. Dark lateral head streak from nostril to temporal region. Pupil vertically oval; iris tan to orange. Tongue and mouth pink. Ventral colouration white to cream. Both sexes reach 60cm TL. **Scalation** MB 19–21 rows, 150–170 VENT, SUB 20–35 and anal scale both single. Scales glossy in appearance and smooth. **Similar species** *S. ordensis* (p. 217).

RANGE Encountered over most of arid QLD, NSW, northern Vic, most of SA and the NT. An isolated population occurs near Lake Argyle, WA.

COMMENTS Nocturnal. Terrestrial. Lives in dry forests, mallee, heaths and deserts. Found actively hunting or crossing roads at night. Shelters under logs, rocks and man-made debris, or in deep soil cracks. Has also been found in trees, hunting small dragons. **Diet** Lizards, frogs, small mammals. **Reproduction** Viviparous. 1–9 per litter. Neonates approximately 16cm TL born in November–April. **Disposition** Generally inoffensive, but readily bites if harassed.

BITE/VENOM DANGEROUSLY VENOMOUS

IUCN LISTING Least Concern.

Gilgandra, NSW, Hal Cogger

60km W Mica Creek, QLD, Scott Eipper

Juvenile, Wilcannia, NSW, Adam Elliott

Mt Isa, QLD, Scott Eipper

Genus *Tropidechis* Günther, 1863

This monotypic, endemic, live-bearing genus is found in two populations about 950km apart. These snakes are predominantly nocturnal, but will bask cryptically. **VENOM** Contains strong procoagulants and neurotoxins; moderately haemolytic, cytotoxic and myotoxic. Envenomations can lead to rapid collapse and death. Tiger snake or Polyvalent antivenom is used to treat envenomations. **Species-level identification difficulty** (within Australia) – 2.

TYPE SPECIES *Hoplocephalus carinatus*.

ETYMOLOGY Keeled adder.

Tropidechis carinatus, Canungra, QLD, Scott Eipper

Tropidechis carinatus, Goomburra, Main Range NP, QLD, Scott Eipper

Rough-scaled Snake *Tropidechis carinatus* (Krefft, 1863)
(Clarence River Snake)

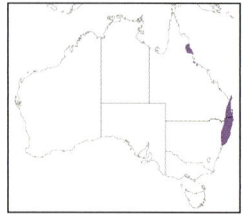

PRONUNCIATION *Tro-pid-deck-iss ca-rin-nah-tuss.*

ETYMOLOGY Keeled adder.

TYPE LOCALITY Near Grafton in the Clarence River district, NSW.

APPEARANCE Medium-sized, robust snake with head distinct from body. Dorsal colouration brown to grey, occasionally with a greenish wash. Some individuals flecked with black. Northern animals can be completely banded. Markings fade with age; however, some animals are patternless. Pupil round; iris tan to orange. Tines of tongue pink, followed by dark band and reddish tongue; mouth lining pink. Ventral colouration yellowish-green. Males and females both reach the same size, 90cm TL. **Scalation** MB 23 rows, 160–180 VENT, SUB 50–60 and anal scale both single. Scales heavily keeled. **Similar species** *Tropidonophis mairii* (p. 73) distinguished by lack of loreal scale and corners of mouth not upturned.

RANGE Encountered in two populations. Southern population occurs from Gosford, NSW, to K'gari (Fraser Island), QLD. Northern population occurs in the wet tropics region of north QLD.

COMMENTS Nocturnal, occasionally basking in the morning or after rain. Predominantly terrestrial but readily climbs. Lives in closed forests, wallum swamps and wetlands. Often found sitting in vegetation or crossing roads at night. Shelters beneath logs, in tree hollows and under rocks or man-made objects. **Diet** Small mammals, frogs, tadpoles, small lizards, occasionally birds. **Reproduction** Viviparous. 5–19 per litter. Neonates approximately 26cm TL born in April–September. **Disposition** Nervous, and inoffensive unless threatened. Will readily bite if harassed.

BITE/VENOM DANGEROUSLY VENOMOUS

IUCN LISTING Least Concern.

Julatten, QLD, Shane Black

Myall Lakes NP, NSW, Jules Farquhar

Mt Glorious, QLD, Scott Eipper

Goomburra, QLD, Scott Eipper

Bandy Bandys, genus *Vermicella* Günther, 1858

This genus currently contains six species, all endemic, and possibly unresolved. Further research is required to determine their species boundaries. They are nocturnal and partly fossorial, and are found across most dry terrestrial habitats, with at least one species present across much of mainland Australia. Some species are very well known, occurring in urban environments, while others are poorly known. These snakes feed on blind snakes and elongated skinks, and are all oviparous. They all have an ingenious defensive strategy involving raising parts of the body to form loops, which, along with thrashing, is employed to deter predators. **VENOM** Toxicity unknown, and no evidence of severe envenomations in humans. One envenomation caused pain and swelling, severe headache and vomiting. Swelling and tenderness of the bite site persisted for two days post-bite. **Species-level identification difficulty** (within Australia) – 3.

TYPE SPECIES: *Calamaria annulata*.
ETYMOLOGY Little worm.

Key to *Vermicella*

1	Internasal scales absent	2
	Internasal scales present	3
2	Fewer than 55 white bands	*V. intermedia* (p. 225)
	More than 60 white bands	*V. multifasciata* (p. 226)
3	Usually less than 286 ventral scales; not found in WA	4
	Usually more than 286 ventral scales; is found in WA	*V. snelli* (p. 228)
4	Less than 230 ventral scales	5
	More than 240 ventral scales	*V. vermiformis* (p. 229)
5	55–94 white bands one to two scales wide; single postocular	*V. parscauda* (p.227)
	36–55 white bands two to four scales wide; usually two postoculars	*V. annulata* (p. 224)

Vermicella annulata, Brisbane Waters NP, NSW, Hal Cogger

Vermicella vermiformis, Ellery Creek, NT, Adam Elliott

BANDY BANDY *Vermicella annulata* (Gray, 1841)

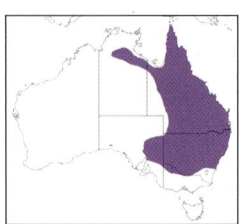

PRONUNCIATION Verm-e-cell-ah an-u-la-tah.
ETYMOLOGY Ringed little-worm.
TYPE LOCALITY Australia.
APPEARANCE Small, moderately robust snake with head indistinct from body. Top and underside of body is a series of alternating black and white rings. Usually 36–39 white rings are present. An unusual, many-banded population is found around Townsville, and has a higher white band count. Pupil vertically oval; iris black. Tongue pinkish; pink mouth lining. Adult females larger than males, reaching 75cm TL. **Scalation** MB 15 rows, 216–220 VENT, SUB 12–30 and anal scale both divided. Scales glossy in appearance and smooth. **Similar species** *V. intermedia* (opposite), *V. parscauda* (p. 227), *V. vermiformis* (p. 229).

RANGE Encountered in eastern Australia, from Cape York, QLD, south to northern Vic, and west to Port Augusta, SA.

COMMENTS Nocturnal. Terrestrial. Lives in forests, open woodland, mallee, brigalow, mulga and rocky areas in arid zones. Shelters under rocks and fallen timber, and beneath leaf litter. **Diet** Blind snakes, elongated skinks. **Reproduction** Oviparous. 4–6 per clutch. Neonates approximately 19cm TL hatching in December–April. **Disposition** Inoffensive, rarely attempting to bite, even if threatened.

BITE/VENOM HARMFUL
IUCN LISTING Least Concern.

Hattah, VIC, Scott Eipper

Narromine, NSW, Scott Eipper

Injune, QLD, Scott Eipper

Mt Nebo, QLD, Scott Eipper

INTERMEDIATE BANDY BANDY *Vermicella intermedia* Keogh & S. A. Smith, 1996

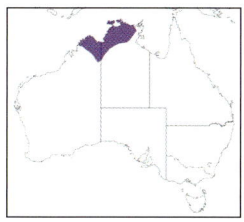

PRONUNCIATION *Verm-e-cell-ah in-ter-meed-e-ah*.

ETYMOLOGY Intermediate little-worm; pertaining to intermediate number of ventral scales compared to other species of *Vermicella* at the time of description.

TYPE LOCALITY Humpty Doo district, NT.

APPEARANCE Small, moderately slender snake with head indistinct from body. Top and underside of body is a series of alternating black and white rings; 50–53 white rings present. Pupil vertically oval; iris black. Tongue pinkish; mouth lining pink. Adult females larger than males, reaching 39cm TL. **Scalation** MB 15 rows, 246–256 VENT, SUB 15–28 and anal scale both divided. Scales glossy in appearance and smooth. **Similar species** *V. multifasciata* (p. 226), *V. vermiformis* (p. 229).

RANGE Encountered from the top end, NT, west across into the Kimberleys of northern WA.

COMMENTS Nocturnal. Terrestrial. Lives in tropical savannah, vine thickets and rocky areas. Shelters under rocks and fallen timber, and beneath leaf litter. **Diet** Probably feeds on blind snakes. **Reproduction** Probably lays eggs. **Disposition** Inoffensive, rarely attempting to bite, even when threatened.

BITE/VENOM HARMFUL

IUCN LISTING Least Concern.

Gubara Track, Kakadu NP, NT, Brad Maryan

Humpty Doo, NT, Paul Horner

Adelaide River, NT, Anders Zimny

Humpty Doo, NT Hal Cogger

NARROW-BANDED BANDY BANDY *Vermicella multifasciata* (Longman, 1915)
(Northern Bandy Bandy)

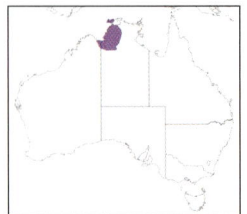

PRONUNCIATION *Verm-e-cell-ah mul-tee-fas-cee-ah-tah*.
ETYMOLOGY Many-banded little-worm.
TYPE LOCALITY Port Darwin, NT.
APPEARANCE Small, moderately slender snake with head indistinct from body. Top of body black to dark brown, with white rings usually formed by white spots arranged in a band; 77–109 white rings present. Pupil vertically oval; iris black. Tongue pinkish; mouth lining pink. Adult females larger than males, reaching 45cm TL. **Scalation** MB 15 rows, 240–296 VENT, SUB 15–25 and anal scale both divided. Scales glossy in appearance and smooth.
Similar species *V. intermedia* (p. 225).

RANGE Encountered from Ord River drainage across into the western 'top end' in the NT and on Melville and Bathurst Islands.

COMMENTS Nocturnal. Terrestrial. Lives in tropical savannah. Shelters under rocks and fallen timber, and beneath leaf litter. **Diet** Probably blind snakes. **Reproduction** Likely to lay eggs. **Disposition** Inoffensive, rarely attempting to bite, even when threatened.

BITE/VENOM HARMFUL

IUCN LISTING Least Concern.

Timber Creek, NT, Allira Costa

Victoria River Downs, NT, Paul Horner

Victoria River Downs, NT, Steve Swanson

CAPE YORK BANDY BANDY *Vermicella parscauda*
Derez, Arbuckle, Ruan, Xie, Huang, Dibben, Shi, Vonk & Fry, 2018

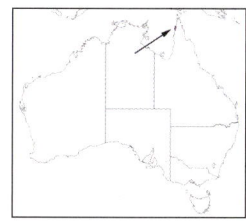

PRONUNCIATION *Verm-e-cell-ah pars-cor-dah*.
ETYMOLOGY Part tail little-worm.
TYPE LOCALITY Weipa, Cape York, QLD.
APPEARANCE Small, moderately slender snake with head indistinct from body. Top and underside of body a series of alternating black and white rings; 51– 94 white rings present. Pupil vertically oval; iris black. Tongue pinkish; pink mouth lining. Adult females larger than males, reaching 39cm TL. **Scalation** MB 15 rows, 213–230 VENT, SUB 27 and anal scale both divided. Scales glossy in appearance and smooth. **Similar species** *V. annulata* (p. 224).
RANGE Encountered on Cape York, between Weipa and Malpoon, QLD.
COMMENTS Nocturnal. Terrestrial. Lives in tropical open woodland on heavy red soils. Poorly known – only described in 2018. **Diet** Probably blind snakes. **Reproduction** Likely to lay eggs. **Disposition** Inoffensive, rarely attempting to bite, even when threatened.
BITE/VENOM HARMFUL
IUCN LISTING Least Concern.

Oyala Thumotang NP, QLD, Anders Zimny

Weipa, QLD, Lauren Dibben

Oyala Thumotang NP, QLD, Anders Zimny

PILBARA BANDY BANDY *Vermicella snelli* Storr, 1968

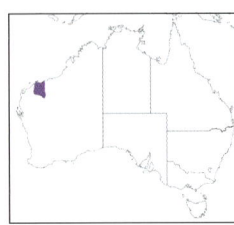

PRONUNCIATION *Verm-e-cell-ah snell-ee.*

ETYMOLOGY Snell's little-worm; pertaining to C. Snell, who collected the holotype.

TYPE LOCALITY Mundiwindi, WA.

APPEARANCE Small, moderately slender snake with head indistinct from body. Top and underside of body a series of alternating black and white rings; 48–64 white rings present. Pupil vertically oval; iris black. Tongue pinkish; pink mouth lining. Adult females larger than males, reaching 41cm TL. **Scalation** MB 15 rows, 262–302 VENT, SUB 12–30 and anal scale both divided. Scales glossy in appearance and smooth.

RANGE Encountered in the Pilbara region, WA.

COMMENTS Nocturnal. Terrestrial. Lives in gorges and rocky areas in arid zones. Shelters under rocks and beneath leaf litter. **Diet** Blind snakes. **Reproduction** Likely to lay eggs. **Disposition** Inoffensive, rarely attempting to bite, even when threatened.

BITE/VENOM HARMFUL

IUCN LISTING Least Concern.

65km S Karratha, WA, Brian Bush

Packsaddle Range, WA, Brad Maryan

65km S Karratha, WA, Brian Bush

85km S Newman, WA, Brian Bush

CENTRALIAN BANDY BANDY *Vermicella vermiformis* Keogh & S. A. Smith, 1996

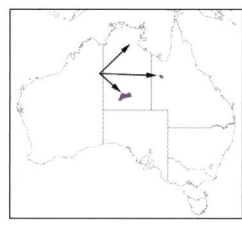

PRONUNCIATION *Verm-e-cell-ah verm-e-form-iss.*
ETYMOLOGY Like worm little-worm.
Type locality Alice Springs, NT.
APPEARANCE Small, moderately slender snake with head indistinct from body. Top and underside of the body a series of alternating black and white rings; 41–45 white rings present. Pupil vertically oval; iris black. Tongue pinkish; pink mouth lining. Adult females larger than males, reaching 32cm TL. **Scalation** MB 15 rows, 263–281 VENT, SUB 12–30 and anal scale both divided. Scales glossy in appearance and smooth. **Similar species** *V. annulata* (p. 224), *V. intermedia* (p. 225).
RANGE Encountered in central Australia surrounding Alice Springs, NT; also found in the southern gulf country across to western QLD.
COMMENTS Nocturnal. Terrestrial. Lives in open woodland, gorges and rocky areas in arid zones. Shelters under rocks and fallen timber, and beneath leaf litter. **Diet** Blind snakes. **Reproduction** Oviparous. Up to nine per clutch. Neonates approximately 16cm TL hatching in January–March. **Disposition** Inoffensive, rarely attempting to bite, even when threatened.
BITE/VENOM HARMFUL
IUCN LISTING Least Concern.

60km W Mica Creek, QLD, Scott Eipper

Alice Springs, NT, Rob Valentic

Lake Moondarra, QLD, Scott Eipper

West Macdonnell Ranges, Jules Farquhar

Family Elapidae (Sea Snakes)

Historically, sea snakes were regarded as two separate families. Further taxonomic work determined that they are better placed in the same family as terrestrial elapids. Some species of sea snake are poorly known, due to the difficulty of identification, presumably low population densities and that they are rarely encountered. As this enigmatic group of snakes can be very difficult to identify, here the sea snakes have been placed together in their own section. Some species included have been historically recorded in Australian waters, but have failed to be detected in the last 30 years. Other species are potentially waifs, which occasionally enter Australian waters but are essentially species from southern Asia or from the Pacific islands. These species have been included in the keys and species accounts, to assist the reader to arrive at an identification. Six species are endemic to Australia.

Key to Australian Marine Elapids

1. Ventral scales large (more than 3 times wider) than adjacent body scales..2
   Ventral scales small (at most twice the width) of adjacent body scales..7

2. Nasal scales in contact..3
   Nasal scales separated by internasal scales..*Laticauda* (p. 274)

3. 5 or more supralabial scales..4
   Less than 4 supralabial scales..*Emydocephalus* (p. 242)

4. Preocular scale present..5
   Preocular scale absent..*Hydrelaps darwiniensis* (p. 249)

5. Ventral scales without notching; posterior chin shields are typical in size..6
   Ventral scales with prominent to slight notching; posterior chin shields are small in size..*Aipysurus* (opposite)

6. More than 30 midbody scale rows, posterior midbody scales smooth..*Parahydrophis mertoni* (p. 280)
   Less than 30 midbody scale rows, posterior midbody scales keeled..*Ephalophis greyae* (p. 247)

7. Posterior ventral scales are either single or paired and overlapping..*Hydrophis* (p. 250)
   Posterior ventral scales are paired and not overlapping..*Microcephalophis gracilis* (p. 278)

Hydrophis stokesii, mating, Port Hedland, WA, Doris Teufel

Genus *Aipysurus* Lacépède, 1804

This genus comprises nine species, all of which occur in Australia, and four of which are endemic. Some species have exhibited severe decline in certain locations for unknown reasons. Further taxonomic work is needed to determine species relationships in some of the wide-ranging taxa. Reports of aggression in sea snakes are for the most part exaggerated. Some species can be curious and this can be misconstrued as aggression. **VENOM** In some species, the venom is strongly neurotoxic, with evidence of myotoxicity. Bites have resulted in, or could cause fatalities. Sea snake or Polyvalent antivenom is used to treat envenomations. In the fish-egg eating specialists, there is evidence of reduced venom toxicity. **Species-level identification difficulty** (within Australia) – 5.
ETYMOLOGY High tail.
TYPE SPECIES *Aipysurus laevis*.

Key to *Aipysurus*

1. 17 midbody scale rows........................2
   More than 17 midbody scale rows........................5

2. Ventral scales with a slight median notch on the posterior edge........................3
   Ventral scales with a deep median notch on the posterior edge........................*A. apraefrontalis* (p. 232)

3. Most head scales including parietals, typical in size and condition........................4
   Some head scales including parietals, fragmented and small........................*A. fuscus* (p. 236) (in part)

4. 29–45 midbody bands........................*A. eydouxii* (p. 234)
   15–21 midbody bands........................*A. mosaicus* (p. 239)

5. Ventral scales with a slight median notch on the posterior edge........................6
   Ventral scales with a deep median notch on the posterior edge........................*A. foliosquama* (p. 235)

6. 19 midbody scale rows........................7
   More than 19 midbody scale rows........................9

7. Some head scales are typically larger than those on the neck........................8
   Most head scales same size or smaller than those on the neck........................*A. duboisii* (p. 233)

8. More than 180 ventral scales........................*A. tenuis* (p. 241)
   Less than 180 ventral scales........................*A. fuscus* (p. 236) (in part)

8. 20–23 midbody scale rows (tuberculate in males),
   found between Carnarvon and Perth, WA........................*A. pooleorum* (p. 240)
   21–25 midbody scale rows (not tuberculate); found in
   WA from Exmouth north; also found NT, QLD and NSW........................*A. laevis* (p. 237)

Aipysurus mosaicus, Tin Can Bay, QLD, Josh Jensen

SHORT-NOSED SEA SNAKE *Aipysurus apraefrontalis* Smith, 1926

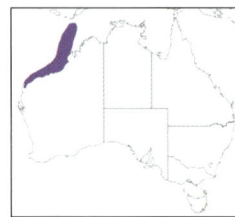

PRONUNCIATION *App-shoe-russ a-pray-fron-tah-liss.*

ETYMOLOGY Lacking prefrontal high tail; pertaining to the species lacking prefrontal scales.

TYPE LOCALITY Ashmore Reef, Timor Sea, WA.

APPEARANCE Medium-sized, heavy bodied snake with head indistinct from body. Dorsal colouration beige to dark brown, with dark brown to purplish-grey cross-bands that taper on flanks. Iris pale. Buccal cavity pale. Ventral colouration pale beige to cream. Both sexes reach a length of 110cm. **Scalation** MB 17 rows, 140–160 VENT, SUB 18–35 all single and anal scale divided. Body scales strongly overlapping, usually with pointed tip and smooth; posteriorly scales can be keeled or with short tubercles. Ventral scales have deep notch with median keel. **Similar species** *A. foliosquama* (p. 235).

RANGE Encountered in waters in WA, from Exmouth to the Ashmore Reef. The northern population experienced a massive decline, but the species is still present at Ashmore Reef and has recently been seen (in 2021) at a depth of 67m.

COMMENTS Cathemeral. Lives in waters over both coral reefs and rocky areas. Throughout the day it has been found beneath lumps of coral and in crevices. **Diet** Gobies, eels. **Reproduction** Thought to be livebearers, with gravid females seen in November and April. **Disposition** Inoffensive unless harassed.

BITE/VENOM VENOMOUS

IUCN LISTING Critically Endangered.

Ashmore Reef, WA, Hal Cogger

Exmouth, WA, Ruchira Somaweera

Ashmore Reef, WA, Hal Cogger

Ashmore Reef, WA, Photo courtesy of Schmidt Ocean Institute

Dubois' Sea Snake *Aipysurus duboisii* Bavay, 1869

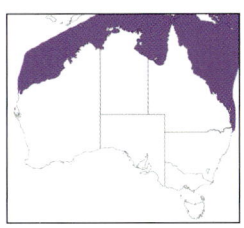

PRONUNCIATION *App-shoe-russ due-boy-see.*

ETYMOLOGY Dubois's high-tail. Pertains to C. F. Dubois, French naturalist.

TYPE LOCALITY Lifou (Loyalty Islands), New Caledonia.

APPEARANCE Large, relatively slender snake with head slightly slightly distinct from body. Dorsal colouration cream to white with grey to black bands. Edges of midbody scales white, forming reticulated pattern on body. Iris pale. Tongue and buccal cavity pale. Ventral colour similar to dorsum. Both sexes reach a length of 116cm. **Scalation** MB 17 rows, 154–181 VENT, SUB 25–30 all divided and anal scale divided. Scales smooth to weakly keeled.

RANGE Encountered in waters around Australia, north of Ballina, NSW, to Exmouth, WA. Also found in the western Pacific to New Caledonia.

COMMENTS Cathemeral. Lives in waters over coral reefs, mudflats and seagrass beds. **Diet** Fish. **Reproduction** Viviparous. 2–7 per litter. Neonate size not recorded. Females gravid in February–May. **Disposition** Usually inoffensive, but may bite if it perceives a threat.

BITE/VENOM DANGEROUSLY VENOMOUS

IUCN LISTING Least Concern.

Noumea, New Caledonia, Claire Goiran

Noumea, New Caledonia, Claire Goiran

Ashmore Reef, WA, Hal Cogger

Bowen, QLD, Steve Swanson

Marbled Sea Snake *Aipysurus eydouxii* (Bibron *in* Gray, 1849)
(Stagger Banded Sea Snake)

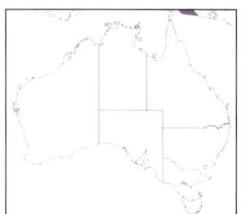

PRONUNCIATION *App-shoe-russ e-you-doox-e.*
ETYMOLOGY Eydoux's high-tail; pertaining to J. F. T. Eydoux, French naturalist.
TYPE LOCALITY Indian Ocean.
APPEARANCE Medium-sized, robust snake with head indistinct from body. Dorsal colouration pale cream to light grey with 29–45 dark bands. Iris dark. Ventral colouration black to cream with dark midline. Both sexes reach a TL of 91cm. **Scalation** MB 17 rows, 127–149 VENT, SUB 21–34 all single and anal scale divided. Scales smooth in appearance. **Similar species** *A. mosaicus* (p. 239).
RANGE Extralimital in Australian waters; known from a single individual listed from Port Moresby, PNG, on the Sahul Shelf and therefore presumed to occur in the Torres Strait, QLD. Occurs extensively throughout SE Asia.
COMMENTS Nocturnal. Found in turbid waters up to 60m deep. **Diet** Fish eggs; this diet specialization has led to an atrophied venom apparatus. **Reproduction** Viviparous. 1–6 per litter born in January–April. Neonate size not recorded. **Disposition** Inoffensive but will bite if harassed.
BITE/VENOM HARMFUL The venom composition has changed due to dietary specialization, and it is far less toxic than in most other members of the genus.
IUCN LISTING Least Concern.

Pulau Tekong, Singapore, Ria Tan

Parit Botak, Malaysia, Harold Voris

Beting Bronok, Singapore, Clay Hoon

Pulau Tekong, Singapore, Ria Tan

LEAF-SCALED SEA SNAKE *Aipysurus foliosquama* Smith, 1926

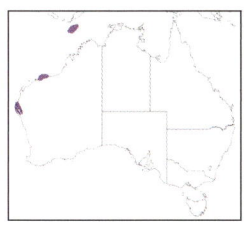

PRONUNCIATION *App-shoe-russ foe-lee-o-squa-mah*.
ETYMOLOGY Leaf-scaled high-tail, pertaining to the leaf-like scalation.
TYPE LOCALITY Ashmore Reef, Timor Sea.
APPEARANCE Small, moderately robust snake with head moderately distinct from body. Dorsal colouration brown to dark brown, sometimes with obscure darker cross-bands that taper on flanks. Iris usually pale. Buccal cavity and tongue pale. Ventral colouration greybrown, sparsely flecked with white. Adult females larger than males, reaching 100cm TL. **Scalation** MB 19–21 rows, 139–153 VENT, SUB 20–29 all single and anal scale divided. Scales smooth but overlap extensively, giving the snake a rough appearance. **Similar species** *A. apraefrontalis* (p. 232).

RANGE Encountered in waters in WA, from the Hibernia and Ashmore Reefs. Also a localized population at Shark Bay.

COMMENTS Nocturnal. Lives in waters over both coral and rocky reefs and in Shark Bay, over seagrass beds. **Diet** Small fish. **Reproduction** One gravid female found in January contained two young. Neonate size not recorded. **Disposition** Inoffensive.

BITE/VENOM VENOMOUS

IUCN LISTING Critically Endangered.

Ashmore Reef, WA, Hal Cogger

Barrow Island, WA, Brad Maryan

Shark Bay, WA, David Robinson

Ashmore Reef, WA, Hal Cogger

DUSKY SEA SNAKE *Aipysurus fuscus* (Tschudi, 1837)

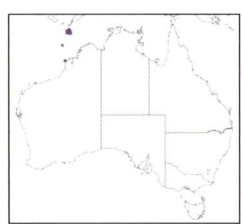

PRONUNCIATION *App-shoe-russ fus-cus*.
ETYMOLOGY Dusky high-tail.
TYPE LOCALITY Celebes (Sulawesi), Indonesia.
APPEARANCE Medium-sized, robust snake with head indistinct from body. Dorsal colouration dark chocolate-brown; some individuals banded entirely or with light bands that extend from ventral surface on to flanks. Ventral colouration similar to upper surface. Iris pale. Both sexes reach a length of 94cm TL. **Scalation** MB 17 rows, 156–172 VENT, SUB 24–37 all single and anal scale divided. Scales smooth and overlapping. **Similar species** *A. laevis* (opposite).
RANGE Encountered in waters in WA, from the Ashmore and Scott Reefs in the Timor Sea.
COMMENTS Predominantly nocturnal. Lives in waters over coral reefs and reef flats. **Diet** Gobies, other fish, fish eggs. **Reproduction** Viviparous. **Disposition** Inoffensive but will bite if harassed.
BITE/VENOM DANGEROUSLY VENOMOUS.
IUCN LISTING Endangered.

Ashmore Reef, WA, Hal Cogger

Ashmore Reef, WA, Mark O'Shea

Ashmore Reef, WA, Mark O'Shea

OLIVE SEA SNAKE *Aipysurus laevis* Lacépède, 1804
(Golden Sea Snake, Olive-brown Sea Snake)

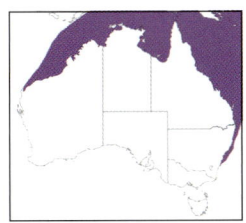

PRONUNCIATION *App-shoe-russ lay-viss*.

ETYMOLOGY Smooth high-tail.

TYPE LOCALITY Locker Island, off Onslow, WA.

APPEARANCE Large, robust snake with head slightly distinct from body. Dorsal colouration grey, yellow, brown to white, with or without darker and lighter blotches and specks; some individuals white with golden, apricot or dark brown head. Juveniles usually have bands that fade as they mature. Iris pale. Buccal cavity and tongue pale. Ventral colouration similar to upper surface. Adult females larger than males, reaching 185cm TL. **Scalation** MB 21–25 rows, 142–156 VENT, SUB 22–30 all single and anal scale divided. Scales usually smooth in females, while those of males can be rugose in some populations. These tubercles can become more prominent during breeding season. **Similar species** *A. fuscus* (opposite), *A. pooleorum* (p. 240), *A. tenuis* (p. 241).

RANGE Encountered in waters around Australia, north of Sydney, NSW, to Exmouth, WA. Also found in New Caledonia and southern PNG.

COMMENTS Diurnal. Lives in waters over both coral reefs and rocky areas. **Diet** Small fish, fish eggs, and occasionally prawns and crabs. **Reproduction** Viviparous. 1–11 per litter. Neonates approximately 37cm TL and born in autumn. Gravid in September–May. **Disposition** Typically inoffensive; misguided males may approach divers in what is perceived to be aggression – this is usually a case of mistaken identity or curiosity.

BITE/VENOM DANGEROUSLY VENOMOUS

IUCN LISTING Least Concern.

Fredrick Reef, QLD, Graham Edgar

Noumea, New Caledonia, Claire Goiran   *Noumea, Claire Goiran*

Barrow Island, WA, Graham Edgar

Ashmore Reef, WA, Hal Cogger

Mosaic Sea Snake *Aipysurus mosaicus*
Sanders, Rasmussen, Elmberg, Mumpuni, Guinea, Blias, Lee & Fry, 2012

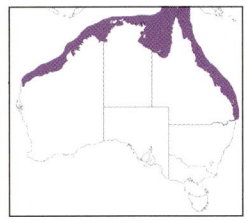

PRONUNCIATION *App-shoe-russ mo-say-e-cuss.*

ETYMOLOGY Mosaic patterned high-tail. Pertaining to the distinctive pattern of the species.

TYPE LOCALITY Gulf of Carpentaria, near Weipa, Australia.

APPEARANCE Medium-sized, robust snake with head indistinct from body. Dorsal colouration pale cream to light grey, with 15–22 dark bands. Iris and tongue pale. Ventral colouration black to cream, with dark midline. Both sexes reach a TL of 110cm. **Scalation** MB 17 rows, 140–157 VENT, SUB 25–38 all single and anal scale divided. Scales smooth in appearance. **Similar species** *A. eydouxii* (p. 2234).

RANGE Encountered from Exmouth, WA, across northern Australia to Brisbane, QLD. Waifs occur in Sydney, NSW. Also recorded off southern PNG.

COMMENTS Nocturnal. Found in turbid waters up to 50m deep. **Diet** Fish eggs; this specialization has led to an atrophied venom apparatus. **Reproduction** Viviparous. Gravid in April, with 1–5 young per litter. Neonates are approximately 26cm TL. **Disposition** Inoffensive but will bite if harassed.

BITE/VENOM HARMFUL The venom composition has changed due to dietary specialization and venom is far less toxic than that of most other members of the genus.

IUCN LISTING Least Concern.

Tin Can Bay, QLD, Josh Jensen

Thursday Island, QLD, Hal Cogger

Bundaberg, QLD, Steve Swanson

Exmouth, WA, Graham Edgar

SHARK BAY SEA SNAKE *Aipysurus pooleorum* L. A. Smith, 1974

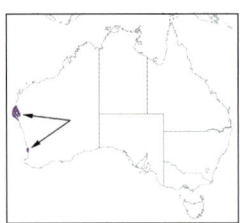

PRONUNCIATION *App-shoe-russ pool-e-or-rum*.

ETYMOLOGY Pooles' high-tail. Pertains to W. & W. Poole, both Fremantle fishermen.

TYPE LOCALITY Shark Bay, WA.

APPEARANCE Medium-sized, robust snake with head slightly distinct from body. Dorsal colouration grey to black with pale grey to cream bands. Iris and tongue pale. Juveniles much brighter than adults, with cleaner markings. Ventral colouration similar to upper surface. Both sexes reach 114cm TL. **Scalation** MB 20–23 rows, 146–159 VENT, SUB 25–33 all single and anal scale divided. Both sexes have overlapping scales. Females have smooth scales, and males generally have tubercles, especially on lower flanks. These can become more prominent during breeding season. **Similar species** *A. laevis* (p. 237).

RANGE Encountered in waters around Shark Bay, occasionally straying to Perth, WA.

COMMENTS Cathemeral. Lives in waters over both coral and rocky reefs, and over seagrass beds in Shark Bay. **Diet** Small fish. **Reproduction** Thought to be livebearers. Neonate size not recorded. **Disposition** Typically inoffensive, but can be defensive when handled.

BITE/VENOM DANGEROUSLY VENOMOUS

IUCN LISTING Vulnerable.

Shark Bay, WA, Scott Eipper

Shark Bay, WA, Scott Eipper

Quobba Beach, WA, Rob Newman

Shark Bay, WA, Sue Blyde

Mjoberg's Sea Snake *Aipysurus tenuis* Lönnberg & Anderson, 1913
(Arafura Sea Snake)

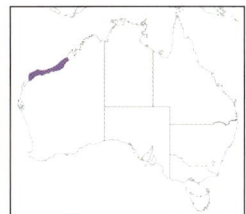

PRONUNCIATION *App-shoe-russ ten-u-iss.*

ETYMOLOGY Slender high-tail. Pertaining to the slender forebody of the species

TYPE LOCALITY Cape Jaubert, Broome, WA

APPEARANCE Large snake with medium build; noticeably lighter forebody in comparison to sympatric *Aipysurus*. Head slightly distinct from body. Dorsal colouration cream to pale grey or brown, sometimes with dark brown speckles forming longitudinal stripes and specks. Golden, apricot or dark brown head. Iris and tongue pale. Ventral colouration similar to upper surface. Both sexes reach 130cm TL. **Scalation** MB 19 rows, 185–194 VENT, SUB 35–40 all single and anal scale divided. Scales smooth in appearance. **Similar species** *A. fuscus* (p. 236), *A. pooleorum* (opposite), *A. laevis* (p. 237).

RANGE Encountered in waters from Broome to the Dampier Archipelago, WA.

COMMENTS Cathemeral. Lives in waters over both coral and rocky reefs. Also hunts in the tidal shallows, investigating burrows. **Diet** Small fish. **Reproduction** Thought to be livebearers. Neonate size not recorded. **Disposition** Typically inoffensive, but can be defensive when handled.

BITE/VENOM DANGEROUSLY VENOMOUS

IUCN LISTING Vulnerable.

Cape Keraudren, WA, Jason Harrison

80 Mile Beach, WA, Lauren Alexander-Kay

Broome, WA, Ruchira Somaweera

Turtle-headed Sea Snakes, genus *Emydocephalus* Krefft, 1869

The genus *Emydocephalus* currently comprises three recognized species, two of which occur in Australia. One species is endemic to Australia. Further work on the populations currently assigned to *E. annulatus* off the NW Australian coast and the Philippines is ongoing. **VENOM** Toxicity is unknown, as there is no evidence of envenomations. Venom apparatus has atrophied due to dietary specialization. **Species-level identification difficulty** (within Australia) – 3.

ETYMOLOGY Turtle head.

TYPE SPECIES *Emydocephalus annulatus*.

Key to *Emydocephalus*

1. Usually less than 144 ventral scales; first supralabial scale not in contact with preocular scale .... *E. annulatus* (opposite)
   144–146 Ventral scales; first supralabial scale in contact with preocular scale .... *E. oriarus* (p. 245)

Emydocephalus annulatus, (pre-ecdysis), male, Noumea, New Caledonia, Claire Goiran

TURTLE-HEADED SEA SNAKE *Emydocephalus annulatus* Krefft, 1869

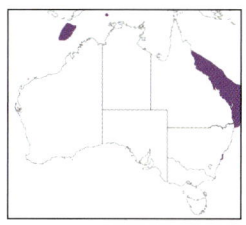

PRONUNCIATION *Em-e-doh-keff-ah-luss an-u-lah-tuss*.
ETYMOLOGY Ringed turtle head.
TYPE LOCALITY Probably the Australian seas.
APPEARANCE Medium-sized snake with moderately robust build. Head indistinct from body. Colouration very variable; some individuals heavily patterned, others plain. Usually dark coloured, with light markings forming bands and rings. Head blunt and rounded. Iris dark. Tongue pale. Both sexes reach 80cm TL. Potentially a species complex.
Scalation MB 15–17 rows, 128–146 VENT, SUB 24–33 all single and anal scale single. Scales smooth to weakly keeled in appearance. Males have a conical-shaped projection on end of rostral scale.
Similar species *E. oriarus* (p. 245).

RANGE Encountered in waters from the Timor Sea in WA, in QLD from Gladstone to Brisbane, and vagrant individuals off Sydney, NSW. Also found in tropical waters between Timor and New Caledonia.

COMMENTS Diurnal. Lives in shallow waters over coral reefs. **Diet** Fish eggs. **Reproduction** Viviparous. 2–8 per litter. Neonate size not recorded. Gravid in October–April. **Disposition** Typically inoffensive; misguided males may approach divers in what is perceived to be aggression, but is usually a case of mistaken identity.

BITE/VENOM HARMFUL
IUCN LISTING Least Concern.

Female, Noumea, New Caledonia, Claire Goiran

Noumea, New Caledonia, Claire Goiran

Noumea, New Caledonia, Claire Goiran

Fredrick Reef, QLD, Graham Edgar

Western Turtle-headed Sea Snake *Emydocephalus oriarus*
Nankivell, Goiran, Hourston, Shine, Rasmussen, Thompson & Sanders, 2020

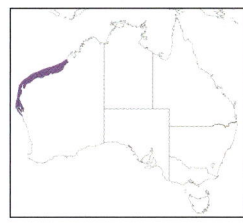

PRONUNCIATION *Em-e-doh-keff-ah-luss or-ree-ah-russ.*
ETYMOLOGY Coastal turtle head.
TYPE LOCALITY Shark Bay, WA.
APPEARANCE Medium-sized snake with moderately robust build. Head indistinct from body. Very variable in colouration and pattern; some individuals banded with black and white to yellow, and others heavily spotted with markings forming coalesced bands. Head blunt and rounded. Males have a conical-shaped projection on end of rostral scale. Iris pale to dark. Tongue pale. Both sexes reach 116cm TL. **Scalation** MB 17 rows, 144–146 VENT, SUB 31–35 all single and anal scale single. Scales smooth in appearance. **Similar species** *E. annulatus* (p. 243).
RANGE Encountered in coastal waters from Broome to Shark Bay, WA.
COMMENTS Diurnal. Lives in shallow waters over soft sandy bottoms. **Diet** Fish eggs. **Reproduction** Viviparous. 2–8 per litter. Neonate size not recorded. **Disposition** Inoffensive, but will bite if threatened.
BITE/VENOM HARMFUL
IUCN LISTING Least Concern.

Exmouth, WA, Alex Hoschke

Exmouth, WA, Alex Hoschke

Exmouth, WA, Ruchira Somaweera

Lighthouse Bay, Exmouth, Alex Hoschke,

Genus *Ephalophis* Smith, 1931

This genus comprises one monotypic, endemic species. It is one of three sea snake species that commonly occur in mangroves and estuaries, and is occasionally seen hunting in shallow pools formed on tidal flats. **VENOM** Toxicity unknown, as there is no evidence of envenomations.
Species-level identification difficulty (within Australia) – 2.

ETYMOLOGY Seaside snake.

TYPE SPECIES *Ephalophis greyi*.

Ephalophis greyae, Port Hedland, WA, Bruce Edley

Ephalophis greyae, Onslow, WA, Ruchira Somaweera

Mangrove Sea Snake *Ephalophis greyae* M. A. Smith, 1931

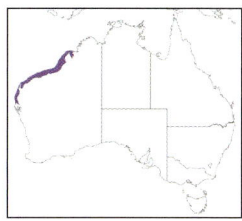

PRONUNCIATION *F-ah-low-fiss grey-e.*

ETYMOLOGY Grey's seaside snake, pertaining to B. Grey, who collected the first specimen.

TYPE LOCALITY Cape Boileau, NW coast of Australia.

APPEARANCE Small snake, with a medium build, the head is slightly distinct from body. Dorsal colouration cream to white with grey to black bands, and skin between scales black, forming reticulated pattern. Juveniles brighter than adults, with distinct markings. Iris and tongue both pale. Ventral colouration similar to dorsum. Both sexes reach 65cm TL. **Scalation** MB 19–21 rows, 151–184 VENT, SUB 24–33 all single and anal scale divided. Scales weakly keeled dorsally, becoming smooth on to flanks.

RANGE Encountered in waters from Shark Bay to Kings Sound, WA.

COMMENTS Activity period dictated by tidal influences. Lives in tidal waters of estuaries and mangrove-lined creeks. **Diet** Small, hole-dwelling fish. **Reproduction** Viviparous. 3–6 per litter. Neonate size not recorded. **Disposition** Inoffensive but will bite if harassed.

BITE/VENOM VENOMOUS

IUCN LISTING Least Concern.

Port Hedland, WA, Doris Teufel

Port Hedland, WA, Doris Teufel

Giralia Bay, WA, Brad Maryan

Genus *Hydrelaps* Boulenger, 1896

Monotypic genus. One of three sea snake species that commonly occur in mangroves and estuaries. Occasionally seen hunting in shallow pools formed on tidal flats. **VENOM** Toxicity unknown, as there is no evidence of envenomations. **Species-level identification difficulty** (within Australia) – 2.

ETYMOLOGY Water elaps – 'Elaps' after genus of African elapid snakes.

TYPE SPECIES *Hydrelaps darwiniensis*.

Hydrelaps darwinensis, Onslow, WA, Tim Karnasuta

Hydrelaps darwiniensis, Weipa, QLD, Hal Cogger

Black-ringed Mangrove Snake *Hydrelaps darwiniensis* Boulenger, 1896
(Port Darwin Sea Snake)

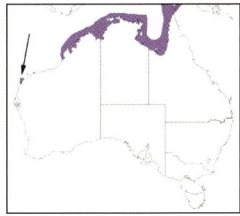

PRONUNCIATION *Hi-dree-laps dar-win-n-siss.*
ETYMOLOGY Darwin's water *Elaps*, in reference to the type locality.
TYPE LOCALITY Port Darwin, NT.
APPEARANCE Small snake with medium build and head slightly distinct from body. Dorsal colouration cream to yellow with black bands or dark polygons; head sometimes grey or black. Ventral colouration similar to upper surface. Iris dark grey. Tongue is pale. Both sexes reach 52cm TL. **Scalation** MB 25–27 rows, 160–179 VENT, SUB 20–36 all single and anal scale divided. Scales smooth in appearance.
RANGE Encountered in waters from Exmouth, WA, across northern Australia to western Cape York Peninsula, QLD. Also found in southern NG.
COMMENTS Nocturnal. Lives in shallow tidal waters of estuaries and mangrove-lined creeks.
Diet Small, hole-dwelling fish. **Reproduction** Viviparous. 3–6 per litter. Neonate size not recorded.
Disposition Typically inoffensive, but can be defensive when handled.
BITE/VENOM VENOMOUS
IUCN LISTING Least Concern.

Milingimbi, NT, Paul Horner

Milingimbi, NT, Paul Horner

Darwin, NT, Ruchira Somaweera

Genus *Hydrophis* Sonnini & Latreille, 1802

The genus *Hydrophis* comprises 48 species, of which 21 occur in Australia and one is endemic. In 2013, a paper looking at the phylogenetic relationships of sea snakes demonstrated that many morphologically divergent species that had been placed into different genera were in fact closely related. This resulted in placing some formerly taxonomically stable genera into synonymy. Genera that were synonymized included *Acalyptophis*, *Astrotia*, *Disteira*, *Enhydrina*, *Lapemis*, *Kerilia*, *Pelamis* and *Thalassophina*. **VENOM** In some species strongly neurotoxic, with evidence of myotoxicity. Bites have resulted in, or could cause fatalities. Sea snake or Polyvalent antivenom is used to treat envenomations. **Species-level identification difficulty** (within Australia) – 5.
ETYMOLOGY Water snake.
TYPE SPECIES *Hydrophis laticauda*

Key to *Hydrophis*

1. Mental groove present 2
   No mental groove .... *H. platurus* (p. 269)

2. Head scales symmetrical and enlarged;
   supraocular scales without spines or tubercles 3
   Head scales asymmetrical and small;
   supraocular scales with spines or tubercles .... *H. peronii* (p. 268)

3. Mental scale broader than long, not obscured by mental groove 4
   Mental scale 4 times longer than broad,
   obscured by mental groove .... *H. zweifeli* (p. 273)

4. Ventral scales not in two overlapping rows 5
   Ventral scales divided into a pair of overlapping scale rows ...... *H. stokesii* (p. 270)

5. Ventral scales do not become obscure posteriorly;
   midbody scales same size, not becoming larger on flanks 6
   Ventral scales become obscure posteriorly;
   lower flank midbody scales larger on upper flanks .... *H. curtus* (p. 256)

6. Body with less than 70 bands around it 7
   Body with more than 70 bands around it;
   fewer than 25 scale rows on neck .... *H. vorisi* (p. 272)

7. No conspicuous white eye-ring 8
   Head black to dark grey with a conspicuous
   white ring around the eye .... *H. kingii* (p. 261)

8. Pattern where present, does not form
   a series of distinct pentagons along the body 9
   Pattern of distinct pentagons along the body .... *H. czeblukovi* (p. 257)

9. Pattern forms a series of blotches or bands with
   spots with blotches or flecks in pale spaces 10
   Pattern where present, forms a series of blotches
   or bands without spots or flecks in pale spaces 13

10. Less than 340 ventral scales 11
    345 ventral scales or more .... *H. elegans* (p. 259)

11. Midbody scale rows 45 or less 12
    Midbody scale rows 50–67 .... *H. ocellatus* (p. 266)

12 Head normal sized, pale yellow to grey with dark flecking.................*H. major* (p. 264)
    Head small compared to the size of the body, and black.................*H. macdowelli* (p. 263)

13 Most midbody scales are heavily keeled.................14
    Most midbody scales smooth.................15

14 Midbody scale rows 33–35; 46–60 bands.................*H. donaldi* (p. 258)
    Midbody scale rows 37–39; 41–43 bands.................*H. caerulescens* (p. 254)

15 More than 42 dark bands or evidence of
    faded bands in older individuals.................16
    40 dark bands or less.................*H. coggeri* (p. 255)

16 Head typically sized or noticeably enlarged.................17
    Head small in size relative to body, and black;
    bands quite broad, fading onto the flanks in adults.................*H. atriceps* (p. 252)

17 Bands when present, dark grey to black or
    faded to plain grey dorsum, with a lighter lower lateral zone.................18
    Up to 60 broad bands; bronze to golden-brown,
    head usually darker, bands darker and more defined
    in juveniles; midbody scales in 32 to 37 rows;.................*H. belcheri* (p. 253)

18 1 or 2 anterior temporal scale(s).................19
    3 anterior temporal scales; 44–46 midbody scale rows.................*H. laboutei* (p. 262)

19 A single anterior temporal scale.................20
    2 anterior temporal scales; 35–48 midbody scale rows.................*H. inornatus* (p. 260)

20 Midbody scale rows 37–43; 50–70 bands.................*H. melanosoma* (p. 265)
    Midbody scale rows 45–49; 45–72 bands (when present).................*H. pacificus* (p. 267)

Hydrophis ocellatus, Fingle Island, Port Stephens, NSW, Hal Cogger

Black-headed Sea Snake Hydrophis atriceps Günther, 1864

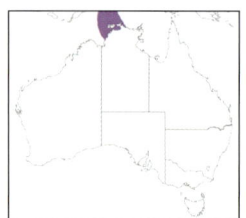

PRONUNCIATION *Hi-dro-fiss at-ree-seps.*

ETYMOLOGY Black-headed water snake; pertains to dark-coloured head of the species.

TYPE LOCALITY Thailand.

APPEARANCE Medium-sized snake with very small head that gradually transitions into the body, which is of medium build. Dorsal colouration cream to white with grey to black bands. Head and nuchal region as the name suggests, black. Juveniles much brighter than adults, with clear black markings. Iris dark. Buccal cavity and tongue both pale. Ventral colouration similar to upper surface. Both sexes reach 100cm TL. **Scalation** MB 39–49 rows, 371–392 VENT, SUB 47–59 all single and anal scale divided. Scales weakly keeled. **Similar species** *H. coggeri* (p. 255), *H. elegans* (p. 259).

RANGE Encountered in waters around the NW of Australia from Darwin, NT. Also found in Indonesia and southern PNG.

COMMENTS Nocturnal. Lives in waters over both coral reefs and rocky areas. **Diet** Small eels, hole-dwelling fish. **Reproduction** Viviparous. 1–10 per litter. Neonate size not recorded. Gravid in July–January. **Disposition** Typically inoffensive, but can be defensive when handled.

BITE/VENOM DANGEROUSLY VENOMOUS

IUCN LISTING Least Concern.

Muar, Malaysia, Harold Voris

Vietnam, Arne Rasmussen

Gulf of Carpentaria, NT, Scott Eipper

Vietnam, Arne Rasmussen

BELCHER'S SEA SNAKE *Hydrophis belcheri* (Gray, 1849)
(Faint-banded Sea Snake)

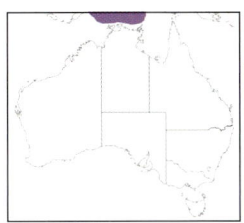

PRONUNCIATION *Hi-dro-fiss bell-cher-e.*
ETYMOLOGY Belcher's water snake; pertains to E. Belcher, English admiral and explorer.
TYPE LOCALITY New Guinea.
APPEARANCE Medium-sized, robust snake with head indistinct from body. Dorsal colouration cream to beige with grey, poorly defined bands. Juveniles much brighter than adults, with clear dark markings. Iris pale. Ventral colouration similar to upper surface. Both sexes reach 100cm TL. **Scalation** MB 32–36 rows, 278–313 VENT, SUB 28–43 all single and anal scale divided. Scales strongly keeled.
RANGE Encountered in waters north of Australia, with waifs entering the Arafura Sea from Indonesia and PNG.
COMMENTS Nocturnal. Lives in waters over coral reefs. **Diet** Small eels. **Reproduction** Viviparous. 2–4 per litter. Neonate size not recorded. **Disposition** Typically inoffensive, but can be defensive when handled.
BITE/VENOM DANGEROUSLY VENOMOUS
IUCN LISTING Least Concern.

New Guinea, Hal Cogger

New Guinea, Hal Cogger

Vietnam, Arne Rasmussen

Vietnam, Arne Rasmussen

DWARF SEA SNAKE *Hydrophis caerulescens* (Shaw, 1802)

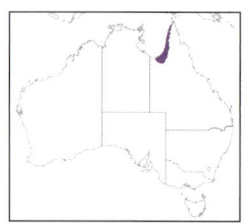

PRONUNCIATION *Hi-dro-fiss say-rule-e-sens.*
ETYMOLOGY Bluish water snake.
TYPE LOCALITY Indian Ocean, Vizagapatam (Visakhapatnam), India.
APPEARANCE Medium-sized snake with smallish head that gradually transitions into body with a medium to heavy build. Dorsal colouration cream to bluish-grey with darker bands. Juveniles brighter than adults, with clear dark markings. Iris dark. Tongue and buccal cavity both pale. Ventral colouration similar to upper surface. Adult males larger than females, reaching 80cm TL. **Scalation** MB 38–54 rows, 253–334 VENT SUB 40–52 all single and anal scale divided. Scales strongly keeled. **Similar species** *H. donaldi* (p. 258).
RANGE Encountered in waters off northern Australia, in the Gulf of Carpentaria, QLD, and a localized record from the Fitzroy River in eastern QLD. Also found throughout SE Asia.
COMMENTS Nocturnal. Lives in river mouths, mangroves and estuaries. **Diet** Small eels, hole–dwelling fish. **Reproduction** Viviparous. 2–15 per litter. Neonate size not recorded. Females gravid in September–June. **Disposition** Typically inoffensive, but can be defensive when handled.
BITE/VENOM DANGEROUSLY VENOMOUS
IUCN LISTING Least Concern.

Gujarat, India, Aadit Patel

Gujarat, India, Aadit Patel

Gujarat, India, Aadit Patel

Juvenile, Gujarat, India, Aadit Patel

Cogger's Sea Snake *Hydrophis coggeri* (Kharin, 1984)
(Slender-necked Sea Snake)

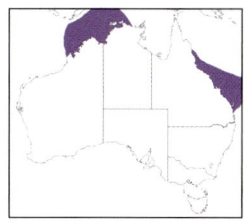

PRONUNCIATION *Hi-dro-fiss cog-err-ee*.

ETYMOLOGY Cogger's water snake; pertains to H. Cogger, Australian herpetologist.

TYPE LOCALITY Port Suva, Fiji.

APPEARANCE Large snake with smallish head that gradually transitions into body with a medium build. Dorsal colouration cream to brown with dark grey to black bands. Juveniles much brighter than adults, with clear black markings. Iris pale. Tongue pale. Ventral colouration similar to upper surface. Adult females larger than males, reaching 120cm TL. **Scalation** MB 34–35 rows, 298–322 VENT, SUB 37 all single and anal scale divided. Scales strongly keeled. **Similar species** *H. atriceps* (p. 252), *H. elegans* (p. 259).

RANGE Encountered in waters off NW Australia, at Ashmore and Scott's Reefs. Also found in the Timor Sea, Vanuatu, New Caledonia and Fiji.

COMMENTS Nocturnal. Lives in waters over both coral reefs and rocky areas. Observed hunting over sand flats. **Diet** Snake eels. **Reproduction** Viviparous. 3–8 per litter. Neonate size not recorded. Females gravid in October. **Disposition** Typically inoffensive, but can be defensive when handled.

BITE/VENOM DANGEROUSLY VENOMOUS

IUCN LISTING Least Concern.

Noumea, New Caledonia, Claire Goiran

Noumea, New Caledonia, Claire Goiran

Ashmore Reef, WA, Hal Cogger

Juvenile, Ashmore Reef, WA, Hal Cogger

SPINE-BELLIED SEA SNAKE *Hydrophis curtus* (Shaw, 1802)

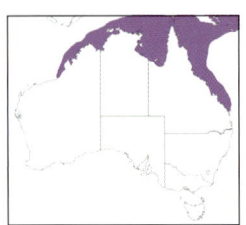

PRONUNCIATION *Hi-dro-fiss kerr-tuss*.
ETYMOLOGY Short water snake.
TYPE LOCALITY East India (by reference in the description).
APPEARANCE Medium-sized, robust snake with head indistinct from body. Dorsal colouration pale brown to dark grey above, with wide cream to white bands. Darker pigments fade with age. Juveniles white or cream with grey bands. Iris dark to greyish-blue. Tongue pale; buccal cavity pink. Ventral colouration similar to upper surface. Adult females larger than males, reaching 100cm TL. **Scalation** MB 25–43 rows, 114–230 VENT, SUB 239–43 all single and anal scale divided. Scales juxtaposed rather than overlapping. This species is sexually dimorphic, with males developing spine-like projections on lower scales.
RANGE Encountered north of Gladstone, QLD, to the Kimberleys, WA. Also occurs across the Indian Ocean into the Arabian Gulf.
COMMENTS Mainly nocturnal. Lives in waters over coral reefs, rocky areas and sandy estuaries. Genetically there is divergence between the Australasian and Indian populations. **Diet** Fish. **Reproduction** Viviparous. 1–15 per litter. Neonates approximately 35cm TL. Gravid in October–April. **Disposition** Typically inoffensive, but can be strongly defensive when handled, readily biting.
BITE/VENOM DANGEROUSLY VENOMOUS
IUCN LISTING Least Concern.

Townsville, QLD, Anders Zimny

Townsville, QLD, Anders Zimny

Juvenile, Fannie Bay, NT, Paul Horner

Weipa, QLD, Hal Cogger

GEOMETRICAL SEA SNAKE *Hydrophis czeblukovi* (Kharin, 1984)
(Fine-spined Sea Snake)

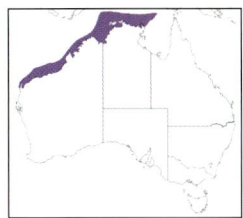

PRONUNCIATION *Hi-dro-fiss zee-blue-kov-ee*.

ETYMOLOGY Czeblukov's water snake; pertains to V. Czeblukov, Russian herpetologist.

TYPE LOCALITY Arafura Sea, NT.

APPEARANCE Large snake with smallish head that gradually transitions into body with a medium build. Dorsal colouration light to dark grey, with large, narrow cream to pale yellow, hexagonal markings. Juveniles thought to be similar to adults. Iris pale. Tongue pale. Ventral colouration usually grey. Both sexes reach 120cm TL. **Scalation** MB 55–56 rows, 315–324 VENT, SUB 48–52 all single and anal scale divided. Scales strongly keeled.

RANGE Encountered in waters from Exmouth to Cable Beach, WA. Sporadic records along the northern coast to Arnhem Land, NT. Also found in Indonesia and PNG.

COMMENTS Nocturnal. Lives in waters over both coral reefs and sandy areas. **Diet** Eels. **Reproduction** Probably gives birth to live young. Neonate size not recorded. **Disposition** Typically inoffensive, but can be defensive when handled.

BITE/VENOM DANGEROUSLY VENOMOUS

IUCN LISTING Least Concern.

Eighty Mile Beach, WA, Ruchira Somaweera

Eighty Mile Beach, WA, Brad Maryan

Rough-scaled Sea Snake *Hydrophis donaldi* Ukuwela, Sanders & Fry, 2012

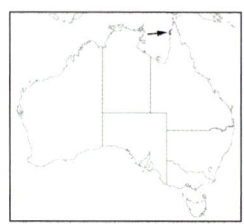

PRONUNCIATION *Hi-dro-fiss do-nald-e*.

ETYMOLOGY Donald's water snake; pertains to D. Donald, who assisted in collecting the first specimens.

TYPE LOCALITY Gulf of Carpentaria, Weipa, QLD.

APPEARANCE Medium-sized snake with smallish head that gradually transitions into body with medium to heavy build. Dorsal colouration light brown to gold, with or without pale grey-brown bands. Iris dark. Tongue and buccal cavity pale. Ventral colouration similar to upper surface. Both sexes reach 90cm TL. **Scalation** MB 33–35 rows, 246–288 VENT, SUB 42–51 all single and anal scale divided. Scales strongly keeled. Sexually dimorphic, with males developing spine-like projections on lower scales. **Similar species** *H. caerulescens* (p. 254).

RANGE Encountered in waters of the Gulf of Carpentaria, QLD.

COMMENTS Nocturnal. Lives in waters over mud flats, estuaries and shallow protected areas, including seagrass beds. **Diet** Likely to be fish. **Reproduction** Probably gives birth to live young. Neonate size not recorded. **Disposition** Typically inoffensive, but can be defensive when handled.

BITE/VENOM DANGEROUSLY VENOMOUS

IUCN LISTING Least Concern.

Weipa, QLD, Scott Eipper

Weipa, QLD, Scott Eipper

Weipa, QLD, Hal Cogger

Elegant Sea Snake *Hydrophis elegans* (Gray, 1842)

PRONUNCIATION *Hi-dro-fiss elle-e-gans.*
ETYMOLOGY Elegant water snake.
TYPE LOCALITY Port Essington, NT.
APPEARANCE Large, moderately slender snake, with head slightly distinct from body. Dorsal colouration cream to white, with both broad grey to black bands and dark colour spots in the pale interspaces. Juveniles much brighter than adults, with clear black markings. Iris, tongue and buccal cavity all pale. Ventral colouration similar to upper surface. Adult females larger than males, reaching 180cm TL. One of the longest sea snakes. **Scalation** MB 39–49 rows, 345–433 VENT, SUB 36–43 all single and anal scale divided. Scales weakly keeled. **Similar species** *H. atriceps* (p. 252), *H. coggeri* (p. 255).
RANGE Encountered in waters around Australia, north of Sydney, NSW, to Exmouth, WA.
COMMENTS Nocturnal. Lives in waters over both coral reefs and rocky areas. Commonly encountered across much of its range. **Diet** Small eels, hole-dwelling fish. **Reproduction** Viviparous. 3–30 per litter. Neonate size 42–47cm TL. Females can be gravid throughout the year. **Disposition** Typically inoffensive, but can be defensive when handled.
BITE/VENOM DANGEROUSLY VENOMOUS
IUCN LISTING Least Concern.

Juvenile, Bundaberg, QLD, Scott Eipper

Juvenile, Bundaberg, QLD, Scott Eipper

Bundaberg, QLD, Steve Swanson

Exmouth, WA, Ruchira Somaweera

Plain Sea Snake *Hydrophis inornatus* (Gray, 1849)

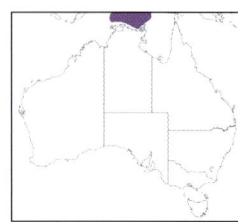

PRONUNCIATION *Hi-dro-fiss in-or-nah-tuss*.

ETYMOLOGY Plain water snake, due to lack of prominent markings on type specimen.

TYPE LOCALITY Indian Ocean.

APPEARANCE Medium-sized, slender snake with head distinct from body. Dorsal colouration pale grey to beige, with sparse pale bands. Juveniles thought to be similar to adults. Ventral colouration similar to upper surface. Both sexes reach 80cm TL. **Scalation** MB 35–48 rows, 195–293 VENT, SUB 36–43 all single and anal scale divided. Scales smooth.

RANGE Encountered in Australian waters, in the Torres Strait off Daru Island and off the NT coastline. Also found off the PNG coast.

COMMENTS Nocturnal. Lives in waters over both coral reefs and rocky areas. It has been suggested that specimens from Australia were incorrectly identified and that the species does not occur in Australian waters. **Diet** Small eels, gobies. **Reproduction** Viviparous. 4–10 per litter. Neonate size not recorded. Females gravid in February–August. **Disposition** Expected to be inoffensive, but defensive when handled.

BITE/VENOM DANGEROUSLY VENOMOUS

IUCN LISTING Least Concern.

Indian Ocean, Hal Cogger

Indian Ocean, Hal Cogger

SPECTACLED SEA SNAKE *Hydrophis kingii* Boulenger, 1896

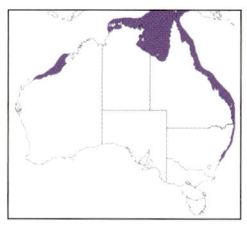

PRONUNCIATION Hi-dro-fiss king-ee.

ETYMOLOGY King's water snake; pertains to P. King, collector of the holotype.

TYPE LOCALITY Northern Australia.

APPEARANCE Large, moderately slender snake, with head indistinct from body. Dorsal colouration from head and first third usually includes broad dark bands, fading to bluish-grey over rear two-thirds. Eye usually has prominent white ring around it. Between bands pale grey to cream. Iris dark. Tongue pale. Ventral colouration similar to upper surface. Both sexes reach 190cm TL. **Scalation** MB 37–39 rows, 324–342 VENT, SUB 36–43 all single and anal scale divided. Scales strongly keeled.

RANGE Encountered in waters north of Coffs Harbour, NSW, across northern Australia to Barrow Island, WA. Also found in Indonesia and PNG.

COMMENTS Nocturnal. Lives in waters over coral reefs and rocky areas, as well as over mudflats and seagrass beds. **Diet** Eels, small fish. **Reproduction** Viviparous. 1–8 per litter. Neonate size not recorded. Gravid in December–March. **Disposition** Typically inoffensive, but can be defensive when handled.

BITE/VENOM DANGEROUSLY VENOMOUS

IUCN LISTING Least Concern.

Hervey Bay, QLD, Scott Eipper

Fannie Bay, NT, Paul Horner

Darwin, NT, Ruchira Somaweera

Eighty Mile Beach, WA, Brad Maryan

LABOUTE'S SEA SNAKE *Hydrophis laboutei* A. R. Rasmussen & Ineich, 2000

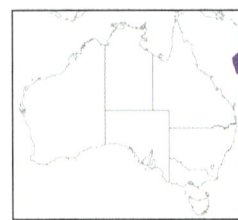

PRONUNCIATION *Hi-dro-fiss la-boo-tey-ee*.

ETYMOLOGY Laboute's water snake pertains to P. Laboute, collector of the holotype.

TYPE LOCALITY Chesterfield Reefs, New Caledonia.

APPEARANCE Medium-sized, moderately robust snake with head distinct from body. Dorsal colouration black with thin cream bands. Ventral colouration similar to upper surface. Both sexes reach 100cm TL. **Scalation** MB 44–46 rows, 265–280 VENT, 35–39 SUB and anal scale divided. Scales smooth.

RANGE Encountered in waters east of Australia at the Chesterfield Reef, and expected to occur in Australian waters. Also found in Noumea.

COMMENTS Nocturnal. Lives in waters with coral reefs and sandy, calcified bottoms with *Halimeda*, to a depth of 62m. Known from a total of three individuals. **Diet** Small eels, hole-dwelling fish. **Reproduction** Presumed to give birth to live young. Neonate size not recorded. **Disposition** Unknown, but expected to be similar to others in the genus.

BITE/VENOM VENOMOUS

IUCN LISTING Data Deficient.

Chesterfield Reef, New Caledonia, Scott Eipper

Chesterfield Reef, New Caledonia, Scott Eipper

Chesterfield Reef, New Caledonia, Scott Eipper

SMALL-HEADED SEA SNAKE *Hydrophis macdowelli* Kharin, 1983

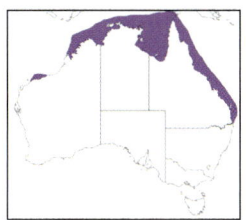

PRONUNCIATION *Hi-dro-fiss   mak-dowel-ee.*
ETYMOLOGY McDowell's water snake; pertains to S. McDowell, American herpetologist.
TYPE LOCALITY North Australian Shelf.
APPEARANCE Small to medium-sized snake, with very small head that gradually transitions into body with medium build. Dorsal colouration pale cream to white, with grey dorsal bands above forming spots on lower flanks. Bands fade with age. Ventral colouration similar to upper surface. Iris pale. Both sexes reach 80cm TL. **Scalation** MB 35–39 rows, 252–274 VENT, SUB 36–44 all single and anal scale divided. Scales weakly keeled.
RANGE Encountered north of Brisbane, QLD, to Barrow Island, WA. Also occurs in PNG and New Caledonia.
COMMENTS Cathemeral. Lives in waters over coral reefs, rocky areas and sandy estuaries. **Diet** Fish. **Reproduction** Viviparous. 2–5 per litter. Neonate size not recorded. Gravid in December–April. **Disposition** Typically inoffensive, but can be defensive when handled.
BITE/VENOM DANGEROUSLY VENOMOUS
IUCN LISTING Least Concern.

Gulf of Carpentaria, NT, Scott Eipper

Noumea, New Caledonia, Claire Goiran

Broome, WA, Ruchira Somaweera

Olive-headed Sea Snake *Hydrophis major* (Shaw, 1802)

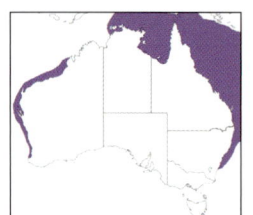

PRONUNCIATION *Hi-dro-fiss may-jor.*

ETYMOLOGY Greater water snake.

TYPE LOCALITY Indian Ocean.

APPEARANCE Large, moderately robust snake with head indistinct from body. Dorsal colouration cream to white, with grey to black, elongated cross-bars in pale interspaces; thin, dark bands present. Juveniles much brighter than adults, with clear black markings. Iris and tongue pale. Buccal cavity pink. Ventral colouration similar to upper surface. Adult females larger than males, reaching 160cm TL. **Scalation** MB 33–43 rows, 197–245 VENT, SUB 39–43 all single and anal scale divided. Scales weakly keeled.

RANGE Encountered in waters around Australia, north of Sydney, NSW, to Bunbury, WA.

COMMENTS Cathemeral. Lives in waters over both coral reefs and rocky areas, as well as sandy inlets, estuaries and seagrass beds. **Diet** Small fish. **Reproduction** Viviparous. 6–12 per litter. Neonate size not recorded. Gravid in October–April. **Disposition** Typically inoffensive, but can be defensive when handled.

BITE/VENOM DANGEROUSLY VENOMOUS

IUCN LISTING Least Concern.

Bribie Island, QLD, Tie Eipper

Noumea, New Caledonia, Claire Goiran

Juvenile, Bribie Isand, QLD, Scott Eipper

Exmouth, WA, Ruchira Somaweera

Black-banded Robust Sea Snake *Hydrophis melanosoma* Günther, 1864

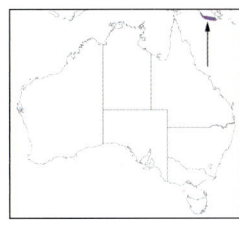

PRONUNCIATION *Hi-dro-fiss mel-an-o-so-ma.*
ETYMOLOGY Black-bodied water snake.
TYPE LOCALITY Unknown.
APPEARANCE Medium-sized, moderately robust snake with head slightly distinct from body. Dorsal colouration black with thin cream to yellow bands. Iris greyish-blue. Ventral colouration similar to upper surface. Both sexes reach 100cm TL. **Scalation** MB 37–43 rows, 260–370 VENT and anal scale divided. Scales weakly keeled.
RANGE Encountered in waters north of Australia, in the Torres Strait. Also found throughout Indonesia and Malaysia.
COMMENTS Nocturnal. Lives in waters over coral reefs, shoals and estuaries. It has been suggested that individuals from Australia were incorrectly identified and that the species does not occur in Australian waters. **Diet** Eels, catfish. **Reproduction** Viviparous. Up to six per litter. Neonate size not recorded. Females gravid in January–April. **Disposition** Typically inoffensive, but can be defensive when handled.
BITE/VENOM DANGEROUSLY VENOMOUS
IUCN LISTING Least Concern.

Malaysia, Scott Eipper

Muar, Malaysia, Harold Voris

Malaysia, Scott Eipper

Muar, Malaysia, Harold Voris

ORNATE SEA SNAKE *Hydrophis ocellatus* Gray, 1849
(Spotted Sea Snake)

PRONUNCIATION *Hi-dro-fiss oss-el-lah-tuss.*
ETYMOLOGY Eye-spotted water snake.
TYPE LOCALITY Australia.
APPEARANCE Large, moderately robust snake with head indistinct from body. Dorsal colouration whitish-cream laterally with pale grey dorsum, and broad grey-brown blotches and markings. Iris dark. Tongue pale. Ventral colouration similar to flanks. Adult males larger than females, reaching 150cm TL. **Scalation** MB 39–59 rows, 240–340 VENT 38–52 single SUB and anal scale divided. Scales smooth in appearance.
RANGE Encountered in waters around Australia, from Barrow Island, WA, across northern and eastern Australia, down to Brisbane, QLD. Waifs occur as far south as the Tas east coast. Also found in waters around Indonesia and PNG.
COMMENTS Nocturnal. Lives in waters over coral reefs and rocky areas, as well as over mudflats and seagrass beds. Formerly thought to be a geographic variant of the wider-ranging species *H. ornatus*. **Diet** Gobies, small fish. **Reproduction** Viviparous. 1–6 per litter. Neonate size not recorded. Gravid in October–January. **Disposition** Typically inoffensive, but can be defensive when handled.
BITE/VENOM DANGEROUSLY VENOMOUS
IUCN LISTING Least Concern.

Fingle Island, Port Stephens, NSW, Hal Cogger

Fingle Island, Port Stephens, NSW, Hal Cogger

Broome, WA, Ruchira Somaweera

Berau Gulf, West Papua, Indonesia, Hal Cogger

LARGE-HEADED SEA SNAKE *Hydrophis pacificus* Boulenger, 1896

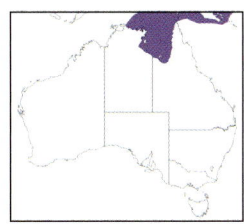

PRONUNCIATION *Hi-dro-fiss pa-sif-e-cuss*
ETYMOLOGY Pacific water snake.
TYPE LOCALITY New Britain, PNG.
APPEARANCE Large, robust snake with head distinct from body. Dorsal colouration grey, usually with dark bands that become obscure with age. Juveniles have clear black markings. Ventral colouration pale yellow to creamy-white. Both sexes reach 150cm TL. **Scalation** MB 45–49 rows, 320–430 VENT and anal scale divided. Scales smooth.
RANGE Encountered in waters in the Gulf of Carpentaria. Also found in waters off PNG.
COMMENTS Nocturnal. Lives in waters over coral reefs and rocky areas. The few Australian records are from prawn trawling and beach-washed individuals. Genetically, this species fits within the Annulated Sea snake *H. cyanocinctus* complex; further work may result in Australian animals being split into more than one taxon. **Diet** Small fish. **Reproduction** Viviparous. Up to 17 per litter. Neonate size not recorded. **Disposition** Typically inoffensive, but can be defensive when handled.
BITE/VENOM DANGEROUSLY VENOMOUS
IUCN LISTING Least Concern.

Vanderlin Island, Gulf of Carpentaria, NT, Scott Eipper

Holotype, New Britain, PNG, Hal Cogger

Mapoon, QLD, Krystal May

Vanderlin Island, Gulf of Carpentaria, NT, Scott Eipper

Horned Sea Snake *Hydrophis peronii* (Duméril, 1853)
(Spiny-headed Sea Snake)

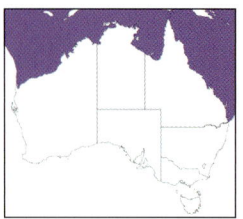

PRONUNCIATION *Hi-dro-fiss pear-ron-ee.*

ETYMOLOGY Peron's water snake; pertaining to F. Peron, French naturalist.

TYPE LOCALITY New Holland (Australia).

APPEARANCE Medium-sized snake with small head that gradually transitions into the body with a medium build. Dorsal colouration brown to grey with whitish blotches and markings; some individuals banded and others plain. Iris and tongue both pale. Buccal cavity pinkish. Ventral colouration similar to upper surface. Adult females larger than males, reaching 99cm TL. **Scalation** MB 21–31 rows, 142–203 VENT, SUB 44 all single and anal scale divided. Head strongly keeled with raised scales above eyes. Body scales strongly keeled.

RANGE Encountered in waters around Australia, north of Brisbane, QLD, to Barrow Island, WA. Also found in Indonesia and the Pacific.

COMMENTS Predominantly nocturnal but has been seen hunting by day. Lives in waters over coral reefs, mudflats and seagrass beds. **Diet** Gobies, small fish. **Reproduction** Viviparous. 1–10 per litter. Neonate size not recorded. Gravid in September–April. **Disposition** Typically inoffensive, but can be aggressively defensive if disturbed.

BITE/VENOM DANGEROUSLY VENOMOUS

IUCN LISTING Least Concern.

Ashmore Reef, WA, Hal Cogger

Darwin, NT, Steve Swanson

Saumarez Reef, QLD, Graham Edgar

Weipa, QLD, Brad Maryan

Yellow-bellied Sea Snake *Hydrophis platurus platurus* (Linnaeus, 1766)

PRONUNCIATION *Hi-dro-fiss pla-tu-russ pla-tu-russ.*
ETYMOLOGY Flat-tailed water snake.
TYPE LOCALITY Unknown.
APPEARANCE Medium-sized, moderately robust snake with slightly distinct head. Dorsal colouration black to dark brown. Bottom half of snake yellow or white. Tail white or yellow with black markings. Iris dark brown to black. Tongue pale; buccal cavity dark. Adult females larger than males, reaching 100cm TL. **Scalation** MB 49–68 rows, 264–408 VENT, SUB 39–51 all single and anal scale divided. Scales smooth and do not overlap.
RANGE Encountered from the east coast of Tas, around the Australian coastline from Port Phillip Bay to Perth, WA. Also found throughout the Indian and Pacific Oceans. An additional subspecies is found off the coast of Costa Rica.
COMMENTS Mainly diurnal. Lives in open water. The world's most widespread snake. **Diet** Small pelagic fish, which it captures while sitting in ambush among floating debris. **Reproduction** Viviparous. 1–6 per litter. Neonates approximately 25–28cm TL. Gravid in March–October. **Disposition** Typically inoffensive, but can be defensive when handled.
BITE/VENOM DANGEROUSLY VENOMOUS
IUCN LISTING Least Concern.

Newcastle, NSW, Scott Eipper

Newcastle, NSW, Scott Eipper

Noumea, New Caledonia, Hal Cogger

Stokes' Sea Snake *Hydrophis stokesii* (Gray, 1846)

PRONUNCIATION *Hi-dro-fiss stohk-see.*

ETYMOLOGY Stokes' water snake; pertains to J. Stokes, who collected the holotype.

TYPE LOCALITY Australian seas.

APPEARANCE Large, very robust snake with large head slightly distinct from body. Dorsal colouration white or cream with grey to black spots. Spots fade with age. Juveniles white or cream with grey to black blotches. Patternless specimens are known. Both iris and tongue pale. Buccal cavity pink. Ventral colouration similar to upper surface but lighter in colour. Adult females larger than males, reaching 160cm TL. **Scalation** MB 54–60 rows, 252–280 VENT, SUB 33–36 all single and anal scale divided. Scales strongly keeled.

RANGE Encountered in waters around Australia, north of Sydney, NSW, to Exmouth, WA. Also occurs across the Indian Ocean to the Arabian Gulf and down the African coast to South Africa.

COMMENTS Mainly nocturnal. Lives in waters over sandy inlets, estuaries and coral reefs. **Diet** Frogfish, stonefish. **Reproduction** Viviparous. 1–14 per litter. Neonates approximately 38–40cm TL. Gravid in December–March. **Disposition** Typically inoffensive, but can be defensive when handled.

BITE/VENOM DANGEROUSLY VENOMOUS

IUCN LISTING Least Concern.

Broome, WA, Ruchira Somaweera

Darwin Harbour, NT, Paul Horner

Darwin Harbour, NT, Hal Cogger

SEA SNAKES 271

Prosperpine, QLD, Steve Swanson

Juvenile, Darwin, NT, Steve Swanson

Darwin Harbour, NT, Paul Horner

PLAIN-BANDED SEA SNAKE *Hydrophis vorisi* Kharin, 1984
(Estuarine Sea Snake)

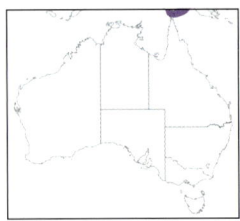

PRONUNCIATION *Hi-dro-fiss vo-riss-e.*

ETYMOLOGY Voris' water snake; pertains to H. Voris, American herpetologist.

TYPE LOCALITY East bank of Fly River, opposite Stuart Island, Western District, PNG.

APPEARANCE Small snake with small head that gradually transitions into body with a medium build. Dorsal colouration cream to grey with dark grey to black bands. Ventral colouration similar to upper surface. Both sexes reach 60cm TL. **Scalation** MB 29–35 rows, 330–350 VENT, 36 single SUB and anal scale divided. Scales smooth.

RANGE Encountered in waters north of Australia, in the Torres Strait and southern NG.

COMMENTS Nocturnal. Known from two specimens that have come from river mouths. **Diet** Small eels, hole-dwelling fish. **Reproduction** Unknown. **Disposition** Unknown; expected to be similar to others in the genus.

BITE/VENOM DANGEROUSLY VENOMOUS

IUCN LISTING Least Concern.

Fly River, PNG, Lauren Vonnahme AMNH

Fly River, PNG, Rynn Dragomirov, AMNH   Fly River, PNG, Rynn Dragomirov, AMNH

Australasian Beaked Sea Snake *Hydrophis zweifeli* (Kharin, 1985)
(Sepik Beaked Sea Snake)

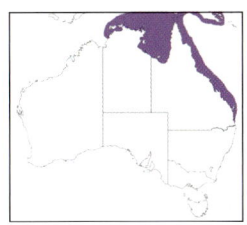

PRONUNCIATION *Hi-dro-fiss zz-why-fell-ee.*

ETYMOLOGY Zweifel's water snake; pertains to R. Zweifel, American herpetologist.

TYPE LOCALITY Off the mouth of Sepik River, PNG.

APPEARANCE Medium-sized, moderately robust snake with head distinct from body. Dorsal colouration white or cream with grey bands. Bands fade with age. Juveniles white or cream with grey to black bands. Loose skin beneath jaw that might be an adaption to enable the snakes to eat larger prey. Iris dark. Tongue pale; buccal cavity pinkish. Ventral colouration similar to upper surface but lighter in colour. Adult females larger than males, reaching 90cm TL. **Scalation** MB 48–55 rows, 272–322 VENT, SUB 52 all single and anal scale divided. Scales strongly keeled.

RANGE Encountered in waters around Australia, north of Brisbane, QLD, around to the west of Darwin, NT. Also found in PNG and Indonesia.

COMMENTS Mainly nocturnal. Lives in waters over sandy inlets, and in estuaries; occasionally found well upriver in fresh water. Formerly thought to occur throughout southern Asia as a part of the wide-ranging species *H. schistosa*; however, the Australasian population was found to be genetically distinctive and assigned to the species *H. zweifeli*. **Diet** Catfish, pufferfish, prawns. **Reproduction** Viviparous. 1–34 per litter. Neonates approximately 28cm TL. Gravid in November–February. **Disposition** Typically inoffensive, but can be very defensive when handled or if trapped in a net inadvertently.

BITE/VENOM DANGEROUSLY VENOMOUS

IUCN LISTING Least Concern.

Darwin, NT, Ruchira Somaweera

Weipa, QLD, Hal Cogger

Weipa, QLD, Hal Cogger

Weipa, QLD, Hal Cogger

Sea Kraits, genus *Laticauda* Laurenti, 1768

Laticauda comprises eight species, two of which occur in Australia as occasional waifs. **VENOM** In some species, strongly neurotoxic, with evidence of myotoxicity. Bites have resulted in, or could cause, fatalities. Sea snake or Polyvalent antivenom is used to treat envenomations. **Species-level identification difficulty** (within Australia) – 3.

ETYMOLOGY Flat tail.

TYPE SPECIES *Laticauda scutatus*.

Key to *Laticauda*

1. Fewer than 20 midbody scale rows;
   supralabials brown to black..*L. laticaudata* (p. 276)
   More than 20 midbody scale rows;
   supralabials yellow to white...*L. colubrina* (opposite)

Laticauda colubrina, Medewei, Bali, Indonesia, Scott Eipper

Laticauda laticaudata, Signal Island, New Caledonia, Hal Cogger

YELLOW-LIPPED SEA KRAIT *Laticauda colubrina* (Schneider, 1799)

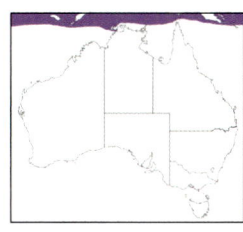

PRONUNCIATION *Lat-e-cor-dah col-u-bree-nah*.
ETYMOLOGY Serpent-like flat tail.
TYPE LOCALITY Unknown.
APPEARANCE Large, robust snake with head slightly distinct from body. Dorsal colouration pale blue to blue-grey with 24– 64 black bands. Lower lateral surfaces lighter before becoming cream to yellow underneath. Iris pale grey. Tongue and buccal cavity pale. Adult females larger than males, reaching 150cm TL. **Scalation** MB 21–25 rows, 210–250 VENT, SUB 25–50 all divided and anal scale divided. Scales shiny in appearance and smooth.
Similar species *L. laticaudata* (p. 276).
RANGE Found across northern Australia, with records from Darwin, NT, and in the Torres Strait. However, all individuals found are best described as waifs. Also found in southern Asia, into the Pacific, across to Tonga.
COMMENTS Nocturnal, but active by day while hunting prey. Aquatic; readily encountered on land outside Australia, entering forests, gorges and caves, and climbing into trees and on to cliff faces. **Diet** Fish, eels. Occasionally eats dead fish found in holes at low tide. **Reproduction** Oviparous. 5–19 per clutch. Neonates approximately 35cm TL. Can be gravid throughout the year. **Disposition** Inoffensive, but can be defensive when handled or if trapped in a net inadvertently.
BITE/VENOM DANGEROUSLY VENOMOUS
IUCN LISTING Least Concern.

Sanur, Bali, Indonesia, Tie Eipper

Mindel Beach, Darwin, Phill Mangion

Medewei, Bali, Indonesia, Scott Eipper

Sanur, Bali, Indonesia, Tie Eipper

Black-lipped Sea Krait *Laticauda laticaudata* (Linnaeus, 1758)

PRONUNCIATION *Lat-e-cor-dah lat-e-cor-dah-tah.*

ETYMOLOGY Flat tail flat tail.

TYPE LOCALITY Indiis (India).

APPEARANCE Medium-sized, moderately robust snake with head distinct from body. Dorsal colouration blue to blue-grey with 26–55 black bands. Lower lateral surfaces lighter, before becoming cream to yellow underneath. Iris pale grey. Tongue and buccal cavity pale. Adult females larger than males, reaching 100cm TL. **Scalation** MB 19 rows, 225–245 VENT, SUB 25–50 all divided and anal scale divided. Scales shiny in appearance and smooth. **Similar species** *L. colubrina* (p. 275).

RANGE Encountered across northern Australia, but all individuals found are best described as waifs. Also found in Asia, into the Pacific, across to Tonga.

COMMENTS Nocturnal, but active by day while hunting food. Aquatic; readily found on land outside Australia, entering forests, rocky seaside shelves and caves. **Diet** Fish, eels. **Reproduction** Oviparous. 5–11 per clutch. Neonates approximately 29cm TL. Can be gravid throughout the year. **Disposition** Inoffensive, but can be defensive when handled or if trapped in a net inadvertently.

BITE/VENOM DANGEROUSLY VENOMOUS

IUCN LISTING Least Concern.

Noumea, New Caledonia, Claire Goiran

Noumea, New Caledonia, Claire Goiran

Noumea, New Caledonia, Hal Cogger

Noumea, New Caledonia, Claire Goiran

Genus *Microcephalophis* Lesson R. P., 1834

Two species in the genus with one occurring in Australian waters. Formerly placed within *Hydrophis* on account of its similar morphology to other microcephalic sea snakes. Phylogenetic studies show this genus to be a sister to *Hydrophis*. **VENOM** In some species, strongly neurotoxic, with evidence of myotoxicity. Bites have resulted in, or could cause, fatalities. Sea snake or Polyvalent antivenom is used to treat envenomations. **Species-level identification difficulty** (within Australia) – 4.
ETYMOLOGY Small-headed snake.
TYPE SPECIES *Hydrus gracilis*.

Microcephalophis gracilis, Sri Lanka, Sanoj Wijayasekara

Microcephalophis gracilis, Sri Lanka, Sanoj Wijayasekara

SMALL-HEADED SEA SNAKE *Microcephalophis gracilis* (Shaw, 1802)

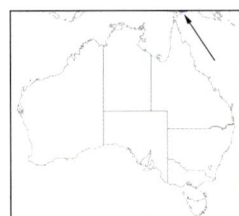

PRONUNCIATION *My-crow-keff-ah-low-fiss gra-sil-us.*
ETYMOLOGY Gracile small-headed snake.
TYPE LOCALITY Unknown.
APPEARANCE Small to medium-sized snake with very small head that gradually transitions into body with a medium build. Dorsal colouration grey, cream or white, with distinct dark grey to black bands. Juveniles brighter than adults, with clear black markings. Iris dark grey. Ventral colouration paler than upper surface. Adult females larger than males, reaching 70cm TL. **Scalation** MB 30–36 rows, 220–287 VENT and anal scale divided. Scales strongly keeled.
RANGE Encountered in waters north of Australia, in the Torres Strait. Also found throughout Indonesia, across Asia to the Persian Gulf.
COMMENTS Nocturnal. Lives in deep, turbid waters. **Diet** Eels. **Reproduction** Viviparous. Gravid in July–December, giving birth to 1–9 in a litter. Neonate size not recorded. **Disposition** Typically inoffensive, but can be defensive when handled.
BITE/VENOM DANGEROUSLY VENOMOUS
IUCN LISTING Least Concern.

Sri Lanka, Sanoj Wijayasekara

Sri Lanka, Sanoj Wijayasekara

Sri Lanka, Sanoj Wijayasekara

Genus *Parahydrophis* Burger & Natsuno, 1974

Monotypic genus. It contains one of three sea snake species that occur in mangroves and estuaries. The genus is significantly less common than *Ephalophis* and *Hydrelaps*. The snakes are occasionally seen hunting in shallow pools formed on tidal flats. **VENOM** Toxicity unknown, and there is no evidence of envenomations. **Species-level identification difficulty** (within Australia) – 3.

ETYMOLOGY False *Hydrophis*.
TYPE SPECIES *Distira mertoni*.

Parahydrophis mertoni, Rapid Creek, Darwin, NT, Hal Cogger

Parahydrophis mertoni, Rapid Creek, Darwin, NT, Hal Cogger

Northern Mangrove Sea Snake *Parahydrophis mertoni* (Roux, 1910)
(Arafura Smooth Sea Snake)

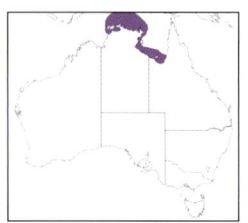

PRONUNCIATION *Para-high-dro-fiss merr-ton-ee.*

ETYMOLOGY Merton's false Hydrophis, pertaining to H. Merton, a German doctor.

TYPE LOCALITY Sungi Waskai, Wokam, SE Maluku, Indonesia.

APPEARANCE Small snake with medium build and head slightly distinct from body. Dorsal colouration greenish-yellow with grey-black bands. Ventral colouration similar to upper surface. Iris dark. Both sexes reach 50cm TL. **Scalation** MB 36–39 rows, 158–160 VENT, SUB 29–35 most single and anal scale divided. Scales smooth.

RANGE Encountered in waters from west of Darwin, NT, across to the Gulf of Carpentaria, QLD. Also found in waters around the Aru islands, south of PNG.

COMMENTS Nocturnal. Lives in shallow tidal waters of estuaries and mangrove-lined creeks.

Diet Small, hole-dwelling fish. **Reproduction** Viviparous. 1–3 per litter. Neonate size not recorded. Gravid in May–September. **Disposition** Strongly defensive when handled.

BITE/VENOM VENOMOUS

IUCN LISTING Least Concern.

Rapid Creek, Darwin, NT, Hal Cogger

Family Homalopsidae (Mangrove Snakes)

The mangrove snakes are a group of mildly venomous, rear-fanged snakes that are either fully or semi-aquatic. They are much more diverse in SE Asia than in Australia. Five of the approximately 57 species in the family occur in northern Australia. For many years, mangrove snakes were treated as a subfamily of the Colubridae. Most Australian species live within mangroves, emerging at night to hunt prey in the intertidal zones and pools, with one species living exclusively in fresh water. One species is endemic to Australia.

Key to Australian Homalopsid snakes

1. Loreal scale absent 2
   Loreal scale present .... *Fordonia leucobalia* (p. 285)

2. Scale smooth or weakly keeled 3
   Scales strongly keeled .... *Cerberus australis* (p. 283)

3. Nasal scales not in contact 4
   Nasal scales in contact .... *Pseudoferania polylepis* (p. 291)

4. 21–23 midbody scale rows .... *Myron richardsonii* (p. 289)
   19 midbody scale rows .... *Myron resetari* (p. 288)

Cerberus australis, Darwin Harbour, NT, Anders Zimny

Genus *Cerberus* Cuvier, 1829

Cerberus comprises five species, represented in Australia by a single species. This was formerly thought to be a monotypic genus. **VENOM** Rear fanged; bites are usually uneventful but can result in stinging and minor swelling. Toxicity is unknown; no records of severe envenomations. **Species-level identification difficulty** (within Australia) – 1.

ETYMOLOGY In reference to the 'dog-like' appearance of the snakes' heads.

TYPE SPECIES *Coluber cerberus*.

Cerberus australis, Darwin NT, Adam Elliott

Cerberus australis, Darwin Harbour, NT, Anders Zimny

Australian Bockadam *Cerberus australis* (Gray, 1842)

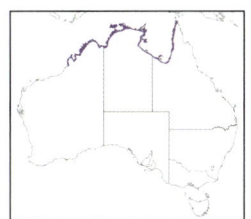

PRONUNCIATION *Sir-bur-russ os-trah-liss.*
ETYMOLOGY Southern dog-like.
TYPE LOCALITY Port Essington, NT.
APPEARANCE Medium-sized, heavy-bodied snake with head distinct from body. Two main colour phases, one reddish-orange, the other grey with variable amounts of black cross-bands and blotches. Usually a black temporal streak. Position of eyes on top of head, and nostril with valves, are adaptations to an aquatic environment. Eyes small with vertically elliptic pupils. Mouth lining and tongue dark. Ventral colouration whitish-cream, with darker spotting. Adult females usually larger than males, reaching 110cm TL. **Scalation** MB 23–25 rows, 140–160 VENT, SUB 45–60 all divided and anal scale divided. Dorsal scales keeled except on head. Scales strongly keeled and dull in appearance.

RANGE Encountered across northern and eastern Australia, from Kalumburu, WA, to the western coast of Cape York Peninsula, QLD. Also found in PNG.

COMMENTS Predominantly nocturnal. Semi-aquatic, living in mangroves, along waterways in brackish and salt water. Shelters in crab holes and among mangrove roots. Recorded basking on mud flats. Lives among mangrove roots and fallen timber. Usually hunts in channels formed on draining mudflats. **Diet** Fish. **Reproduction** Viviparous. 6–9 per litter. Neonates approximately 19cm TL born in February–May. **Disposition** Generally defensive, and will readily bite.
BITE/VENOM HARMFUL
IUCN LISTING Least Concern.

Darwin Harbour, NT, Paul Horner

Darwin Harbour, NT, Anders Zimny

Buffalo Creek, NT, Paul Horner

Darwin, NT, Jules Farquhar

Genus *Fordonia* Gray, 1842

Monotypic genus. **VENOM** Rear fanged, and bites can result in stinging. Toxicity unknown, and no evidence of severe envenomations recorded. **Species-level identification difficulty** (within Australia) – 1.
ETYMOLOGY Possibly after J. Ford.
TYPE SPECIES *Homalopsis leucobalia*.

Fordonia leucobalia, Darwin, NT, Jesse Campbell

Fordonia leucobalia, Darwin, NT, Jesse Campbell

Fordonia leucobalia, Darwin, NT, Jesse Campbell

White-bellied Mangrove Snake Fordonia leucobalia (Schlegel, 1837)
(Crab-eating Snake)

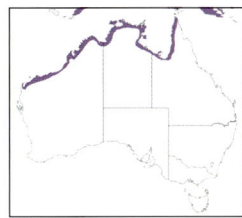

PRONUNCIATION For-doh-nee-ah luke-o-bail-lee-ah.
ETYMOLOGY White dappled.
TYPE LOCALITY Timor, Indonesia.
APPEARANCE Medium-sized, heavy-bodied snake with head indistinct from body. Colouration and pattern very variable, with combination of whites, yellows, browns, reds, blacks, greys and oranges. Can be patterned with dots, stripes or blotches, or plain. Lower flanks and ventral surface usually white, as the common name suggests; beneath tail white with dark medial streak. Eyes small with vertically elliptic pupils. Mouth lining and tongue pink. Dentition unique among Australian snakes, with short, robust teeth designed to crack open crustacean exoskeletons. Adult females usually larger than males, reaching 90cm TL. **Scalation** MB 23–29 rows, 130–160 VENT, SUB 25–45 all divided and anal scale divided. No loreal scale. Scales dull in appearance and smooth.

RANGE Encountered across northern and eastern Australia, from Nichol Bay, WA, to the western coast of Cape York Peninsula and Torres Strait Islands of QLD. Also found in PNG and SE Asia.

COMMENTS Predominantly nocturnal to cathemeral, depending on temperature and tides. Semi-aquatic, living in mangroves, along waterways in brackish and salt water. Lives among mangrove roots and fallen timber, sheltering in crab holes. **Diet** Emerges on to flats to hunt crabs and small lobsters. Usually hunts in tidal pools formed on mudflats. One of two Australian snakes that dismember their prey, removing and consuming the legs before eating the rest. **Reproduction** Viviparous. 2–17 per litter. Neonates approximately 19cm TL born in February–March. **Disposition** Inoffensive but will bite if threatened.

BITE/VENOM HARMFUL

IUCN LISTING Least Concern.

Darwin Harbour, NT, Anders Zimny

Darwin Harbour, NT, Anders Zimny

Pajinka, QLD, Ryan Francis

Juvenile, Darwin Harbour, NT, Anders Zimny

Broome, WA, Anders Zimny

Genus *Myron* Gray, 1849

The *Myron* genus comprises three species, and is represented in Australia by two. The genus was formerly thought to be monotypic. **VENOM** Rear fanged, and bites can result in minor stinging. Toxicity unknown, and no evidence of severe envenomations recorded. **Species-level identification difficulty** (within Australia) – 3.

ETYMOLOGY Probably meaningless.

TYPE SPECIES *Myron richardsonii*.

Myron richardsonii, Darwin Harbour, NT, Anders Zimny

Myron richardsonii, Darwin Harbour, NT, Paul Horner

ROEBUCK BAY MANGROVE SNAKE *Myron resetari* Murphy, 2011

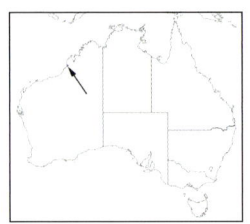

PRONUNCIATION *My-ron rez-ah-tah-ree.*
ETYMOLOGY Pertains to A. Resetar, American herpetologist.
TYPE LOCALITY Broome, WA.
APPEARANCE Small, heavy-bodied snake with head distinct from body. Dorsal colouration silver-grey, brown to olive, with series of black cross-bands and blotches. Usually a black temporal streak. Eyes small with vertically elliptic pupils. Mouth lining and tongue pink. Ventral colouration whitish-cream to yellow. Adult females usually larger than males, reaching 40cm TL. **Scalation** MB 19 rows, 137–145 VENT, SUB 30–40 all divided and anal scale divided. Scales dull in appearance and weakly keeled. **Similar species** *M. richardsonii* (opposite).

RANGE Encountered only around Roebuck Bay and Broome, WA. Expected to occur over a larger expanse of the WA coastline.

COMMENTS Predominantly nocturnal. Semi-aquatic, living in mangroves, along waterways in brackish and salt water. Lives among mangrove roots and in burrows, emerging on to flats to hunt in small pools. **Diet** Gobies. **Reproduction** Thought to give birth to live young. **Disposition** Inoffensive but will bite if threatened.

BITE/VENOM HARMFUL
IUCN LISTING Vulnerable.

Roebuck Bay, WA, Ian 'Bushrat' Bool

Roebuck Bay, WA, Ian 'Bushrat' Bool

Broome, WA, Jake Meney

Richardson's Mangrove Snake *Myron richardsonii* Gray, 1849

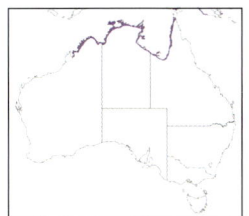

PRONUNCIATION *My-ron rich-hard-son-ee.*
ETYMOLOGY Pertains to J. Richardson, collector of the type specimen.
TYPE LOCALITY NW Australia.
APPEARANCE Small, heavy-bodied snake with head distinct from body. Dorsal colouration silver-grey, brown to olive-green, with irregular black cross-bands and blotches. Usually a black temporal streak. Some individuals in the Torres Strait reddish-orange or black. Eyes small with vertically elliptic pupils. Mouth lining and tongue dark. Ventral colouration whitish-cream or yellow. Adult females usually larger than males, reaching 50cm TL. **Scalation** MB 21–23 rows, 130–147 VENT, SUB 30–40 all divided and anal scale divided. Scales dull in appearance and weakly keeled. **Similar species** *M. resetari* (opposite).

RANGE Encountered across northern and eastern Australia, from Derby, WA, to the western coast of Cape York Peninsula and Torres Strait Islands of QLD. Also found in PNG and Indonesia.

COMMENTS Predominantly nocturnal. Semi-aquatic, living in mangroves, along waterways in brackish and salt water. Shelters in crab holes and among mangrove roots. Usually hunts in channels formed on draining mudflats. **Diet** Gobies, mudskippers. May also feed on molluscs. **Reproduction** Viviparous. 6–9 per litter. Neonates approximately 14cm TL born in February–March. **Disposition** Inoffensive but may bite if threatened

BITE/VENOM HARMFUL

IUCN LISTING Least Concern.

Darwin, NT, Ryan Francis

Darwin Harbour, NT, Anders Zimny

Sigabaduru, Western Province, PNG, David Williams

Genus *Pseudoferania* Fitzinger, 1849

This monotypic genus is represented in Australia by a single species. The eastern (inclusive of the Papuan animals) and western animals form two distinct groups that have been suggested as being worthy of taxonomic recognition based on morphology. **VENOM** Rear fanged, and bites can result in minor stinging and mild swelling. Toxicity unknown, and there is no recorded evidence of severe envenomations. **Species-level identification difficulty** (within Australia) – 1.

ETYMOLOGY False *Ferania* (a genus of Asian aquatic snakes).

TYPE SPECIES *Pseudoferania macleayi*.

Pseudoferania polylepis, Emu Creek, QLD, Scott Eipper

Pseudoferania polylepis, Marreba Wetlands, QLD, Scott Eipper

MACLEAY'S WATER SNAKE *Pseudoferania polylepis* (J. G. Fischer, 1886)

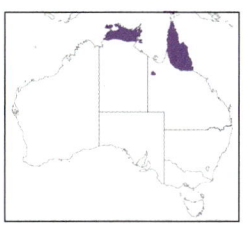

PRONUNCIATION Sue-doh-fur-ran-ee-ah poly-lep-iss.
ETYMOLOGY Many scales.
TYPE LOCALITY Fly River, Western Province, PNG.
APPEARANCE Medium-sized, heavy-bodied snake with head distinct from body. Dorsal colouration dark grey to pale brown, usually striped with darker tones. Lower flanks cream to yellow. Many individuals are flecked with dark speckling. Ventral colouration cream to bright yellow with heavy dark flecking. Adult females usually larger than males, reaching 110cm TL. **Scalation** MB 21–23 rows, 140–165 VENT, SUB 35–50 all divided and anal scale divided. Scales smooth and glossy and have an iridescent sheen.

RANGE Two separate Australian populations. One extends through the northern NT, across to Riversleigh, QLD. The second occurs from Georgetown, across to Ingham, QLD, north across the Torres Strait. Also occurs in PNG and neighbouring Indonesia.

COMMENTS Predominantly nocturnal. Aquatic, living in freshwater rivers, swamps and billabongs. Rarely found on land. Lives amongst submerged tree roots and vegetation. **Diet** Fish, eels, frogs. **Reproduction** Viviparous. 4–27 young per litter. Neonates approximately 18cm TL born in November–April. **Disposition** Inoffensive but may bite if threatened.

BITE/VENOM HARMFUL

IUCN LISTING Least Concern.

Marreba wetlands, QLD, Tie Eipper

Julatten, QLD, Shane Black

Adel Grove, QLD, Scott Eipper

Emu Creek, QLD, Scott Eipper

Family Typhlopidae (Blind Snakes)

Blind snakes are some of the world's oldest snakes and also one of the least-known snake groups. There are three reasons for this. Firstly, they are fossorial and rarely seen on the surface. Secondly, they are morphologically conservative, with small rudimentary eyes and smooth scales, and have a similar body shape. Lastly, they are hard to find – some species dwell in very deep underground caverns and are rarely encountered. There are about 270 described species worldwide. However, according to current phylogenetic assessments, this number could increase dramatically. Many of the Australian species are known from less than 10 specimens.

Blind snakes are often mistaken for earthworms. They are often small, worm-like burrowing snakes, pink, brown to grey in colour, and with smooth scales around their bodies that allow them to travel through soil easily.

Due to their secretive nature, the basic ecology of blind snakes is poorly known at a species level. They are usually encountered when people dig them up, when they shift cover and when crossing roads at night after rain.

Their eyes are vestigial, appearing as small, dark spots under the scales of the head. They do appear sensitive to light, often moving away from artificial light sources at night if disturbed. The mouth is well behind and below the tip of the snout – probably as an adaptation for moving through the soil to minimize the amount of matter entering by accident. The body is uniform in thickness along its length, with a noticeably short tail usually ending in a conical spine.

To find their food, blind snakes flick their short, pale tongue to smell the trail of ants and termites, which they follow to the nest. They rake ants and termites into the mouth with their top jaw, usually swallowing their food whole.

Blind snakes are non-venomous and harmless. They cannot bite and have limited defensive capabilities. They can produce a pungent odour from the anal glands, regurgitate their last meal or probe with their tail spine, producing an unpleasant prickling sensation. If picked up some species thrash about, smearing the anal gland secretion on a predator. Some species will also squeak when grasped. Some exhibit fluorescence when exposed to UVA light sources. It is not clear why this is the case, but it may be an anti-predation strategy.

Because blind snakes are seemingly very similar in ecology and utilize the same defensive strategies, the following species descriptions can seem repetitious. Unlike in the other species descriptions, due to the vestigial nature of the eyes, the pupil shape and iris colour are not described. The mouth lining colouration is also omitted. In all the species the authors have examined, the tongue has been pale.

Due to the small size and fossorial nature of blind snakes, they can prove a challenge to photograph. This can also lead to difficulties in identification. The characteristics of these snakes can be almost too difficult to easily see without the aid of a jeweller's loupe or microscope.

Blind snakes are Australia's second most diverse group of snakes. As currently placed, all Australian species are represented by three genera. Forty-five species are endemic to Australia. The key below applies to all currently recognized taxa.

Key to Australian Blind Snakes

1. Terminal spine present...3
   Terminal spine absent..2

2. Occurs in the Kimberley region of WA..*Anilios zonula* (p. 342)
   Occurs in western QLD..*Anilios aspina* (p. 298)

3. Fewer than 23 midbody scale rows...8
   24 midbody scale rows..4

4. Head rounded in profile...5
   Head angular in profile..7

5. Colouration without sharp distinction on lower flank..6
   Colouration with sharp distinction on lower flank.............................*Anilios ligatus* (p. 318)

6. Found in the Pilbara..*Anilios ganei* (p. 309)
   Found in the Northern Territory...*Anilios yirrikalae* (p. 341)

7. Found in northern Australia; fewer than 20 subcaudal scales.............*Anilios unguirostris* (p. 336)

	Found in southern NSW; more than 20 subcaudal scales	*Anilios batillus* (p. 300)
8	18 midbody scale rows or more	12
	16 midbody scale rows	9
9	Colouration without stripes and a dark coloured tail	10
	Colouration with stripes and a dark-coloured tail	*Anilios minimus* (p. 322)
10	Body thread-like; more than 700 dorsal scale rows	*Anilios longissimus* (p. 319)
	Body slender, usually less than 700 dorsal scale rows	11
11	More than 15 subcaudal scales	12
	Subcaudal scales 9–14, found in northern NT	*Anilios nema* (p. 323)
12	Found in south-east QLD; subcaudal scales 19	*Anilios insperatus* (p. 314)
	Found in coastal WA; more than 15 subcaudal scales; more than 575 ventrals	*Anilios leptosoma* (p. 316)
13	18 midbody scale rows	14
	20–22 midbody scale rows	24
14	Snout rounded when viewed from above	15
	Snout not rounded when viewed from above	19
15	Snout rounded when viewed in profile	16
	Snout not rounded when viewed in profile	*Anilios affinis* (p. 296)
16	Body without stripes	17
	Body with stripes and a dark-coloured tail	*Anilios chamodracaena* (p. 305)
17	Colouration the same dorsally without a dark-coloured tail	18
	Dark-coloured tail	*Anilios guentheri* (p. 311)
18	No conspicuous glands on the head, ventral scales more than 400; restricted to northern WA	19
	White to cream gular region; ventral scales less than 400	*Indotyphlops braminus* (p. 343)
19	Nasal cleft not visible from above	*Anilios howi* (p. 313)
	Nasal cleft visible from above	*Anilios micromma* (p. 321)
20	Colouration the same dorsally, without a dark tail	20
	Dark-coloured tail; rostral scale hooked in profile	*Anilios grypus* (p. 310)
21	Snout weakly trilobed when viewed from above	*Anilios margaretae* (p. 320)
	Snout bluntly pointed when viewed from above	22
22	Fewer than 12 subcaudal scales; found in northern WA	*Anilios yampiensis* (p. 340)
	More than 12 subcaudal scales; found in mid-coastal WA	23
23	Fewer than 17 subcaudal scales; fewer than 595 ventral scales	*Anilios obtusifrons* (p. 325)
	More than 17 subcaudal scales; more than 595 ventral scales	*Anilios systenos* (p. 332)
24	20 midbody scale rows	35
	22 midbody scale rows	25
25	Snout rounded when viewed from above	26
	Snout not rounded when viewed from above	31
26	Snout rounded when viewed in profile	27
	Snout not rounded when viewed in profile	*Anilios australis* (p. 299)
27	Head not dorsally depressed; snout rounded when viewed in profile	28
	Head strongly depressed dorsally, snout rounded when viewed in profile	*Anilios troglodytes* (p. 335)
28	No sharp edge between dorsal and ventral colouration; fewer than 550 ventral scales	29

	Sharp edge between dorsal & ventral colouration; more than 550 ventral scales	*Anilios robertsi* (p. 329)
29	More than 400 ventral scales	**30**
	Fewer than 400 ventral scales	*Anilios torresianus* (p. 333)
30	Found in northern Australia; nasal cleft not joining the first supralabial scale; head slightly flattened	*Anilios kimberleyensis* (p. 315)
	Found in eastern Australia; nasal cleft joining the first supralabial scale; head not flattened	*Anilios nigrescens* (p. 324)
31	Snout weakly trilobed when viewed from above	**33**
	Snout moderately trilobed when viewed from above	**32**
32	Snout moderately angular when viewed in profile	*Anilios endoterus* (p. 307)
	Snout bluntly rounded when viewed in profile	*Anilios vagurima* (p. 337)
33	Snout angular when viewed in profile	**34**
	Snout strongly angular when viewed in profile	*Anilios pilbarensis* (p. 326)
34	Dark grey to brown, rostral scale not hooked in profile; terminal spine dark	*Anilios bicolor* (p. 301)
	Rostral hooked in profile; terminal spine same colour as body	*Anilios hamatus* (p. 312)
35	Snout not strongly trilobed when viewed from above	**36**
	Snout strongly trilobed when viewed from above	*Anilios bituberculatus* (p. 302)
36	Snout weakly trilobed when viewed from above	**37**
	Snout rounded when viewed from above	**40**
37	Snout weakly angular in profile; robust; restricted to coastal WA	*Anilios splendidus* (p. 331)
	Snout angular or bluntly angular in profile	**38**
38	Body robust to heavily robust	**39**
	Body slender	*Anilios waitii* (p. 338)
39	Body heavily robust; snout bluntly angular in profile	*Anilios pinguis* (p. 327)
	Body robust; snout angular in profile	*Anilios proximus* (p. 328)
40	Snout rounded when viewed in profile	**41**
	Snout hook-shaped when viewed in profile	*Anilios centralis* (p. 304)
41	Body with a pattern of clear stripes	**42**
	Body without a pattern of clear stripes	**44**
42	Dark-bodied with faint stripes; sharp colour delineation between dorsal and ventral surfaces; south-east QLD	*Anilios silvia* (p. 330)
	Striped; a gradual colour shift from dorsal to ventral surfaces	**43**
43	Found in north QLD	*Anilios broomi* (p. 303)
	Found on Christmas Island	*Ramphotyphlops exoceti* (p. 344)
44	Ventral scales more than 300	**45**
	Ventral scales fewer than 280	*Anilios tovelli* (p. 334)
45	Ventral scales are paler than dorsum	**46**
	Ventral scales are black like dorsum	*Anilios leucoproctus* (p. 317)
46	Nasal cleft visible from above; fewer than 500 dorsal scale rows	**47**
	Nasal cleft not visible from above; more than 500 dorsal scale rows	*Anilios fossor* (p. 308)
47	Rostral scale without dark brown streak	**48**
	Rostral scale with dark brown streak	*Anilios wiedii* (p. 339)
48	Nasal cleft divides nasal scale	*Anilios diversus* (p. 306)
	Nasal cleft does not divide the nasal scale	*Anilios ammodytes* (p. 297)

Genus *Anilios* Gray, 1845

This genus currently contains 48 species, with the exception of *A. erycinus* all occurring in Australia. Forty-four species are endemic to Australia. Many are incredibly similar in appearance. Very little is known about this group of snakes. This is probably due to their small size, fossorial lifestyle and difficult detectability. Preliminary phylogenetic assessments suggest that only half of the species are described. Therefore taxonomic change will be unavoidable as more species groups and their relationships to each other are examined. Until a revision in 2014, most Australian species were placed within *Ramphotyphlops*. **BITE/VENOM** Blind snakes are unable to bite humans due to their mouth shape, and are harmless to humans. **Species-level identification difficulty** (within Australia) – 5.
ETYMOLOGY Probably meaningless.
TYPE SPECIES *Anilios australis*.

Anilios ligatus, Condamine, QLD, Scott Eipper

Anilios chamodracaena, Western Cape York Peninsula, QLD, Eric Vanderduys

SMALL-HEADED BLIND SNAKE *Anilios affinis* (Boulenger, 1889)

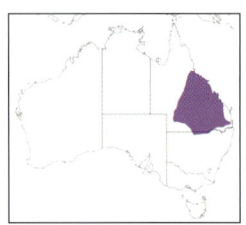

PRONUNCIATION *Ah-nil-e-oss ah-fin-iss.*
ETYMOLOGY Affinity with (*A. unguirostris*).
TYPE LOCALITY QLD.
APPEARANCE Small, slender blind snake. Dorsal colouration pale brown to pinkish, lighter on head. Ventral colouration white, cream to yellowish. Head rounded when viewed from above and angular in profile. Tail quite short and rounded at the end, with a short, blunt conical spine. Reaches 23cm TL. **Scalation** MB 18 rows, 278–357 DSR and SUB 10–19. Scales small, smooth and glossy.
RANGE Encountered in eastern QLD, from Charters Towers, to just over the NSW border. Also recorded in the Solomon Islands, although this population is likely to represent a morphologically similar but different species.
COMMENTS Nocturnal. Fossorial, living in open woodland of clay and loamy soils. Shelters in termite mounds and beneath fallen timber. **Diet** Ant larvae. **Reproduction** Oviparous. Recorded as laying three eggs. Neonates are approximately 11 cm TL. **Disposition** Would rather retreat than have contact. When threatened can excrete a pungent musk.
BITE/VENOM HARMLESS
IUCN LISTING Least Concern.

Injune, QLD, Scott Eipper

Yelarbon, QLD, Rob Valentic, specimen courtesy of Jamie Gover

Jimna, QLD, Steve Swanson

Middleton, QLD, Rob Valentic

AMMODYTE BLIND SNAKE *Anilios ammodytes* (Montague, 1914)

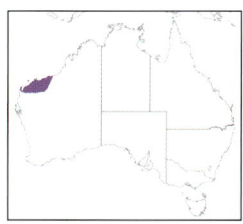

PRONUNCIATION *Ah-nil-e-oss am-mo-die-tees*.
ETYMOLOGY Sand dwelling.
TYPE LOCALITY Hermite Island, Monte Bello Islands, WA.
APPEARANCE Slender, medium-sized snake. Dorsal colouration dark purplish-brown to pinkish-brown, darker on head. Ventral colouration lighter than dorsum. Head rounded when viewed from above and in profile. Tail short and rounded at the end, with a short, blunt conical spine. Reaches 35cm TL. **Scalation** MB 20 rows, 389–498 VENT, SUB 8–18. Scales small, smooth and glossy. **Similar species** *A. diversus* (p. 306).
RANGE Encountered in the Pilbara region of WA, from the Northwest Cape to Eighty Mile Beach.
COMMENTS Nocturnal. Fossorial, living in open woodland, grassland and coastal dunes with spinifex and mulga, on both stony and sandy soil types. Shelters beneath logs and rocks. **Diet** Ant larvae. **Reproduction** Probably oviparous. **Disposition** Would rather retreat than have contact. When threatened can excrete a pungent musk.
BITE/VENOM HARMLESS
IUCN LISTING Least Concern.

Shay Gap, WA, Brian Bush

Burrup Peninsula, WA, Brad Maryan

Shay Gap, WA, Brian Bush

Shay Gap, WA, Brian Bush

ROUND-TAILED BLIND SNAKE *Anilios aspina*
(Couper, Covacevich & S. Wilson, 1998)

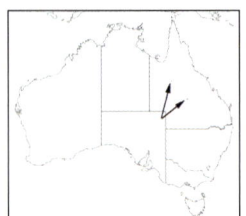

PRONUNCIATION *Ah-nil-e-oss as-spine-ah.*
ETYMOLOGY Without a spine; reference to lacking an apical spine.
TYPE LOCALITY Margot Station, 20km N Barcaldine, QLD.
APPEARANCE Small, slender snake. Dorsal colouration brown to pink. Posterior edges of midbody scales can be darker. Ventral colouration lighter than dorsum. One of two Australian blind snakes that lack a terminal (apical) spine. Head rounded when viewed from above and in profile. Reaches 28cm TL. **Scalation** MB 18 rows, 437 VENT, SUB 14. Scales small, smooth and glossy.

RANGE Encountered in central QLD, from Julia Creek to Barcaldine.

COMMENTS Nocturnal. Fossorial, living in Mitchell grasslands on heavy clay soils and black soil plains. Shelters under rocks. Poorly known from a few individuals. **Diet** Ant larvae. **Reproduction** Probably oviparous. **Disposition** Would rather retreat than have contact. When threatened, recorded as poking the antagonizer as if it had a tail spine. Unusually not recorded as producing a defensive odour.

BITE/VENOM HARMLESS
IUCN LISTING Least Concern.

Julia Creek, QLD, Eric Vanderduys

Julia Creek, QLD, Eric Vanderduys

Julia Creek, QLD, Eric Vanderduys

Southern Blind Snake *Anilios australis* Gray, 1845

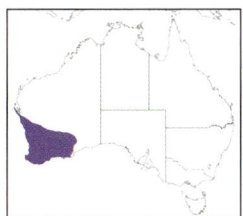

PRONUNCIATION *Ah-nil-e-oss os-trah-liss.*

ETYMOLOGY Southern.

TYPE LOCALITY WA.

APPEARANCE Large, robust snake. Dorsal colouration dark purplish-brown to pinkish-brown, lighter on head. Ventral colouration lighter than dorsum. Head rounded when viewed from above and angular in profile. The northernmost specimens often have pale collar on nape and may require further examination to determine relationships within the species. Tail quite short and rounded at the end, with short, blunt conical spine. Females larger than males, reaching 43cm TL. **Scalation** MB 22 rows, 278–357 VENT, SUB 10–18. Scales small, smooth and glossy. **Similar species** *A. bituberculatus* (p. 302), *A. hamatus* (p. 312), *A. splendidus* (p. 331), *A. bicolor* (p. 301).

RANGE Encountered in SW WA, from Cape Arid to Shark Bay.

COMMENTS Nocturnal. Fossorial, living in open woodland, mallee, heaths, grassland and coastal dunes with spinifex and mulga. Predominantly on sandy soil types. Shelters beneath stumps, rocks, and man-made debris, and in stick-ant nests. **Diet** Ant larvae. **Reproduction** Oviparous. 2–11 per clutch. Neonates approximately 10cm TL hatching in February–April. **Disposition** Would rather retreat than have contact. When threatened, can excrete a pungent musk or poke an antagonizer with the tail spine. Also recorded as being able to produce a squeak when grasped.

BITE/VENOM HARMLESS

IUCN LISTING Least Concern.

Wellington NP, WA, Steve Swanson

City Beach, WA, Danny Melville

Juvenile, Bibra Lake, WA, Danny Melville

Bluff Noll, WA, Brian Bush

Shovel-snouted Blind Snake *Anilios batillus* (Waite, 1894)

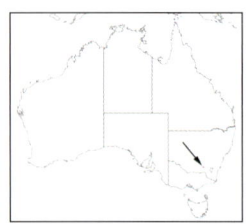

PRONUNCIATION Ah-nil-e-oss bah-till-us.
ETYMOLOGY Shovel – in reference to head shape.
TYPE LOCALITY Wagga Wagga, NSW.
APPEARANCE Medium-sized, robust snake. Colouration based on a preserved specimen. Dorsum colouration yellowish-brown with evidence of dark pigment forming dark longitudinal stripes. Underside paler, being yellow to cream. Head bluntly pointed when viewed from above and angular in profile. Tail quite short and rounded at the end, with short, blunt conical spine. Reaches 32cm TL. **Scalation** MB 24 rows, 557 DSR, SUB 21. Scales small, smooth and glossy.
RANGE Known from a single specimen from Wagga Wagga, NSW, described in 1894.
COMMENTS Presumed to be nocturnal. Fossorial, living in open woodland. Collection location is possibly an error, and this may not be an Australian species. **Diet** Ant larvae. **Reproduction** Presumed to be oviparous. **Disposition** Unknown, but presumed to be the same as the rest of the genus.
BITE/VENOM HARMLESS
IUCN LISTING Data Deficient.

Wagga Wagga, NSW, Scott Eipper

Wagga Wagga, NSW, Scott Eipper

Wagga Wagga, NSW, Scott Eipper

DARK-SPINED BLIND SNAKE *Anilios bicolor* (P. Schmidt, in Peters, 1858)

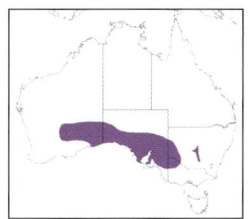

PRONUNCIATION *Ah-nil-e-oss bye-co-lore*.

ETYMOLOGY Two coloured – in reference to the dorsal and ventral colouration.

TYPE LOCALITY Adelaide, SA.

APPEARANCE Large, robust snake. Dorsal colouration dark purplish-brown to brown. Colour lightens on lower flanks, merging to white underneath. Head weakly trilobed when viewed from above and angular in profile. Tail quite short and rounded at the end, with short, blunt conical spine. Females larger than males, reaching 42cm TL. **Scalation** MB 22 rows, 357 VENT, SUB 14. Scales small, smooth and glossy. **Similar species** *A. bituberculatus* (p. 302), *A. australis* (p. 299).

RANGE Encountered in southern WA, from Cocklebiddy, across southern Australia, to Lake Cargelligo, NSW; also found in NW Vic.

COMMENTS Nocturnal. Fossorial, living in open woodland and mallee with spinifex and mulga. Predominantly on sandy soil types. Shelters beneath stumps, rocks and accumulated leaf litter. Can make an audible squeak if grasped. **Diet** Ant larvae and pupae. **Reproduction** Oviparous. Up to six per clutch. Neonates approximately 11cm TL. **Disposition** Would rather retreat than have contact. When threatened, can excrete a pungent musk or poke an antagonizer with the tail spine.

BITE/VENOM HARMLESS

IUCN LISTING Least Concern.

Hattah, VIC, Scott Eipper

Round Hill, NSW, Hal Cogger

Hattah, VIC, Scott Eipper

Ilkurlka, WA, Brian Bush

Prong-snouted Blind Snake *Anilios bituberculatus* (W. C. H. Peters, 1863)

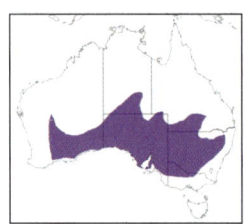

PRONUNCIATION *Ah-nil-e-oss bye-to-bur-cue-lah-tuss*.
ETYMOLOGY Two tubercled.
TYPE LOCALITY Loos, 4.5km west of Gawler, SA.
APPEARANCE Large, slender snake. Dorsal colouration dark purplish-black to pink. Ventral colouration lighter than dorsum. Head strongly trilobed when viewed from above and angular in profile. Tail quite short and rounded at the end, with short, blunt conical spine. Females larger than males, reaching 46cm TL. **Scalation** MB 20 rows, 414–485 VENT, SUB 11–18. Scales small, smooth and glossy. **Similar species** *A. endoterus* (p. 307), *A. bicolor* (p. 301).

RANGE Encountered across arid and semi-arid southern Australia, from the WA coastline, through SA, into western Vic, NSW and southern QLD.

COMMENTS Nocturnal. Fossorial, living in open woodland, mallee, heaths and coastal dunes with spinifex and mulga. Predominantly on sandy soil types. Shelters beneath stumps, rocks and leaf litter, and inside termite nests. **Diet** Ant pupae and larvae. **Reproduction** Oviparous. 2–9 per clutch. **Disposition** Would rather retreat than have contact. When threatened, can excrete a pungent musk or poke an antagonizer with the tail spine.

BITE/VENOM HARMLESS
IUCN LISTING Least Concern.

Lake Cargelligo, NSW, Scott Eipper

W of Coolgardie, WA, Brian Bush

Hattah, VIC, Scott Eipper

Quilpe, QLD, Rob Valentic

Lake Cargelligo, NSW, Scott Eipper

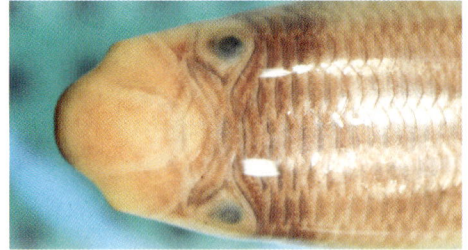

Currawinya NP, QLD, Scott Eipper.

Faint-striped Blind Snake *Anilios broomi* (Boulenger, 1898)

PRONUNCIATION *Ah-nil-e-oss broom-ee.*
ETYMOLOGY Pertains to R. Broom, South African zoologist.
TYPE LOCALITY Muldiva, QLD.
APPEARANCE Small, slender snake. Dorsal colouration dark pinkish-brown. Scales with darker centres align, forming thin dorsal stripes running the length of body. Usually darker on head. Ventral colouration white. Head rounded when viewed from above and in profile. Tail quite short and rounded at the end, with a short, blunt conical spine. Reaches 26cm TL. **Scalation** MB 20 rows, 456–460 VENT, SUB 15–16. Scales small, smooth and glossy.
RANGE Encountered from Cooktown to Innot Hot Springs, along the Einasleigh Uplands, QLD.
COMMENTS Nocturnal. Fossorial, living in open woodland. Shelters beneath rocks and fallen timber. **Diet** Ant larvae. **Reproduction** Presumed oviparous. **Disposition** Would rather retreat than have contact. When threatened can excrete a pungent musk or poke an antagonizer with the tail spine.
BITE/VENOM HARMLESS
IUCN LISTING Least Concern.

Brooklyn Station, QLD, Eric Vanderduys

Near Laura, QLD, Eric Vanderduys

Laura, QLD, Anders Zimny

Centralian Blind Snake *Anilios centralis* (Storr, 1984)

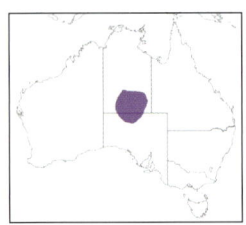

PRONUNCIATION *Ah-nil-e-oss sen-trah-liss.*
ETYMOLOGY In reference to where it is found in Australia.
TYPE LOCALITY Alice Springs, NT.
APPEARANCE Medium-sized, slender snake. Dorsal colouration dark pinkish-brown to yellowish-brown. Ventral colouration paler. Head rounded when viewed from above and pointed, with a weakly recurved hook-shaped rostral in profile. Tail quite short and rounded at the end, with a short, blunt conical spine. Reaches 30cm TL. **Scalation** MB 20 rows, 417–502 VENT, SUB 12–20. Scales small, smooth and glossy.
RANGE Encountered in central Australia around Alice Springs, and the MacDonnell Ranges of the NT.
COMMENTS Nocturnal. Fossorial, living in rocky gorges with spinifex and low woodland. Predominantly on sandy soil types. Shelters beneath leaf litter and fallen timber. **Diet** Ant larvae. **Reproduction** Oviparous. 1–5 per clutch. **Disposition** Would rather retreat than have contact. When threatened, can excrete a pungent musk or poke an antagonizer with the tail spine.

BITE/VENOM HARMLESS
IUCN LISTING Least Concern.

Alice Springs, NT, Paul Horner

Alice Springs, NT, Paul Horner

Alice Springs, NT, Matt Summerville

CAPE YORK STRIPED BLIND SNAKE *Anilios chamodracaena*
(Ingram & Covacevich, 1993)

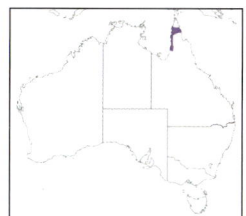

PRONUNCIATION *Ah-nil-e-oss cam-o-drah-cee-nah*.
ETYMOLOGY 'Earth-Snake' – one of the 12½ names of the female demon Gello.
TYPE LOCALITY 40km N Aurukun, QLD.
APPEARANCE Small, slender snake. Dorsal colouration dark pinkish-brown. Scales with darker centres align, forming thin dorsal stripes running along length of body. Usually darker on head. Ventral colouration white. Head rounded when viewed from above and in profile. Tail quite short and rounded at the end, with a short, blunt conical spine. Reaches 21cm TL.
Scalation MB 18 rows, 464–523 VENT, SUB 14–16. Scales small, smooth and glossy.
RANGE Encountered from Weipa to Inkerman Station on Cape York Peninsula, QLD.
COMMENTS Nocturnal. Fossorial, living in open woodland, tussock grassland and suburban environments. Shelters beneath rocks and inside rotten timber. **Diet** Ant larvae. **Reproduction** Presumed oviparous. **Disposition** Would rather retreat than have contact. When threatened, can excrete a pungent musk or poke an antagonizer with the tail spine.
BITE/VENOM HARMLESS
IUCN LISTING Least Concern.

Western Cape York Peninsula, QLD, Eric Vanderduys

Western Cape York Peninsula, QLD, Eric Vanderduys

Archer River, QLD, Anders Zimny

Archer River, QLD, Anders Zimny

Northern Blind Snake *Anilios diversus* (Waite, 1894)

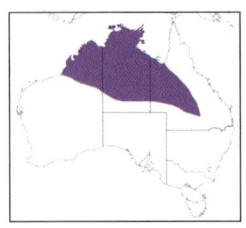

PRONUNCIATION *Ah-nil-e-oss die-ver-suss*.
ETYMOLOGY Separated.
TYPE LOCALITY Morven, QLD.
APPEARANCE Slender, medium-sized snake. Dorsal colouration dark purplish-brown to pinkish-brown. Ventral colouration paler. Head rounded when viewed from above and in profile. Tail quite short and rounded at end, with a short, blunt conical spine. Reaches 36cm TL.
Scalation MB 20 rows, 384–457 VENT, SUB 8–18. Scales small, smooth and glossy. **Similar species** *A. ammodytes* (p. 297).
RANGE Encountered from Morven, QLD, through the NT, to the Kimberley region of WA.
COMMENTS Nocturnal. Fossorial, living in open woodland, black soil grassland and coastal savannah, predominantly on sandy soil types. Shelters beneath rocks and fallen timber. **Diet** Ant larvae. **Reproduction** Oviparous. 5–8 per clutch. **Disposition** Would rather retreat than have contact. When threatened, can excrete a pungent musk or poke an antagonizer with the tail spine.
BITE/VENOM HARMLESS
IUCN LISTING Least Concern.

Weddell, NT, Anders Zimny

Weddell, NT, Anders Zimny

Gove, NT, Anders Zimny

Mica Creek, QLD, Scott Eipper

DESERT BLIND SNAKE *Anilios endoterus* (Waite, 1918)
(Interior Blind Snake)

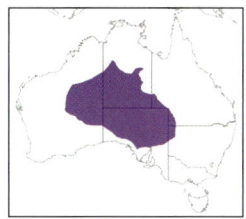

PRONUNCIATION *Ah-nil-e-oss end-o-tear-russ.*

ETYMOLOGY In the earth – in reference to fossorial nature of the species.

TYPE LOCALITY Hermannsburg, NT.

APPEARANCE Medium-sized, slender snake. Dorsal colouration dark purplish-brown to pinkish-brown, sometimes reddish. Ventral colouration cream to white. Head moderately trilobed when viewed from above, and gently angular in profile. Tail quite short and rounded at end, with a short, blunt conical spine. Females larger than males, reaching 36cm TL. **Scalation** MB 22 rows, 406–438 VENT, SUB 9–16. Scales small, smooth and glossy. **Similar species** *A. bituberculatus* (p. 302).

RANGE Encountered in central Australia from the top of the Eyre Peninsula, SA, through western NSW and QLD, southern NT and much of the interior of WA. Isolated population at Mt Isa, QLD.

COMMENTS Nocturnal. Fossorial, living in open woodland, mallee, grassland and deserts, predominantly on sandy soil types. Shelters beneath fallen timber and rocks. **Diet** Ant larvae. **Reproduction** Oviparous. Recorded mating in late October. **Disposition** Would rather retreat than have contact. When threatened, can excrete a pungent musk or poke an antagonizer with the tail spine.

BITE/VENOM HARMLESS

IUCN LISTING Least Concern.

Mt Isa, QLD, Ryan Francis

Pernatty, SA, Wes Read

Ethabuka, QLD, Anders Zimny

Ilkurlka, WA, Brian Bush

Ruby Gap Blind Snake *Anilios fossor* Shea, 2015
(Miner Blind Snake)

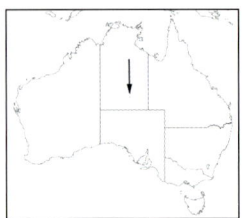

PRONUNCIATION *Ah-nil-e-oss foss-or.*
ETYMOLOGY Digging – in reference to the habits and the type locality history of mining.
TYPE LOCALITY Ruby Gap NP, NT.
APPEARANCE Small, slender snake. Dorsal colouration pale yellowish-brown. Ventral colouration paler. Colouration in life may be pink. Head rounded when viewed from above and in profile. Tail quite short and rounded at end, with a short, blunt conical spine. Reaches 25cm TL.
Scalation MB 20 rows, 514 DSR, SUB 11. Scales small, smooth and glossy.
RANGE Known from a single specimen found at Glen Annie in Ruby Gap Nature Reserve in the upper reaches of the Hale River in the southern NT, about 100km east of Alice Springs.
COMMENTS Nocturnal. Fossorial, living in open river gum woodland with *Acacia* on sandy soil types. **Diet** Ant larvae. **Reproduction** Presumed oviparous. **Disposition** Unknown, but presumed to be the same as the rest of the genus.
BITE/VENOM HARMLESS
IUCN LISTING Data Deficient.

Ruby Gap NR, NT, Glenn Shea

GANE'S BLIND SNAKE *Anilios ganei* (Aplin, 1998)

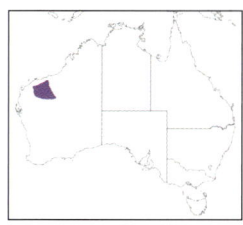

PRONUNCIATION *Ah-nil-e-oss gain-e.*
ETYMOLOGY Pertains to L. Gane, collector of the type specimen.
TYPE LOCALITY Cathedral Gorge, 30km west of Newman, WA.
APPEARANCE Medium-sized, moderately robust snake. Dorsal colouration dark greyish-brown to brown above. Ventral colouration white to cream. Head rounded when viewed from above and in profile. Tail quite short and rounded at the end, with a short, blunt conical spine. Reaches 34cm TL. **Scalation** MB 24 rows, 406–448 VENT, SUB 12–19. Scales small, smooth and glossy.
RANGE Encountered in WA in the Pilbara region between Newman, Millstream and Pannawonica.
COMMENTS Nocturnal. Fossorial, living in moist, shady gorges. Has been encountered in townships and while crossing roads. Shelters beneath fallen timber and rocks. **Diet** Ant larvae. **Reproduction** Oviparous. **Disposition** Would rather retreat than have contact. When threatened, can excrete a pungent musk or poke an antagonizer with the tail spine.
BITE/VENOM HARMLESS
IUCN LISTING Least Concern.

Duck Creek, Mt Stuart Station, WA, Brad Maryan

Cathedral Gorge, WA, Brian Bush

Northern Beaked Blind Snake *Anilios grypus* (Waite, 1918)
(Long-beaked Blind Snake)

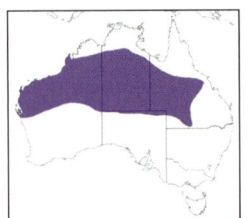

PRONUNCIATION *Ah-nil-e-oss gree-pus.*
ETYMOLOGY Hook nosed.
TYPE LOCALITY Australia; paratype from Marble Bar, WA.
APPEARANCE Very slender, large snake. Dorsal colouration pale brown to brown, with black tail. Snout pale in most populations; head usually black. Ventral colouration paler than dorsum. Head angular when viewed from above and deeply hooked in profile. Tail quite short and rounded at the end, with a short, blunt conical spine. Reaches 53cm TL.
Scalation MB 18 rows, 526–677 VENT, SUB 13–36. Scales small, smooth and glossy.
RANGE Encountered in WA, from Shark Bay to the Kimberley, across Australia into central QLD.
COMMENTS Nocturnal. Fossorial, living in open woodland, grassland, sand ridge deserts and coastal dunes with spinifex and mulga. Shelters beneath rocks, termite nests, fallen timber and leaf litter. Genetics and variation in morphology suggest there is more than the one species currently recognized within the taxon. **Diet** Ant larvae. **Reproduction** Oviparous. Lays three eggs in a clutch.
Disposition Would rather retreat than have contact. When threatened, can excrete a pungent musk or poke an antagonizer with the tail spine.

BITE/VENOM HARMLESS

IUCN LISTING Least Concern.

Tom Price, WA, Jules Farquhar

Indee Station, WA, Brian Bush

Coonarie Creek, WA, Brian Bush

Ophthalmia Range, WA, Anders Zimny

BLACK-TAILED BLIND SNAKE *Anilios guentheri* (Peters, 1865)

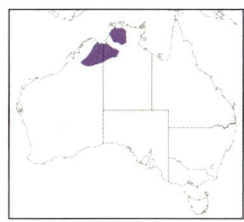

PRONUNCIATION Ah-nil-e-oss goon-ther-e.

ETYMOLOGY Pertains to A. Günther, German herpetologist at the British Museum.

TYPE LOCALITY North Australia.

APPEARANCE Medium-sized, relatively slender snake. Dorsum colouration dark purplish-brown to pinkish-brown, lighter on head. Ventral colouration paler. Head rounded when viewed from above and in profile. Tail quite short and rounded at the end, with a short, blunt conical spine. Reaches 30cm TL. **Scalation** MB 18 rows, 564–610 VENT (western) 464–547 (eastern), SUB 10–15. Scales small, smooth and glossy.

RANGE Encountered in WA, from the southern Kimberley across into the northwestern part of the NT.

COMMENTS Nocturnal. Fossorial, living in tropical open woodland, grassland and escarpment margins, predominantly on sandy soil types. Shelters beneath fallen timber and under leaf litter. Eastern (*nigricauda*) and western (nominate) populations may prove to be different taxa, with consistent differences in ventral counts. More research is required. **Diet** Ant larvae. **Reproduction** Oviparous. 2–4 per clutch. **Disposition** Would rather retreat than have contact. When threatened, can excrete a pungent musk or poke an antagonizer with the tail spine.

BITE/VENOM HARMLESS

IUCN LISTING Least Concern.

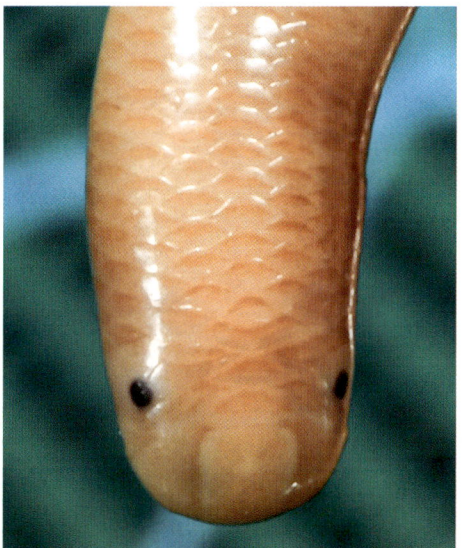

Kununurra Caravan Park, WA, Scott Eipper

Kununurra Caravan Park, WA, Scott Eipper

Top Springs, NT, Paul Horner

Edith Falls, NT, Rob Valentic

Gordon Downs, WA, Hal Cogger

Northern Hook-snouted Blind Snake *Anilios hamatus* (Storr, 1981)
(Pale-headed Blind Snake)

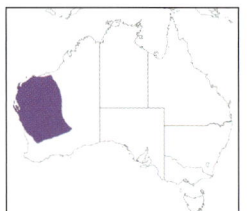

PRONUNCIATION *Ah-nil-e-oss ha-mah-tuss*.

ETYMOLOGY Hooked – in relation to the snout.

TYPE LOCALITY Marandoo, WA.

APPEARANCE Large, moderately robust snake. Dorsal colouration blackish-brown to pinkish-brown; rostral lighter in colour. Some individuals can have pale flecks on forebody. Ventral colouration paler than dorsum. Head weakly trilobed when viewed from above and angular in profile, terminating in a hooked rostral scale. Tail quite short and rounded at the end, with a short, blunt conical spine. Adult females larger than males, reaching 43cm TL. **Scalation** MB 22 rows, 330–396 VENT, SUB 11–22. Scales small, smooth and glossy. **Similar species** *A. australis* (p. 299).

RANGE Encountered in central western WA, from the Hamersley Range, south to Merredin and Shark Bay, and east across to the Kalgoorlie region.

COMMENTS Nocturnal. Fossorial, living in open woodland, mallee, heath, saltbush with spinifex and mulga, predominantly on compacting loam and stony soil types. Shelters beneath rocks and in stick-ant nests. **Diet** Ant eggs, larvae and pupae. **Reproduction** Oviparous. 2–9 per clutch. **Disposition** Would rather retreat than have contact. When threatened, can excrete a pungent musk or poke an antagonizer with the tail spine.

BITE/VENOM HARMLESS

IUCN LISTING Least Concern.

Newman, WA, Anders Zimny

Pindar, WA, Steve Swanson

Laverton, WA, Brian Bush

Newman, WA, Anders Zimny

How's Blind Snake *Anilios howi* (Storr, 1983)

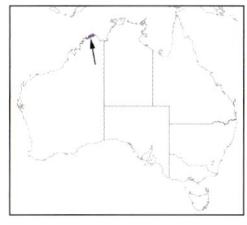

PRONUNCIATION *Ah-nil-e-oss how-ee.*
ETYMOLOGY Pertains to R. How of the Western Australian Museum.
TYPE LOCALITY Walsh Point, Port Warrender/Mitchell Plateau, WA.
APPEARANCE Small, relatively slender snake. Dorsum colouration purplish-brown to blackish-brown. Ventral colouration paler brown. Head rounded when viewed from above and in profile. Tail quite short and rounded at the end, with a short, blunt conical spine. Reaches 23cm TL. **Scalation** MB 18 rows, 430–533 VENT, SUB 10–16. Scales small, smooth and glossy.
RANGE Found in the Kimberley region of WA, between Kalumburu and Walsh Point in the Admiralty Gulf.
COMMENTS Nocturnal. Fossorial, living in tropical woodland and coastal plains. Found on damp reddish clay and stony soils. **Diet** Ant larvae. **Reproduction** Presumed to be oviparous. **Disposition** Would rather retreat than have contact. When threatened, can excrete a pungent musk or poke an antagonizer with the tail spine.
BITE/VENOM HARMLESS
IUCN LISTING Data Deficient.

Bachsten Creek, WA, Scott Eipper

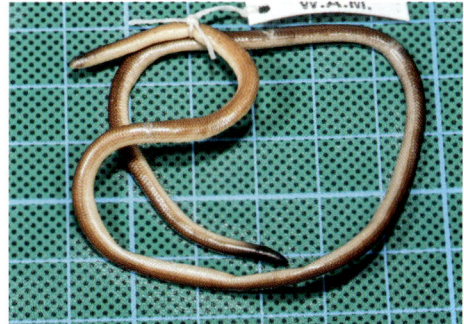
Bachsten Creek, WA, Scott Eipper

Mitchell River Plateau, WA, Philip Griffin

Fassifern Blind Snake *Anilios insperatus* Venchi, S. Wilson & Borsboom, 2015

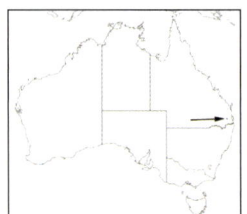

PRONUNCIATION *Ah-nil-e-oss in-spur-rah-tuss.*

ETYMOLOGY Unexpected; pertaining to how close to a major city this new species turned up.

TYPE LOCALITY Warrill View, QLD.

APPEARANCE Currently Australia's smallest known snake. Pale yellow in preservative, without evidence of pigmentation on the body. This snake was light pink in life, likely due to oxygenation of the blood. Body type slender with head weakly trilobed when viewed from above and bluntly angular in profile. Tail quite short and rounded at the end, with a short, blunt conical spine. Reaches 9.6cm TL. **Scalation** MB 16 rows, 442 DSR and SUB 19. Scales small, smooth and glossy.

RANGE Encountered in SE QLD at Warrill View.

COMMENTS Known from a single specimen, found beneath a small stone on a cleared hillside that was previously open woodland of Lemon-scented gum, Red ironbark and Grey box. Soil type was shallow, hard-setting clay. **Diet** Ant larvae. **Reproduction** Presumed oviparous. **Disposition** Unknown but presumed to be the same as for the rest of the genus.

BITE/VENOM HARMLESS

IUCN LISTING Critically Endangered.

Warrill View, QLD, Scott Eipper

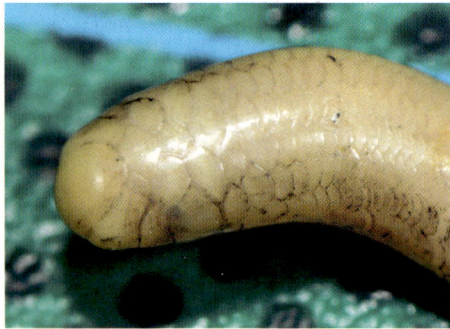

Warrill View, QLD, Scott Eipper

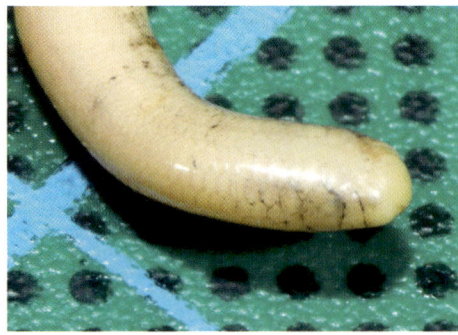

Warrill View, QLD, Scott Eipper

KIMBERLEY BLIND SNAKE *Anilios kimberleyensis* (Storr, 1981)

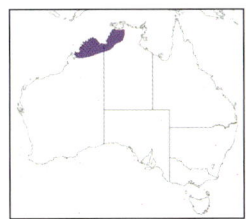

PRONUNCIATION *Ah-nil-e-oss kim-burr-lee-n-siss*.
ETYMOLOGY In reference to the Kimberley region of northern WA.
TYPE LOCALITY Biggie Island, WA.
APPEARANCE Medium-sized, relatively slender snake. Dorsal colouration dark brown to grey-brown, merging from lower flanks to creamish in colour below. Head slightly flattened and lighter in colour. Head rounded when viewed from above and in profile. Tail quite short and rounded at the end, with a short, blunt conical spine. Reaches 30cm TL. **Scalation** MB 22 rows, 405–504 VENT, SUB 10–16. Scales small, smooth and glossy.
RANGE Encountered in the Kimberley region of WA, from Koolan Island to Kalumburu. Also occurs in the northwestern NT, from the border region across to Litchfield NP.
COMMENTS Nocturnal. Fossorial, living in tropical woodland and moist vine thickets in gorges. Has been found in leaf litter and beneath rocks. **Diet** Ant larvae. **Reproduction** Presumed oviparous. **Disposition** Would rather retreat than have contact. When threatened, can excrete a pungent musk or poke an antagonizer with the tail spine.
BITE/VENOM HARMLESS
IUCN LISTING Least Concern.

Doongan Station, WA, Anders Zimny

Litchfield NP, NT, Paul Horner

Doongan Station, WA, Anders Zimny

MURCHISON BLIND SNAKE *Anilios leptosoma* (Robb, 1972)

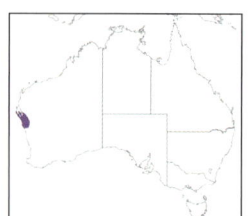

PRONUNCIATION *Ah-nil-e-oss lep-toe-so-ma.*
ETYMOLOGY Thin bodied.
TYPE LOCALITY 'The Loop', lower Murchison River, WA.
APPEARANCE Medium-sized, relatively slender snake. Dorsal colouration purplish-brown to pink. Ventral colouration paler than dorsum. Head weakly trilobed when viewed from above and bluntly angular in profile, the rostral forming a weak hook. Tail quite short and rounded at the end, with a short, blunt conical spine. Reaches 39cm TL.
Scalation MB 16 rows, 583–781 VENT, SUB 30 with 4 anal scales. Scales small, smooth and glossy.
Similar species *A. obtusifrons* (p. 325), *A. systenos* (p. 332).
RANGE Encountered in WA, from Northampton to Wooramel, in the Murchison region.
COMMENTS Nocturnal. Fossorial, living in open woodland, mallee, heaths, suburban gardens and coastal dunes, predominantly on sandy soil types. Shelters beneath fallen timber and rocks. Also found in stick-ant nests and garden beds. **Diet** Ant larvae. **Reproduction** Presumed oviparous. **Disposition** Would rather retreat than have contact. When threatened, can excrete a pungent musk or poke an antagonizer with the tail spine.
BITE/VENOM HARMLESS
IUCN LISTING Least Concern.

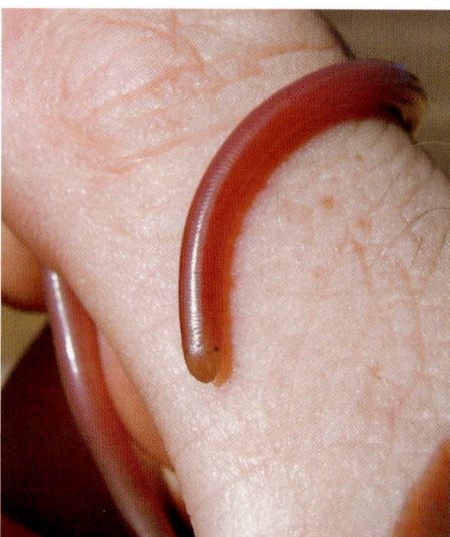

Kalbarri, WA, Nick Cairns

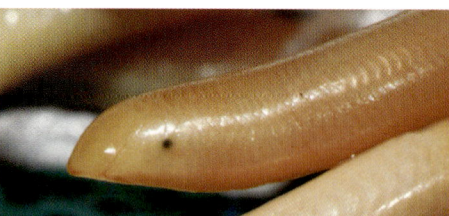

Kalbarri, WA, Scott Eipper

Kalbarri, WA, Scott Eipper

Kalbarri, WA, Nick Cairns

CAPE YORK BLIND SNAKE *Anilios leucoproctus* (Boulenger, 1889)
(White-tailed Blind Snake)

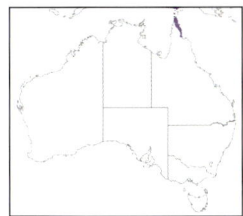

PRONUNCIATION *Ah-nil-e-oss luke-o-proc-tuss*.

ETYMOLOGY White-vented.

TYPE LOCALITY Fly River, PNG.

APPEARANCE Relatively slender, small species. Australian individuals dark purplish-black all over. De Rooij refers to yellow labials, yellowish anal region and a noticeably lighter ventral colouration in animals from PNG. This is not seen in Australian individuals. Head rounded when viewed from above and in profile. Tail quite short and rounded at the end, with a short, blunt conical spine. Reaches 25cm TL. **Scalation** MB 20 rows, 386–426 DSR, 14–17 SUB. Scales small, smooth and glossy. **Similar species** *Indotyphlops braminus* (p. 343).

RANGE Encountered in QLD, from Mt Tozer to the tip of Cape York and some islands of the Torres Strait. Also found in southern PNG.

COMMENTS Nocturnal. Fossorial, living in lowland forests, often with termite colonies, predominantly on sandy soil types. Shelters beneath logs and rocks. **Diet** Ant larvae. **Reproduction** Presumed oviparous. **Disposition** Unknown but presumed to be the same as for the rest of the genus.

BITE/VENOM HARMLESS

IUCN LISTING Least Concern.

Badu Island, QLD, Scott Eipper

Badu Island, QLD, Scott Eipper

Mt Tozer, QLD, Hal Cogger

Punsand, QLD, Mark Simpson

ROBUST BLIND SNAKE *Anilios ligatus* (W. C. H. Peters, 1879)

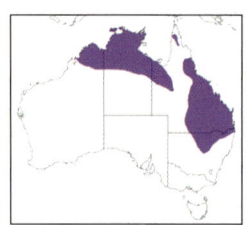

PRONUNCIATION *Ah-nil-e-oss le-gah-tus*.
ETYMOLOGY Bound.
TYPE LOCALITY Port Mackay, QLD.
APPEARANCE Large, very robust snake. Dorsal colouration dark purplish-brown to blackish-brown; slightly lighter on head in some individuals. Hatchlings pinkish-purple. Ventral colouration yellowish to white. Head rounded when viewed from above and in profile. Western population has a higher average ventral count than eastern animals. Tail quite short and rounded at the end, with a short, blunt conical spine. Adult females larger than males, reaching 41cm TL. **Scalation** MB 24 rows, 296–355 VENT, SUB 11–17. Scales small, smooth and glossy.

RANGE Encountered in two separated populations. Western population extends from the Kimberley region of WA, across the NT, and into northwestern QLD to about Mt Isa. Eastern population extends through eastern QLD, from Weipa, into northern NSW to about Gunnedah.

COMMENTS Nocturnal. Has been encountered diurnally active on the surface after rain. Fossorial, living in both closed and open woodland, grassland and brigalow, predominantly on sandy soil types. Shelters beneath fallen timber and rocks, and in ant nests. **Diet** Ant larvae. **Reproduction** Oviparous. 2–13 per clutch that took 37 days to hatch. Neonates approximately 9.8cm TL, born in March. **Disposition** Would rather retreat than have contact. When threatened, can excrete a pungent musk or poke an antagonizer with the tail spine.

BITE/VENOM HARMLESS
IUCN LISTING Least Concern.

Charleville, QLD, Rob Valentic

Darwin, NT, Ryan Francis

Basalt, QLD, Scott Eipper

Weipa, QLD, Scott Eipper

Condamine, QLD, Scott Eipper

Basalt, QLD, Scott Eipper

BARROW ISLAND BLIND SNAKE *Anilios longissimus* (Aplin, 1998)

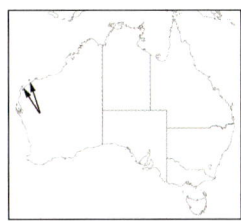

PRONUNCIATION *Ah-nil-e-oss lon-gis-im-uss.*
ETYMOLOGY Extremely long.
TYPE LOCALITY Bandicoot Bay, Barrow Island, WA.
APPEARANCE Small, very slender snake. Dorsal colouration pale yellow in preservative. The specimen was semi-translucent and appeared light pink in life due to oxygenation of the blood. Head strongly depressed, squarish when viewed from above and rounded in profile. Tail quite short and rounded at the end, with a short, blunt conical spine. The sole specimen measured is 27.8cm TL. **Scalation** MB 16 rows, 750 vertebral scales, SUB 15. Scales small, smooth and glossy.
RANGE Encountered at Bandicoot Bay, Barrow Island, WA.
COMMENTS Habits largely unknown. Potentially a true troglodyte living in reddish soil in subterranean limestone caverns. Two specimens are known, one escaping below the surface while the other was collected and sent to the Western Australian Museum. It was found on the outer casing of a bore, removed from deep below the surface. A similar snake was found in Exmouth and was thought to be a second specimen, but it is an undescribed taxon. **Diet** Ant larvae. **Reproduction** Presumed oviparous. **Disposition** Unknown but presumed to be the same as for the rest of the genus.
BITE/VENOM HARMLESS
IUCN LISTING Data Deficient.

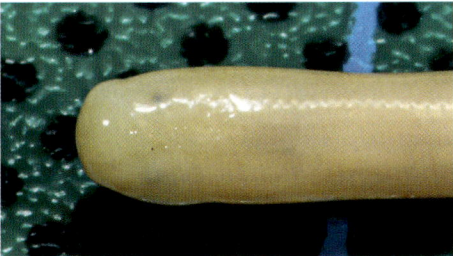
Bandicoot Bay, Barrow Island, WA, Scott Eipper

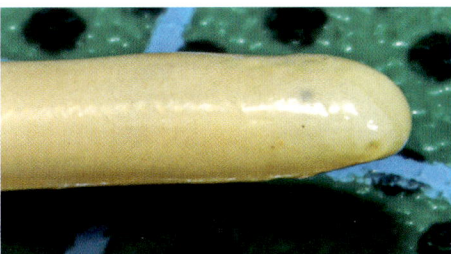
Bandicoot Bay, Barrow Island, WA, Scott Eipper

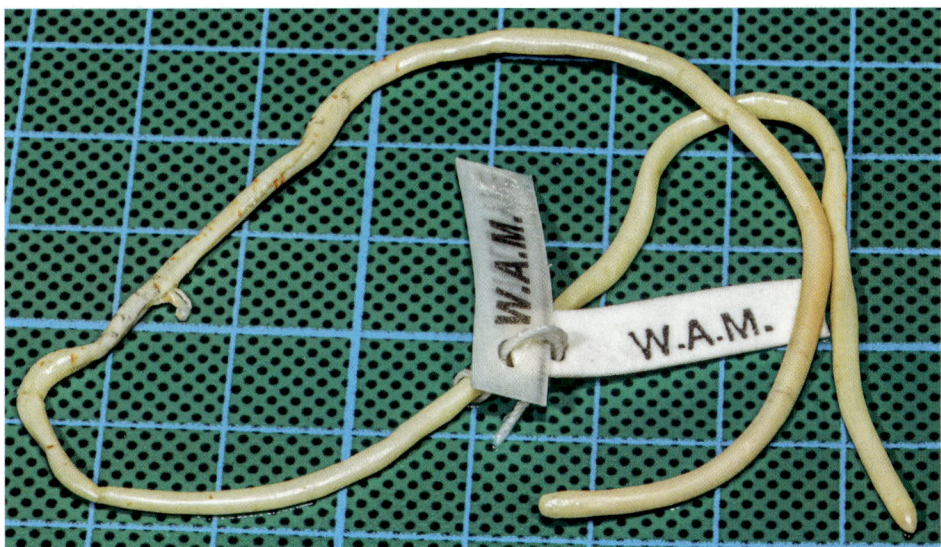
Bandicoot Bay, Barrow Island, WA, Scott Eipper

Buff-snouted Blind Snake *Anilios margaretae* (Storr, 1981)

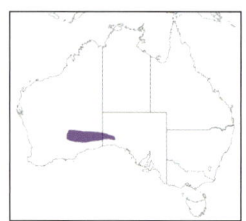

PRONUNCIATION *Ah-nil-e-oss mar-grat-ee.*
ETYMOLOGY Pertains to Margaret Butler, Australian naturalist.
TYPE LOCALITY Lake Throssell, WA.
APPEARANCE Relatively slender, medium-sized snake. Dorsal colouration dark purplish-brown to pinkish-brown, lighter on head. Ventral colouration pale grey. Head trilobed when viewed from above and angular in profile. Tail quite short and rounded at the end, with a short, blunt conical spine. Reaches 30cm TL. **Scalation** MB 18 rows, 559 VENT, SUB 12. Scales small, smooth and glossy.
RANGE Encountered in southern central WA at Lake Throssell, and near Maralinga, SA.
COMMENTS Nocturnal. Fossorial, living at salt-lake margins with reddish sandy loams. **Diet** Ant larvae. **Reproduction** Presumed oviparous. **Disposition** Unknown but presumed to be the same as for the rest of the genus.
BITE/VENOM HARMLESS
IUCN LISTING Data Deficient.

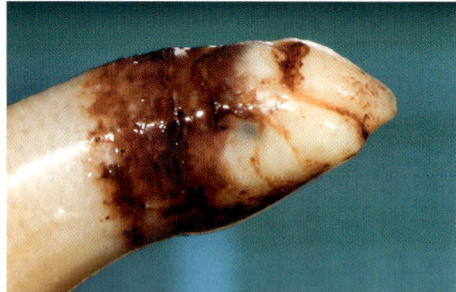
Lake Throssell, WA, Scott Eipper

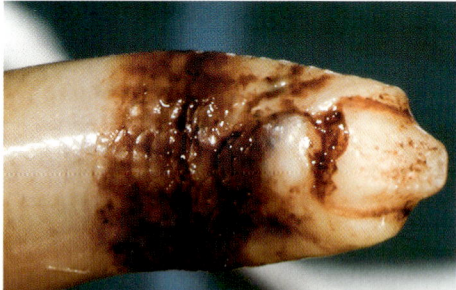
Lake Throssell, WA, Scott Eipper

40km S Neale Junction Nature Reserve, WA, Glen Gaikhorst

SMALL-EYED BLIND SNAKE *Anilios micromma* (Storr, 1981)

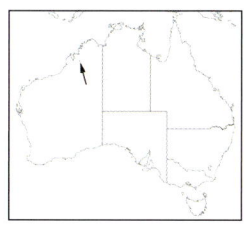

PRONUNCIATION *Ah-nil-e-oss my-crow-mah*.

ETYMOLOGY Small eyed.

TYPE LOCALITY Leopold Downs, WA.

APPEARANCE Relatively slender, small snake. Pale yellow in colour in preservative – may have appeared light pink in life due to oxygenation of the blood. Head rounded when viewed from above and in profile. Eyes very small. It has been suggested that the eyes are normal in size and that the lack of pigment surrounding them has led to the assumption of smaller eye size. Tail quite short and rounded at the end, with a short, blunt conical spine. The sole specimen measured is 21cm TL. **Scalation** MB 18 rows, 478 VENT, SUB 15. Scales small, smooth and glossy.

RANGE Encountered in the southern Kimberley region of WA from Leopold Downs Station.

COMMENTS Largely unknown. Currently only known from a single specimen collected in 1924. The habitat on the station is typical tropical open woodland. **Diet** Ant larvae. **Reproduction** Presumed oviparous. **Disposition** Unknown but presumed to be the same as for the rest of the genus.

BITE/VENOM HARMLESS

IUCN LISTING Data Deficient.

Leopold Downs Station, WA, Scott Eipper

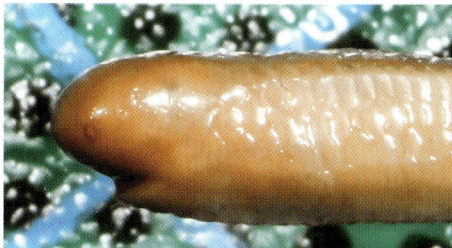

Leopold Downs Station, WA, Scott Eipper

Leopold Downs Station, WA, Scott Eipper

GROOTE EYLANDT BLIND SNAKE *Anilios minimus* (Kinghorn, 1929)

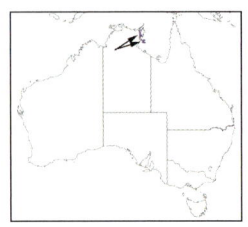

PRONUNCIATION *Ah-nil-e-oss min-e-muss.*

ETYMOLOGY The smallest.

TYPE LOCALITY Groote Eylandt, Gulf of Carpentaria, NT.

APPEARANCE Slender, small species that is dark pinkish-brown with dark longitudinal streaks. Head and tail dark in colour. Pale pinkish-white below. Head rounded when viewed from above and in profile. Tail quite short and rounded at the end, with a short, blunt conical spine. Reaches 22cm TL. **Scalation** MB 16 rows, 381–457 VENT, SUB 9–17. Scales small, smooth and glossy.

RANGE Found on Groote Eylandt and adjacent areas of the NT, to Lake Evella.

COMMENTS Nocturnal. Fossorial, living in tropical woodland and coastal plains. Found on sandy soils; one individual encountered in a coastal dune, beneath an embedded log on moist sandy soil. Known from very few individuals. **Diet** Ant larvae. **Reproduction** Presumed oviparous. **Disposition** Unknown but presumed to be the same as for the rest of the genus.

BITE/VENOM HARMLESS

IUCN LISTING Data Deficient.

Groote Eylandt, NT, Scott Eipper

Groote Eylandt, NT, Scott Eipper

Groote Eylandt, NT, Scott Eipper

SLENDER BLIND SNAKE *Anilios nema* (Shea & Horner, 1997)

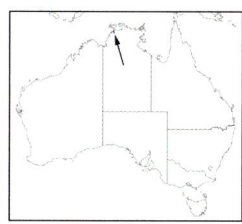

PRONUNCIATION *Ah-nil-e-oss nee-mah.*
ETYMOLOGY 'A thread'.
TYPE LOCALITY Fannie Bay, Darwin, NT.
APPEARANCE Medium-sized, relatively slender snake. Dorsal colouration dark purplish-brown to pale pink. Ventral colouration pale pinkish-white. Head rounded when viewed from above and in profile. Tail quite short and rounded at the end, with a short, blunt conical spine. Reaches 28cm TL. **Scalation** MB 16 rows, 520–589 VENT, SUB 9–14. Scales small, smooth and glossy. **Similar species** *A. tovelli* (p. 334).
RANGE Encountered in and around Darwin, NT.
COMMENTS Nocturnal. Fossorial, living in tropical woodland and heavily vegetated gardens. Found beneath rocks, fallen timber and household building materials. **Diet** Ant larvae. **Reproduction** Presumed oviparous. **Disposition** Unknown but presumed to be the same as for the rest of the genus.
BITE/VENOM HARMLESS
IUCN LISTING Data Deficient.

Rum Jungle, NT, Paul Horner

Rum Jungle, NT, Scott Eipper

Rum Jungle, NT, Scott Eipper

Blackish Blind Snake *Anilios nigrescens* Gray, 1845

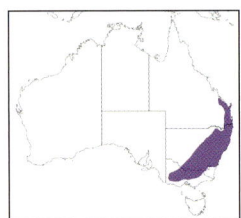

PRONUNCIATION *Ah-nil-e-oss nigh-gres-sens*.

ETYMOLOGY Blackish.

TYPE LOCALITY Parramatta, NSW.

APPEARANCE Large, robust snake. Dorsal colouration black, dark purplish-brown to pinkish-brown; juveniles pink. Edges of scales whitish, forming fine, netted pattern. Ventral colouration white. Head rounded when viewed from above and in profile. Tail quite short and rounded at the end with a short, blunt conical spine. Adult females larger than males, reaching 57cm TL. **Scalation** MB 22 rows, 420–457 VENT, SUB 13–19. Scales small, smooth and glossy. **Similar species** *A. proximus* (p. 328).

RANGE Encountered in southern QLD, from Kroombit Tops, south through eastern NSW, to Seymour, Vic.

COMMENTS Nocturnal. Fossorial, living in open and closed woodland, rocky outcrops, heaths, grassland and rainforest, predominantly on sandy soil types. Shelters beneath stumps, rocks and leaf litter, sometimes forming aggregations of multiple individuals. Frequently encountered in suburban gardens, in swimming pools, or crossing roads on warm, wet nights. Has also been found in rotten standing trees to a height of 5m. **Diet** Ant larvae. Anecdotal reports of eating worms and leeches. **Reproduction** Oviparous. 5–20 per clutch. Neonates approximately 12cm TL hatching in December–March. **Disposition** Would rather retreat than have contact. When threatened, can excrete a pungent musk or poke an antagonizer with the tail spine. Will also produce a squeak when grasped.

BITE/VENOM HARMLESS

IUCN LISTING Least Concern.

Springbrook, QLD, Scott Eipper

Springbrook, QLD, Scott Eipper

Kroombit Tops, QLD, Scott Eipper

Mt Glorious, QLD, Scott Eipper

BLUNT-SNOUTED BLIND SNAKE *Anilios obtusifrons*
Ellis, Doughty, Donnellan, Marin & Vidal, 2017

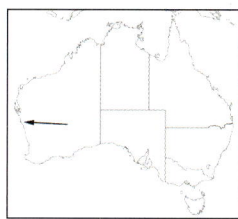

PRONUNCIATION *Ah-nil-e-oss ob-tus-e-frons*.
ETYMOLOGY Blunt snouted.
TYPE LOCALITY 23km south of Kalbarri, WA.
APPEARANCE Small, slender snake. Dorsal colouration pink, becoming lighter on lower flanks. Ventral colouration yellow to cream. Head bluntly pointed when viewed from above and angular in profile. Tail quite short and rounded at the end, with a short, blunt conical spine. Reaches 22.5cm TL. **Scalation** MB 18 rows, 581–590 VENT, SUB 15 and anal scale 4. Scales small, smooth and glossy. **Similar species** *A. leptosoma* (p. 316), *A. systenos* (p. 332).
RANGE Known from three specimens found in two locations south of Kalbarri, WA.
COMMENTS Presumed to be nocturnal. Fossorial, living in *Acacia* woodland, heath and mallee with brown sandy soil. **Diet** Ant larvae. **Reproduction** Presumed oviparous. **Disposition** Unknown but presumed to be the same as for the rest of the genus.
BITE/VENOM HARMLESS
IUCN LISTING Vulnerable.

Near Kalbarri, WA, Brad Maryan

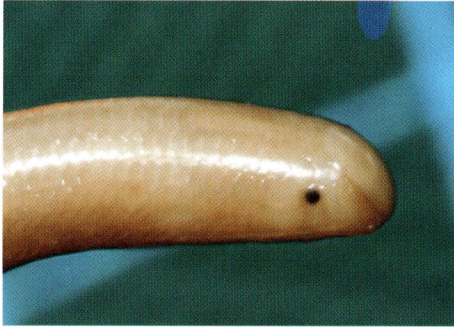

Near Kalbarri, WA, Scott Eipper

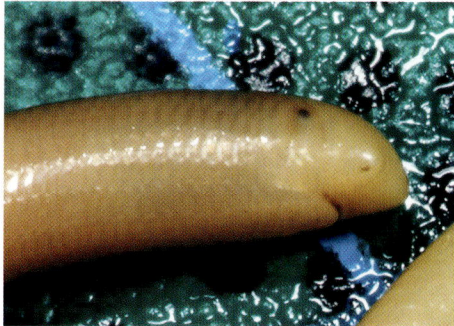

Near Kalbarri, WA, Scott Eipper

Pilbara Hook-snouted Blind Snake *Anilios pilbarensis*
(Aplin & Donnellan, 1993)

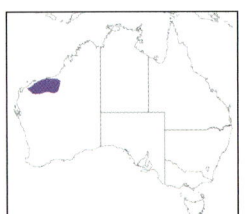

PRONUNCIATION *Ah-nil-e-oss pill-bra-n-siss*.

ETYMOLOGY In reference to the Pilbara region.

TYPE LOCALITY Woodstock Station, Pilbara, WA.

APPEARANCE Large, moderately slender snake. Dorsal colouration brown to purplish-brown, or grey gradually merging into whitish ventral surface. Rostral lighter in colour. Head weakly trilobed when viewed from above and strongly angular in profile. Tail quite short and rounded at the end, with a short, blunt conical spine. Females larger than males, reaching 38cm TL. **Scalation** MB 22 rows, 363–425 VENT, SUB 15–22. Scales small, smooth and glossy.

RANGE Encountered in central western WA, from Muccan to Balfour Downs, across to Chichester Range NP.

COMMENTS Nocturnal. Fossorial, living in open woodland with spinifex and tussock grassland. Also found in areas with *Acacia* with granite outcrops. Predominantly on compacting loam soil types. Shelters beneath rocks and fallen timber. **Diet** Ant larvae. **Reproduction** Presumed oviparous. **Disposition** Would rather retreat than have contact. When threatened, can excrete a pungent musk or poke an antagonizer with the tail spine.

BITE/VENOM HARMLESS

IUCN LISTING Data Deficient.

Woodie Woodie Mine, WA, Brad Maryan

Marble Bar, WA, Jesse Campbell

Marble Bar, WA, Jesse Campbell

FAT BLIND SNAKE *Anilios pinguis* (Waite, 1897)

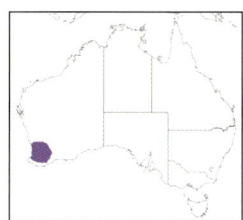

PRONUNCIATION *Ah-nil-e-oss ping-wiss.*
ETYMOLOGY Stout.
TYPE LOCALITY SA.
APPEARANCE Large, very robust snake. Dorsal colouration dark purplish-brown to black, lighter on head. Lower flanks become gradually lighter in colour, turning pale grey. Head weakly trilobed when viewed from above and bluntly angular in profile. Tail quite short and rounded at the end, with a short, blunt conical spine. Reaches 50cm TL. **Scalation** MB 20 rows, 280–377 VENT, SUB 10–18. Scales small, smooth and glossy.
RANGE Encountered in SW WA, from Gingin to Bruce Rock, across to Bunbury.
COMMENTS Nocturnal. Fossorial, living in open woodland, mallee, heaths and rock outcrops, predominantly on heavily stony soil types. Shelters beneath stumps and rocks, and inside stick-ant nests. **Diet** Ant larvae. **Reproduction** Oviparous. 1–5 per clutch. Neonates approximately 10cm TL. **Disposition** Would rather retreat than have contact. When threatened, can excrete a pungent musk or poke an antagonizer with the tail spine.

BITE/VENOM HARMLESS
IUCN LISTING Least Concern.

East of Forrestania, WA, Brian Bush

John Forrest NP, WA, Robert Audcent

East of Forrestania, WA, Brian Bush

WOODLAND BLIND SNAKE *Anilios proximus* (Waite, 1893)
(Proximus Blind Snake)

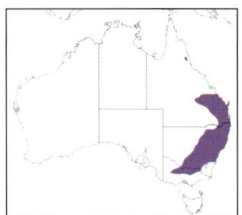

PRONUNCIATION *Ah-nil-e-oss prox-im-us*.
ETYMOLOGY Nearest.
Type locality NSW.
APPEARANCE Large, very robust snake. Dorsal colouration dark purplish-brown to brown; hatchlings pinkish-purple. Ventral colouration yellowish to white. Head slightly trilobed when viewed from above and angular in profile. Atherton population has a higher DSR count (390–392 v 326–378) than southern population. Tail quite short and rounded at the end, with a short, blunt conical spine. Adult females larger than males, reaching 62cm TL.
Scalation MB 220 rows, DSR 326–378, SUB 12–23. Scales small, smooth and glossy. **Similar species** *A. nigrescens* (p. 324), *A. torresianus* (p. 333).

RANGE Two separated populations. Northern population centres around the Atherton tablelands, QLD. Main population extends through eastern QLD, from Rockhampton through NSW, to Macedon, Vic, and across to the SA border.

COMMENTS Nocturnal but has been encountered diurnally active on the surface after rain. Fossorial, living in both closed and open woodland, grassland and brigalow, predominantly on sandy soil types. Known to shelter beneath fallen timber and rocks, and in ant nests. **Diet** Ant larvae. **Reproduction** Oviparous. 3–34 per clutch. Neonates approximately 13cm TL. **Disposition** Would rather retreat than have contact. When threatened, can excrete a pungent musk or poke an antagonizer with the tail spine. Can also produce a squeak when grasped.

BITE/VENOM HARMLESS
IUCN LISTING Least Concern.

Beechmont, QLD, Scott Eipper

Miles, QLD, Scott Eipper

Beechmont, QLD, Scott Eipper

ROBERTS' BLIND SNAKE *Anilios robertsi* (Couper, Covacevich & Wilson, 1998)

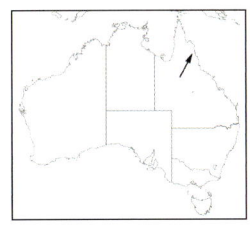

PRONUNCIATION *Ah-nil-e-oss robert-see.*
ETYMOLOGY Pertains to L. Roberts, collector of the holotype.
TYPE LOCALITY Romeo Creek, via Shipton's Flat, QLD.
APPEARANCE Medium-sized, slender snake. Dorsal colouration dark purplish-black with sharp, straight-edged delineation to stark white underside. Reaches 29cm TL. Head rounded when viewed from above and in profile. Tail quite short and rounded at the end, with a short, blunt conical spine. **Scalation** MB 22 rows, 556 VENT, SUB 12. Scales small, smooth and glossy.
RANGE Known only from Romeo Creek near Shipton's Flat in north QLD.
COMMENTS Thought to be nocturnal. Known from a single individual found in February 1983 coming out of a recently fallen hollow log in open woodland. **Diet** Ant larvae. **Reproduction** Presumed oviparous. **Disposition** Unknown, but presumed to be the same as for the rest of the genus.
BITE/VENOM HARMLESS
IUCN LISTING Data Deficient.

Romeo Creek, Shipton's Flat, QLD, Scott Eipper

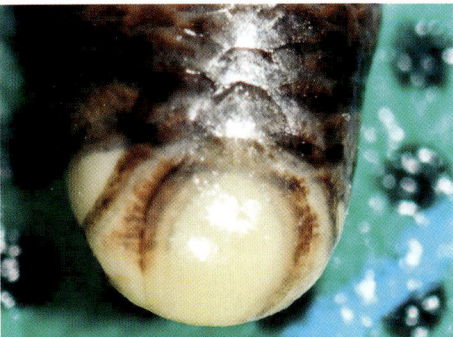

Romeo Creek, Shipton's Flat, QLD, Scott Eipper

Romeo Creek, Shipton's Flat, QLD, Scott Eipper

COOLOOLA BLIND SNAKE *Anilios silvia* (Ingram & Covacevich, 1993)

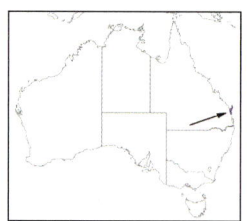

PRONUNCIATION *Ah-nil-e-oss sil-vee-ah.*

ETYMOLOGY Pertains to R. Sylvia.

TYPE LOCALITY Seary's Scrub, Cooloola, QLD.

APPEARANCE Small, slender snake. Broadly striped with dark purplish-black on yellowish-cream body; stripes usually coalesce, giving the appearance of a whole dark, purplish-black snake above, with sharp, variable-edged delineation to stark white underside. Head rounded when viewed from above and in profile. Tail quite short and rounded at the end, with a short, blunt conical spine. Reaches 21cm TL. **Scalation** MB 20 rows, 272–320 VENT, SUB 14–21. Scales small, smooth and glossy.

RANGE Encountered in SE QLD, from Palmwoods to K'gari (Fraser Island), QLD.

COMMENTS Nocturnal. Fossorial, living in closed woodland, leafy suburban gardens and rainforests, predominantly on sandy soil types. Known to shelter beneath fallen logs, rocks and rotten timber. **Diet** Ant larvae. **Reproduction** Presumed oviparous. **Disposition** Would rather retreat than have contact. When threatened, can excrete a pungent musk or poke an antagonizer with the tail spine.

BITE/VENOM HARMLESS

IUCN LISTING Least Concern.

Weyba Nature Reserve, QLD, Scott Eipper

Weyba Nature Reserve, QLD, Scott Eipper

Weyba Nature Reserve, QLD, Scott Eipper

Noosa, QLD, Scott Eipper

SPLENDID BLIND SNAKE *Anilios splendidus* (Aplin, 1998)

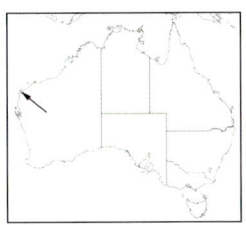

PRONUNCIATION *Ah-nil-e-oss splen-did-us.*
ETYMOLOGY Magnificent.
TYPE LOCALITY Milyering Well, Cape Range NP, WA.
APPEARANCE Large, moderately robust snake. Dorsal colouration silver purplish-grey with randomly placed white scales, lighter on front half of head. Ventral colouration paler than dorsum. Head sub-rectangular and weakly trilobed when viewed from above and weakly angular in profile. Tail quite short and rounded at the end, with a short, blunt conical spine. Reaches 51cm TL. **Scalation** MB 20 rows, 377 VENT, SUB 13. Scales small, smooth and glossy. **Similar species** *A. australis* (p. 299).

RANGE Found at the Ranger's residence (Milyering Well) in Cape Range NP, WA.

COMMENTS Nocturnal. Fossorial; known from two specimens, one found in 1995 and another in 2023. Habitat in the area of collection was sand over coral limestone with Tangling Melaleuca, small shrubs and *Triodia*. Suggested to be a junior synonym of the Fat Blind Snake *A. pinguis* (p. 327.) **Diet** Ant larvae. **Reproduction** Presumed oviparous. **Disposition** Unknown, but presumed to be the same as for the rest of the genus.

BITE/VENOM HARMLESS
IUCN LISTING Data Deficient.

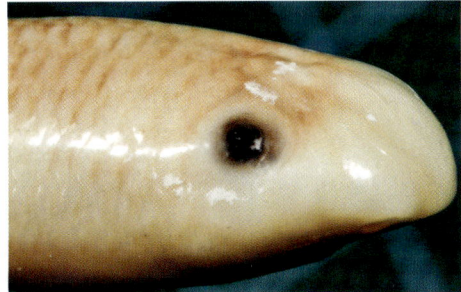

Milyering Well, Cape Range, WA, Scott Eipper

Milyering Well, Cape Range, WA, Scott Eipper

Milyering Well, Cape Range, WA, Mark Cowan

Yardie Creek, WA, Robert Audcent

Sharp-snouted Blind Snake *Anilios systenos*
Ellis, Doughty, Donnellan, Marin & Vidal, 2017

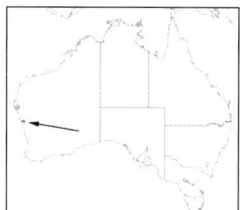

PRONUNCIATION *Ah-nil-e-oss sis-ten-oss.*
ETYMOLOGY Tapering to a point.
TYPE LOCALITY 15km east of Geraldton, WA.
APPEARANCE Small, slender snake. Dorsal colouration yellowish-brown with evidence of dark pigment forming dark longitudinal stripes. Ventral colouration paler, yellow to cream. Head bluntly pointed when viewed from above and angular in profile. Tail quite short and rounded at the end, with a short, blunt conical spine. Reaches 27cm TL. **Scalation** MB 18 rows, 598–621 VENT, SUB 30 and 4 anal scales. Scales small, smooth and glossy. **Similar species** *A. leptosoma* (p. 316), *A. obtusifrons* (p. 325).
RANGE Known from a few specimens just east of Geraldton, WA.
COMMENTS Presumed to be nocturnal. Fossorial, living in open woodland. **Diet** Ant larvae. **Reproduction** Presumed oviparous. **Disposition** Unknown, but presumed to be the same as for the rest of the genus.
BITE/VENOM HARMLESS
IUCN LISTING Data Deficient.

15km E of Geraldton, WA, Scott Eipper

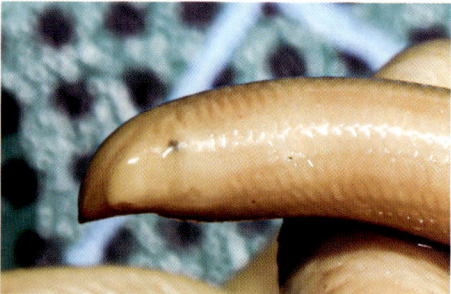

15km E of Geraldton, WA, Scott Eipper

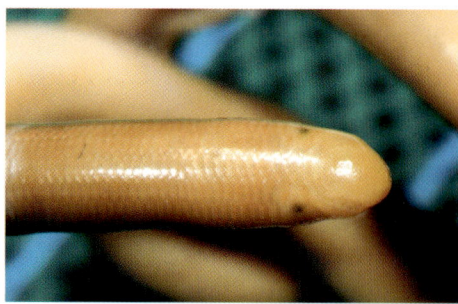

15km E of Geraldton, WA, Scott Eipper

NORTH-EAST BLIND SNAKE *Anilios torresianus* (Boulenger, 1889)

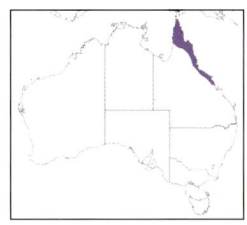

PRONUNCIATION *Ah-nil-e-oss tor-reez-e-ah-nuss*.

ETYMOLOGY In reference to the Torres Strait, which contains the type locality.

TYPE LOCALITY Murray Island, Torres Strait, QLD.

APPEARANCE Large, very robust snake. Dorsal colouration dark purplish-brown to pinkish-brown, lighter on head. Ventral colouration paler. Head rounded when viewed from above and in profile. Tail quite short and rounded at the end, with a short, blunt conical spine. Adult females larger than males, reaching 40cm TL. **Scalation** MB 22 rows, 348–387 VENT, SUB 14–22. Scales small, smooth and glossy. **Similar species** *A. proximus* (p. 328).

RANGE Encountered in NE QLD, from Mackay to the tip of Cape York. Also found in southern PNG.

COMMENTS Nocturnal. Fossorial, living in open and closed woodland, vine thickets, rainforests and moist gardens. One individual was found at 6m on a tree trunk during a shower. Shelters beneath stumps, rocks and man-made debris, and in termite nests. **Diet** Ant larvae. **Reproduction** Oviparous. 2–7 per clutch. Neonates approximatley 11cm TL. **Disposition** Would rather retreat than have contact. When threatened, can excrete a pungent musk or poke an antagonizer with the tail spine.

BITE/VENOM HARMLESS

IUCN LISTING Least Concern.

Lake Morris, QLD, Scott Eipper

Lake Morris, QLD, Scott Eipper

Archer River, QLD, Anders Zimny

Lake Eacham, QLD, Shane Black

Darwin Blind Snake *Anilios tovelli* (Loveridge, 1945)

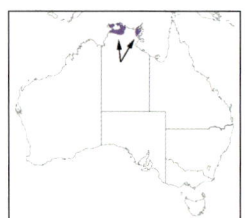

PRONUNCIATION *Ah-nil-e-oss to-vell-ee.*

ETYMOLOGY Pertains to G. Tovell, who collected the holotype.

TYPE LOCALITY Koonawarra Sports Ground, about 5m south of Darwin, NT.

APPEARANCE Small, slender snake. Dorsal colouration dark brown to pinkish-brown, with dark medial pigment on each scale, which can form faint stripes. Tail dark in colour. Ventral colouration pinkish-grey. Head rounded when viewed from above and in profile. Tail quite short and rounded at the end, with a short, blunt conical spine. Reaches 13cm TL. **Scalation** MB 20 rows, 240–262 VENT, SUB 13–16. Scales small, smooth and glossy. **Similar species** *A. nema* (p. 323).

RANGE Encountered in Darwin and surrounding areas, including the Cobourg Peninsula, NT.

COMMENTS Nocturnal. Fossorial, living in open woodland, vine thickets and floodplains. Shelters beneath stumps and rocks. **Diet** Ant larvae. **Reproduction** Presumed oviparous. **Disposition** Unknown, but presumed to be the same as for the rest of the genus.

BITE/VENOM HARMLESS

IUCN LISTING Data Deficient.

Darwin, NT, Scott Eipper

Darwin, NT, Scott Eipper

Berry Springs, NT, Steve Swanson

Sandamara Blind Snake *Anilios troglodytes* (Storr, 1981)

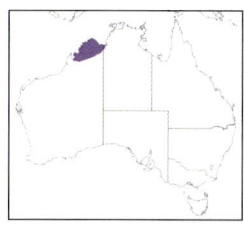

PRONUNCIATION Ah-nil-e-oss trog-low-dy-tees.
ETYMOLOGY Hole dwelling.
TYPE LOCALITY Tunnel Cave, Napier, Distribution, WA.
APPEARANCE Medium-sized, slender snake. Dorsal colouration pale brown to pinkish, and lighter on head. Ventral colouration white or cream to yellowish. Head flattened and rounded when viewed from above and in profile. Tail quite short and rounded at the end, with a short, blunt conical spine. Reaches 40cm TL. **Scalation** MB 22 rows, 587–641 VENT, SUB 12–14. Scales small, smooth and glossy.
RANGE Encountered in northwestern WA, from El Questro Station, across to Tunnel Cave in the Napier Range.
COMMENTS Nocturnal. Fossorial, living in open woodland. Shelters beneath stumps and rocks. **Diet** Ant larvae. **Reproduction** Presumed oviparous. **Disposition** Would rather retreat than have contact. When threatened, can excrete a pungent musk or poke an antagonizer with the tail spine.
BITE/VENOM HARMLESS
IUCN LISTING Least Concern.

Mornington Station, WA, Ian 'Bushrat' Bool

Mornington Station, WA, Ian 'Bushrat' Bool

Bachsten Creek, WA, Scott Eipper

Claw-snouted Blind Snake *Anilios unguirostris* (W. C. H. Peters, 1867)

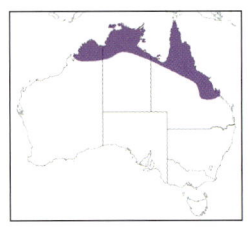

PRONUNCIATION *Ah-nil-e-oss un-jer-ros-tris*.
ETYMOLOGY Claw beaked.
TYPE LOCALITY Rockhampton, QLD.
APPEARANCE Large, moderately slender snake. Dorsal colouration dark purplish-brown to pinkish-brown, darker on head. Ventral colouration lighter. Head poorly trilobed when viewed from above and strongly hook-like and angular in profile. Tail quite short and rounded at the end, with a short, blunt conical spine. Adult females larger than males, reaching 61cm TL. **Scalation** MB 24 rows, 387–474 VENT, SUB 11–16. Scales small, smooth and glossy.
RANGE Encountered in the Kimberley region of WA, across northern Australia to central QLD.
COMMENTS Nocturnal. Fossorial, living in open woodland, grassland, and mulga lands, on both stony and sandy soil types. Shelters beneath logs and rocks. **Diet** Ant larvae. **Reproduction** Presumed oviparous. **Disposition** Would rather retreat than have contact. When threatened, can excrete a pungent musk or poke an antagonizer with the tail spine.

BITE/VENOM HARMLESS
IUCN LISTING Least Concern.

Palmer River, QLD, Shane Black

Palmer River, QLD, Shane Black

Daly Waters, NT, Paul Horner

MORNINGTON BLIND SNAKE *Anilios vagurima* Ellis, 2019

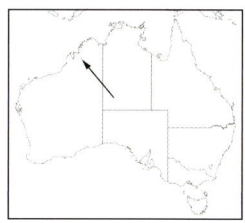

PRONUNCIATION *Ah-nil-e-oss vah-goo-ree-mah.*
ETYMOLOGY Wandering cleft – in reference to the termination point of the nasal cleft.
TYPE LOCALITY Mornington Sanctuary, WA.
APPEARANCE Moderately slender snake. Dorsal colouration light brown; scales have a darker posterior edge. Ventral colouration pale tan, light brown. Colouration described following preservation. Head moderately trilobed when viewed from above and bluntly rounded in profile. Tail quite short and rounded at the end, with a short, blunt conical spine. Reaches 33cm TL.
Scalation MB 22 rows, 542 dorsal body scales, SUB 14. Scales small, smooth and glossy.
RANGE Known only from the Mornington Wildlife Sanctuary in the Kimberley region of WA, from a single specimen.
COMMENTS Nocturnal. Fossorial, living in tropical woodland. Found on reddish clay loam. **Diet** Ant larvae. **Reproduction** Presumed oviparous. **Disposition** Would rather retreat than have contact. When threatened, can excrete a pungent musk or poke an antagonizer with the tail spine.
BITE/VENOM HARMLESS
IUCN LISTING Least Concern.

Mornington Station, WA, Ryan Ellis

SOUTHERN BEAKED BLIND SNAKE *Anilios waitii* (Boulenger, 1895)

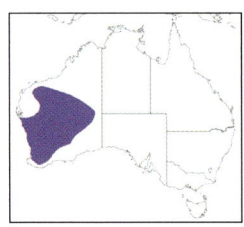

PRONUNCIATION *Ah-nil-e-oss wait-ee.*
ETYMOLOGY Pertains to E. Waite, Australian zoologist.
TYPE LOCALITY NW Australia.
APPEARANCE Large, moderately slender snake. Dorsal colouration pale yellowish-brown to dark brown. Brown merges into pale greyish-cream below. Rostral usually pale. Head weakly trilobed when viewed from above and strongly angular in profile. Reaches 62cm TL. **Scalation** MB 20 rows, 535–667 VENT, SUB 13–26. Scales small, smooth and glossy.
RANGE Encountered in southern WA, from the Warburton Range, through the goldfields and wheatbelt regions, to Perth.
COMMENTS Nocturnal. Fossorial, living in mallee, sand ridge deserts, grassland on red loams and pale sands. There is a record of 10 sub-adults found under a rock. Has been seen entering a meat-ant nest. **Diet** Ant eggs and pupae. **Reproduction** Presumed oviparous. **Disposition** Would rather retreat than have contact. When threatened, can excrete a pungent musk or poke an antagonizer with the tail spine.
BITE/VENOM HARMLESS
IUCN LISTING Least Concern.

20km S Menzies, WA, Brian Bush

Leinster, WA, Brian Bush

Pinnacles Station, WA, Brad Maryan

Red Hill, WA, Brad Maryan

Brown-snouted Blind Snake *Anilios wiedii* (Peters, 1867)

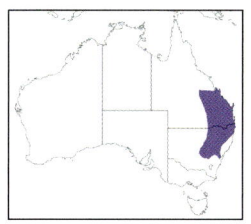

PRONUNCIATION *Ah-nil-e-oss weed-ee.*
ETYMOLOGY Pertains to M. Wied-Neuwied, German zoologist.
TYPE LOCALITY Brisbane, QLD.
APPEARANCE Small, moderately slender snake. Dorsal colouration dark purplish-brown to pinkish-brown. Rostral scale usually brown. Ventral colouration paler than dorsum. Head rounded when viewed from above and in profile. Tail quite short and rounded at the end, with a short, blunt conical spine. Adult females larger than males, reaching 30cm TL. **Scalation** MB 20 rows, 390–410 VENT, SUB 20. Scales small, smooth and glossy. **Similar species** *A. nigrescens* (p. 324).
RANGE Encountered in eastern QLD, from Mackay south into NSW, to the Hunter Valley and through to the plains of central NSW.
COMMENTS Nocturnal. Fossorial, living in open woodland, brigalow, grassland and mulga, predominantly on sandy soil types. Shelters beneath stumps and rocks, and in leaf litter. **Diet** Ant larvae. **Reproduction** Oviparous. 1–8 per clutch. Neonates approximately 11cm TL. **Disposition** Would rather retreat than have contact. When threatened, can excrete a pungent musk or poke an antagonizer with the tail spine.
BITE/VENOM HARMLESS
IUCN LISTING Least Concern.

Chambers Flat, QLD, Scott Eipper

Forest Lake, QLD, Scott Eipper

Chambers Flat, QLD, Scott Eipper

Armidale, NSW, Scott Eipper

YAMPI BLIND SNAKE *Anilios yampiensis* (Storr, 1981)

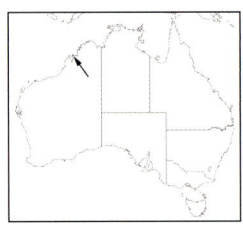

PRONUNCIATION *Ah-nil-e-oss yam-pee-n-siss.*
ETYMOLOGY After Yampi Sound, WA.
TYPE LOCALITY Koolan Island, WA.
APPEARANCE Small, slender snake that in preservative is yellowish-brown with evidence of dark pigment on head and tail. Underside paler, being yellow to cream. Head bluntly pointed when viewed from above and angular in profile. Tail quite short and rounded at the end, with a short, blunt conical spine. The sole specimen is 13cm TL. **Scalation** MB 18 rows, 480 VENT, SUB 11. Scales small, smooth and glossy.
RANGE Known from a single specimen collected on Koolan Island off the Yampi Peninsula in the Kimberley region of WA in March 1966.
COMMENTS Presumed to be nocturnal and fossorial; probably lives in gorges made up of quartzite sandstone with vine forests. **Diet** Ant larvae. **Reproduction** Presumed oviparous. **Disposition** Unknown but presumed to be the same as for the rest of the genus.
BITE/VENOM HARMLESS
IUCN LISTING Data Deficient.

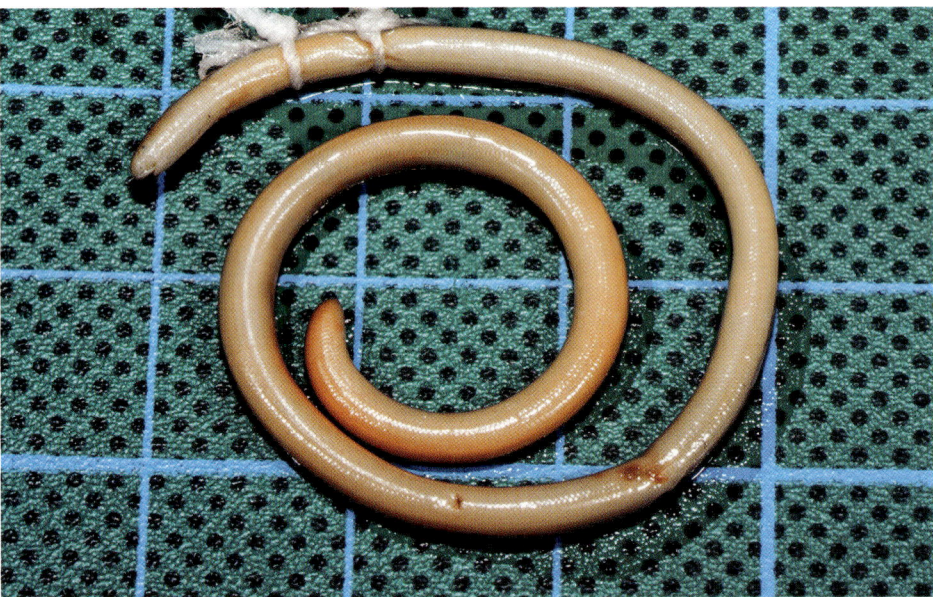

Koolan Island, WA, Scott Eipper

Koolan Island, WA, Scott Eipper

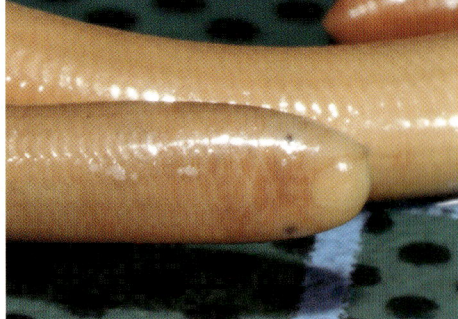

Koolan Island, WA, Scott Eipper

Yirrkala Blind Snake *Anilios yirrikalae* (Kinghorn, 1942)

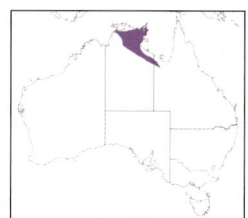

PRONUNCIATION *Ah-nil-e-oss yee-ree-car-lee*.
ETYMOLOGY In reference to Yirrkala, the type locality.
TYPE LOCALITY Yirrkala Mission Station, near Caledon Bay, Arnhem Land, NT.
APPEARANCE Small, slender snake. Dorsal colouration dark purplish-brown to pinkish-brown. Ventral colouration paler than dorsum. Head rounded when viewed from above and in profile. Tail quite short and rounded at the end, with a short, blunt conical spine. Reaches 22cm TL.
Scalation MB 24 rows, VENT 434, SUB 13. Scales small, smooth and glossy.
RANGE Encountered from Riversleigh, QLD, through the top end region of the NT, to near Darwin.
COMMENTS Nocturnal. Fossorial, living in open tropical woodland, predominantly on sandy soil types. Known to shelter beneath fallen timber and rocks. **Diet** Ant larvae. **Reproduction** Presumed oviparous. **Disposition** Would rather retreat than have contact. When threatened, can excrete a pungent musk or poke an antagonizer with the tail spine.
BITE/VENOM HARMLESS
IUCN LISTING Least Concern.

Yirrkala Mission, Caledon Bay, NT, Scott Eipper

Yirrkala Mission, Caledon Bay, NT, Scott Eipper

Yirrkala Mission, Caledon Bay, NT, Scott Eipper

West Kimberley Blind Snake *Anilios zonula* Ellis, 2016

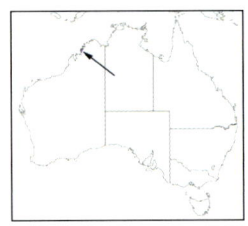

PRONUNCIATION *Ah-nil-e-oss zon-u-lah.*
ETYMOLOGY Little belt – in reference to the narrow, slender appearance of the species.
TYPE LOCALITY Storr Island, WA.
APPEARANCE Small, slender snake. Dorsal colouration dark purplish-brown to pinkish-brown, slightly darker towards head. Ventral colouration paler than dorsum. Head rounded when viewed from above and in profile. Tail quite short and rounded at the end, with a short, blunt conical spine. Reaches 18cm TL. **Scalation** MB 18 rows, 446–482 DSR, SUB 11–15. Scales small, smooth and glossy.

RANGE Presently found on Augustus and Storr Islands in the western Kimberley, WA, but is likely to occur on the adjacent mainland.

COMMENTS Nocturnal. Fossorial, living in vine thickets surrounded by open woodland. Shelters beneath sandstone rock slabs. **Diet** Ant larvae. **Reproduction** Presumed oviparous. **Disposition** Unknown, but presumed to be the same as the rest of the genus.

BITE/VENOM HARMLESS
IUCN LISTING Data Deficient.

Storr Island, WA, Scott Eipper

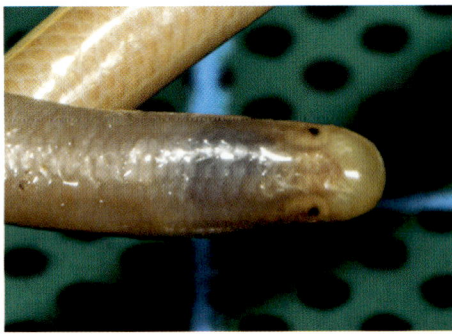

Storr Island, WA, Scott Eipper

Storr Island, WA, Michael Lyons

BLIND SNAKES

> ## Genus *Indotyphlops* Hedges, Marion, Lipp, Marin & Vidal, 2014
> This genus currently contains 20 species, and is represented in Australia by a single species, an accidental introduction into Australia and its territories. **BITE/VENOM** Blind snakes are unable to bite humans due to the way their mouth is shaped. **Species-level identification difficulty** (within Australia) – 5.
> **ETYMOLOGY** Indian/Indonesian blind eye.
> **TYPE SPECIES** *Typhlops pammeces*.

FLOWERPOT SNAKE *Indotyphlops braminus* (Daudin, 1803)
(Brahminy Blind Snake, Bootlace Snake)

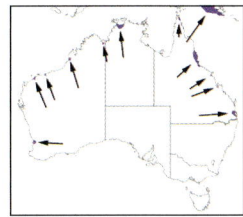

PRONUNCIATION *In-doe-tif-lops bra-min-uss*.

ETYMOLOGY Indian blind-eye caste – pertains to caste (of priests) among Hindu society.

TYPE LOCALITY Vizagapatam, coast of Coromandel, India.

APPEARANCE Small, slender snake. Dorsal colouration brown to purplish-black. Snout and cloaca lighter in colour. Underside of head and neck cream to white. Head rounded when viewed from above and in profile. Conspicuous glands on head form faint, scribbly pale lines. Tail quite short and rounded, with a short, blunt conical spine. Capable of reaching a TL of 19cm. All specimens are female. **Scalation** MB 18 rows, 261–358 DSR, 300–331 VENT, 8–15 SUB. Scales small, smooth and glossy.

RANGE Introduced, arriving in Australia in the 1960s. Currently found in scattered locations across northern Australia, including Brisbane, Townsville, Cairns, Torres Strait, Darwin, Derby, the Pilbara coast, Perth and Christmas Island. Found worldwide in tropical areas.

COMMENTS Nocturnal. Fossorial, living in gardens and forests. Populations in the Pacific also arboreal. Shelters in decaying logs, loose soil and flowerpots, and under leaf litter and man-made debris. Has been distributed worldwide in pot-plant soil. **Diet** Ants, termites and their larvae. **Reproduction** Parthenogenetic. 1–10 per clutch. These hatch after 39 days; hatchlings are about 6cm TL. **Disposition** Would rather retreat than have contact. When threatened, can excrete a pungent musk or poke an antagonizer with the tail spine.

BITE/VENOM HARMLESS

IUCN LISTING Least Concern.

Apia, Samoa, Scott Eipper

Apia, Samoa, Scott Eipper

Marsden, QLD, Scott Eipper

Marsden, QLD, Scott Eipper

Genus *Ramphotyphlops* Fitzinger, 1843

This genus currently contains 21 species, represented in Australia by a single species occurring only on Christmas Island. Until a revision in 2014, most Australian species were placed within *Ramphotyphlops*. **BITE/VENOM** Blind snakes are unable to bite humans due to their gape ability and mouth size. **Species-level identification difficulty** (within Australia) – 4.
ETYMOLOGY Beaked blind eye.
TYPE SPECIES *Typhlops multilineatus*.

CHRISTMAS ISLAND BLIND SNAKE *Ramphotyphlops exocerti* (Boulenger, 1887)

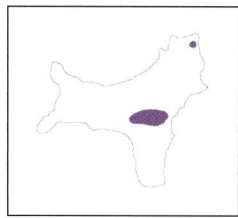

PRONUNCIATION *Ram-fo-tif-lops x-o-sir-tee*.
ETYMOLOGY Outside sleeping beaked blind-eye.
TYPE LOCALITY Christmas Island, Indian Ocean.
APPEARANCE Medium-sized, slender snake. Dorsal colouration pink. Each scale has brown centre, giving the appearance of variegations or fine longitudinal stripes. Ventral colouration cream to white. Head rounded when viewed from above and in profile. Tail quite short and rounded, with short, blunt conical spine. Reaches 35cm TL. **Scalation** MB 20 rows, 354 VENT, SUB 21. Scales small, smooth and glossy.
RANGE Encountered only on Christmas Island in the tropical Indian Ocean.
COMMENTS Nocturnal. Fossorial, living in rainforests among limestone casts. Most individuals have been found under fallen timber after rain. The species is at risk of extinction due to the inadvertent introduction of multiple exotic predators on the island. **Diet** Ant larvae. **Reproduction** Presumed oviparous. **Disposition** Would rather retreat than have contact. When threatened, can excrete a pungent musk or poke an antagonizer with the tail spine.
BITE/VENOM HARMLESS
IUCN LISTING Endangered.

Christmas Island, Dave Hunter

Christmas Island, Scott Eipper

Christmas Island, Scott Eipper

Tropidonophis mairii, Marsden, QLD, Tie Eipper

Encountering a Snake in a House or Garden

Snakes move through Australian suburban properties with a much higher frequency than most people realize. They are excellent at remaining undetected and are usually transient at a property, moving off on their own accord and searching for food and mates. They often consume pest species such as rodents and introduced geckos. As such, they provide a free form of biological pest control to home owners. Snakes, particularly harmless species, should whenever possible remain on site, which gives them a much higher chance of survival compared with translocated individuals. The maximum distance translocated snakes should be moved is ideally less than 1km from the point of capture.

Snakes injured by household pets should be checked over by a wildlife rescuer and/or a vet. Pets should be monitored for signs of envenomation in this instance.

Snake caught in bird netting, Scott Eipper

If you notice a snake in your house or garden the following is recommended:
1. Remain calm.
2. Keep any children and/or pets away from the snake; a distance of more than 10m is ideal.
3. Do not approach the snake, try to pick it up, capture it or hit it with an object – this could cause a defensive response.
4. If possible, take a picture of the snake from a distance of more than 5m from it. Mobile phones these days have amazing cameras, so there is no need to get close to get a photo. Use the zoom function to stay a safe distance away.
5. If the snake is indoors, contact a snake catcher for further advice; if it is outdoors and you are uncomfortable, call a snake catcher for advice. Many ethical snake catchers provide free identifications that may help you to decide whether to have the animal relocated. There is usually a fee for removing snakes from properties.

Snake on a property, Morelia spilota spilota, Tie Eipper

Snake on a property, Morelia spilota spilota, Tie Eipper

Encountering a Stranded Sea Snake

Apart from a few specialist species that hunt in the intertidal zone, most Australian sea snakes do not come ashore wilfully. When they are found on a beach, this is usually as the result of an injury or rough weather. Like all snakes, they are sensitive to temperature but unlike terrestrial snakes, their physiology is vastly different.

Many species are incapable of sustained movement on land. This is due to their musculature and scalation, which is designed for life below the water's surface. They are extremely sensitive to handling – pinning the head and handling by the tail can cause injuries that can lead to the death of an otherwise healthy snake. If you encounter a live, beach-washed sea snake, alert a local snake catcher or wildlife carer. Sometimes determining if a snake is alive is difficult as stranded individuals are often exhausted and very lethargic.

If you encounter a live stranded sea snake the following is recommended:
1. Remain calm.
2. Keep any children and/or pets away from the snake – ideally at a distance of more than 10m.
3. Do not try to pick up the snake or hit it with an object – this could cause a defensive response or injure the snake.
4. If possible, take a picture of the snake from a distance of more than 5m from it.
5. Contact a snake catcher or marine wildlife carer for advice. There are also excellent resources available on social media, such as pages dedicated to sea snakes. Many ethical snake catchers will assist with stranded sea snakes.
6. Avoid trying to put the snake back in the water; usually, this will only result in further injuries being sustained by the snake.

Stranded sea snake, Scott Eipper

Snakes must *only* be handled by appropriately experienced and licenced people, in this case using methods appropriate to working with sea snakes. These include:
a. Avoiding head restraint, and never allowing the body to be unsupported, even for brief periods. This is due to the snakes' adaptation for life in the water.
b. Avoiding placing sea snakes in bags for transport; the dry fabric can cause issues with the scales.
c. Using a plastic container with a soaking wet towel (utilizing fresh water or sea water) for cushioning and hydration, for transporting the snake to a carer; bringing the container to the snake, and without allowing the body to be unsupported, transferring it to the container supporting the snake's body appropriately.
d. When transporting the snake in a vehicle, avoiding excessive vibration and sun exposure; keeping the snake cool while transferring it to an appropriate person or facility.
e. Releasing the snake as close to the point of capture as practicable.

Stranded sea snake transport, Scott Eipper

Stranded sea snake handling, supporting body during assessment, Tie Eipper

The World's Most Venomous Snakes

It has been widely reported that Australian venomous snakes are significantly more toxic than other venomous snakes across the world. This has come from the incorrect interpretation of a paper on the treatment of snake bite in Australia, which included the toxicity of 20 Australian species, with three exotic species as controls. Since then, many snake venoms have been tested in the same fashion, and this has resulted in the table below, which ranks the species in order of toxicity, from the most to least toxic species tested. Australian species are shaded.

Rank	Scientific name	Common name	LD 50 value	Continent
1	Oxyuranus microlepidotus	Inland taipan	0.01	Australia
2	Aipysurus duboisii	Dubois' sea snake	0.032	Australia/Asia
3	Pseudonaja textilis	Eastern brown snake	0.041	Australia
4	Crotalus durissus	South American rattlesnake	0.048	South America
5	Oxyuranus scutellatus	Coastal taipan	0.05	Australia
6	Hydrophis platurus	Yellow-bellied sea snake	0.055	Australia
7	Dendroaspis polylepis	Black mamba	0.055	Africa
8	Hydrophis peronii	Horned sea snake	0.062	Australia/Asia
9	Aipysurus laevis	Olive sea snake	0.069	Australia/Asia
10	Notechis scutatus niger	Peninsula tiger snake	0.074	Australia/Asia/Africa/North & South America
11	Notechis scutatus	Common tiger snake	0.099	Australia
12	Notechis scutatus ater	Krefft's tiger snake	0.101	Australia
13	Hydrophis melanocephalus	Black-headed sea snake	0.111	Asia
14	Notechis scutatus humphreysi	Tasmanian tiger snake	0.129	Australia
15	Bungarus multicinctus	Many-banded krait	0.16	Asia
16	Hydrophis stricticollis	Yellow sea snake	0.164	Asia
17	Hydrophis schistosus	Beaked sea snake	0.164	Asia
18	Laticauda laticaudata	Blue-banded sea krait	0.179	Australia/Asia
19	Hydrophis coggeri	Cogger's sea snake	0.18	Australia/Asia
20	Notechis scutatus occidentalis	Western tiger snake	0.194	Australia
21	Hydrophis major	Olive-headed sea snake	0.194	Australia/Asia
22	Naja phillippinensis	North-Phillippine cobra	0.2	Asia
23	Laticauda semifasciata	Chinese sea krait	0.2	Asia
24	Naja samarensis	South-Phillippine cobra	0.21	Asia
25	Crotalus tigris	Tiger rattlesnake	0.21	North America
26	Naja melanoleuca	Forest cobra	0.225	Africa
27	Hydrophis belcheri	Belcher's sea snake	0.24	Australia/Asia
28	Naja kaouthia	Monocled cobra	0.24	Asia
29	Pseudonaja inframacula	Peninsula brown snake	0.25	Australia
30	Hydrophis stokesii	Stoke's sea snake	0.26	Australia/Asia
31	Hydrophis elegans	Elegant sea snake	0.262	Australia/Asia
32	Naja naja	Indian cobra	0.28	Asia
33	Naja atra	Chinese cobra	0.29	Asia
34	Crotalus scutulatus	Mohave rattlesnake	0.31	North America
35	Notechis scutatus serventyi	Chappell Island tiger snake	0.338	Australia
36	Pseudonaja aspidorhyncha	Strap-snouted brown snake	0.338	Australia
37	Acanthophis antarcticus	Common death adder	0.338	Australia
38	Hydrophis nigrocinctus	Black-banded sea snake	0.344	Asia
39	Pseudonaja guttata	Speckled brown snake	0.36	Australia
40	Walterinnesia aegyptia	Desert black cobra	0.4	Africa
41	Naja nivea	Cape cobra	0.4	Africa
42	Naja oxiana	Central Asian cobra	0.45	Asia
43	Laticauda colubrina	Yellow-lipped sea krait	0.45	Australia/Asia
44	Bungarus caeruleus	Blue krait	0.45	Asia

45	Hydrophis cyanocinctus	Annulated sea snake	0.465	Asia
46	Cerastes cerastes	Saharan horned viper	0.475	Africa
47	Microcephalophis gracilis	Slender sea snake	0.48	Australia/Asia
48	Austrelaps superbus	Lowland copperhead	0.5	Australia
49	Hydrophis curtus	Spine-bellied sea snake	0.541	Australia/Asia
50	Pseudonaja affinis	Dugite	0.56	Australia
51	Tropidechis carinatus	Rough-scaled snake	0.58	Australia
52	Austrelaps ramsayi	Highland copperhead	0.6	Australia
53	Bitis arietans	Puff adder	0.6	Africa
54	Agkistrodon contortrix	Copperhead	0.6	North America
55	Dendroaspis viridis	West African green mamba	0.79	Africa
56	Vipera berus	Common adder	1	Europe
57	Vipera aspis	Asp	1	Europe
58	Hemachatus haemachatus	Rinkhal	1	Africa
59	Dendroaspis jamesoni	Jameson's mamba	1.02	Africa
60	Pseudechis papuanus	Papuan black snake	1.09	Australia
61	Ophiophagus hannah	King cobra	1.091	Asia
62	Agkistrodon laticinctus	Southern copperhead	1.118	North America
63	Micrurus fulvius	Eastern coral snake	1.3	North America
64	Austrelaps labialis	Pygmy copperhead	1.3	Australia
65	Hoplocephalus stephensii	Stephens' banded snake	1.36	Australia
66	Daboia siamensis	Eastern Russell's viper	1.37	Asia
67	Pseudechis guttatus	Spotted black snake	1.53	Australia
68	Micrurus nigrocinctus	Central American coral snake	1.7	South America
69	Naja annulifera	Snouted cobra	1.75	Africa
70	Pseudechis australis	Mulga snake	1.91	Australia
71	Vipera ammodytes	Nosed-horned viper	2	Europe
72	Naja nigricollis	Black-necked spitting cobra	2	Africa
73	Micrurus frontalis	Southern coral snake	2.04	South America
74	Agkistrodon piscivorus	Cottonmouth	2.044	North America
75	Crotalus horridus	Timber rattlesnake	2.25	North America
76	Daboia palaestinae	Palestine viper	2.333	Asia
77	Pseudechis colletti	Collett's snake	2.38	Australia
78	Pseudechis porphyriacus	Red-bellied black snake	2.52	Australia
79	Cryptophis nigrescens	Eastern small-eyed snake	2.67	Australia
80	Macrovipera lebetinus	Blunt-nosed viper	2.72	Europe
81	Crotalus basiliscus	Mexican west-coast rattlesnake	2.78	North America
82	Dendroaspis angusticeps	East African green mamba	3.32	Africa
83	Bungarus fasciatus	Banded krait	3.58	Asia
84	Glyphodon tristis	Brown headed snake	3.61	Australia
85	Trimeresurus gramineus	Indian green pit viper	4	Asia
86	Naja haje	Egyptian cobra	4	Africa
87	Gloydius halys	Siberian pit viper	4	Asia
88	Protobothrops flavoviridis	Habu	4.3	Asia
89	Bothrops jararaca	Jararaca	4.32	South America
90	Micrurus tener	Texan coral snake	4.4	North America
91	Montivipera latifii	Latifi's viper	4.61	Asia
92	Tropidolaemus wagleri	Wagler's pitviper	4.625	Asia
93	Bothrops jararacussu	Jararacussu	4.92	South America
94	Bitis gabonica	East African gaboon viper	5	Africa
95	Echis coloratus	Painted saw-scaled viper	5.167	Africa
96	Echis carinatus	Saw-scaled viper	5.5	Africa
97	Crotalus cerastes	Sidewinder	5.5	North America
98	Lachesis muta	South American bushmaster	6	South America
99	Montivipera bornmuelleri	Lebanon viper	6.25	Asia
100	Sistrurus catenatus	Eastern massauga	6.8	North America

Checklist of the Snakes of Australia

Scientific name	Common Name	Seen ✓	Notes
ACROCHORDIDAE	**FILE SNAKES**		2
Acrochordus arafurae	Arafura file snake		
Acrochordus granulatus	Little file snake		
PYTHONIDAE	**PYTHONS**		17 + 4 subspecies
Antaresia childreni childreni	Children's python		
Antaresia childreni stimsoni	Western children's python		
Antaresia maculosa maculosa	Southern spotted python		
Antaresia maculosa peninsularis	Cape York spotted python		
Antaresia papuensis	Papuan spotted python		
Antaresia perthensis	Pygmy python		
Aspidites melanocephalus	Black-headed python		
Aspidites ramsayi	Woma		
Leiopython fredparkeri	Southern white-lipped python		
Liasis fuscus	Water python		
Liasis olivaceus barroni	Pilbara olive python		
Liasis olivaceus olivaceus	Olive python		
Morelia bredli	Centralian carpet python		
Morelia carinata	Rough-scaled python		
Morelia imbricata	Western carpet python		
Morelia spilota metcalfei	Inland carpet python		
Morelia spilota spilota	Diamond python		
Morelia viridis	Green python		
Nyctophilopython oenpelliensis	Oenpelli python		
Simalia amethistina	Amethystine python		
Simalia kinghorni	Scrub python		
COLUBRIDAE	**COLUBRID SNAKES**		6 species
Boiga irregularis	Brown tree snake		
Dendrelaphis calligastra	Northern tree snake		
Dendrelaphis punctulatus	Common tree snake		
Lycodon capucinus	Common wolf snake*		
Pantherophis guttatus	Corn snake*		
Stegonotus australis	Slaty-grey snake		
NATRICIDAE	**WATER SNAKES**		1 species
Tropidonophis mairii	Keelback		
ELAPIDAE	**ELAPIDS**		109 species + 9 subspecies
Acanthophis antarcticus	Common death adder		
Acanthophis cryptamydros	North-western death adder		
Acanthophis hawkei	Barkly Tableland death adder		
Acanthophis laevis	Eastern smooth-scaled death adder		
Acanthophis praelongus	Northern death adder		
Acanthophis pyrrhus	Desert death adder		
Acanthophis rugosus	Papuan death adder		
Acanthophis wellsei	Pilbara death adder		
Antaioserpens albiceps	North-eastern plain-nosed burrowing snake		
Antaioserpens warro	Robust burrowing snake		
Austrelaps labialis	Pygmy copperhead		
Austrelaps ramsayi	Highlands copperhead		
Austrelaps superbus	Lowlands copperhead		
Brachyurophis approximans	North-western shovel-nosed snake		
Brachyurophis australis	Australian coral snake		
Brachyurophis campbelli	Einsleigh shovel-nosed snake		
Brachyurophis fasciolatus fasciatus	Eastern narrow-banded shovel-nosed snake		
Brachyurophis fasciolatus fasciolatus	Western narrow-banded shovel-nosed snake		
Brachyurophis incinctus	Unbanded shovel-nosed snake		
Brachyurophis morrisi	Arnhem shovel-nosed snake		
Brachyurophis roperi	Northern shovel-nosed snake		
Brachyurophis semifasciatus	Southern shovel-nosed snake		
Cacophis churchilli	Northern crowned snake		
Cacophis harriettae	White-crowned snake		

CHECKLIST OF THE SNAKES OF AUSTRALIA

Scientific name	Common Name	Seen ✓	Notes
Cacophis krefftii	Southern dwarf crowned snake		
Cacophis squamulosus	Golden-crowned snake		
Cryptophis boschmai	Carpentaria snake		
Cryptophis incredibilis	Pink snake		
Cryptophis nigrescens	Eastern small-eyed snake		
Cryptophis nigrostriatus	Black-striped snake		
Cryptophis pallidiceps	Northern small-eyed snake		
Demansia angusticeps	Narrow-headed whipsnake		
Demansia calodera	Black-necked whipsnake		
Demansia cyanochasma	Desert whipsnake		
Demansia flagelliato	Carpentarian whipsnake		
Demansia olivacea	Olive whipsnake		
Demansia papuensis	Greater black whipsnake		
Demansia psammophis	Yellow-faced whipsnake		
Demansia quaesitor	Sombre whipsnake		
Demansia reticulata	Reticulated whipsnake		
Demansia rimicola	Crack-dwelling whipsnake		
Demansia rufescens	Rufous whipsnake		
Demansia shinei	Shine's whipsnake		
Demansia simplex	Grey whipsnake		
Demansia torquata	Collared whipsnake		
Demansia vestigiata	Lesser black whipsnake		
Denisonia devisi	De Vis' banded snake		
Denisonia maculata	Ornamental snake		
Drysdalia coronoides	White-lipped snake		
Drysdalia mastersii	Masters' snake		
Drysdalia rhodogaster	Mustard-bellied snake		
Echiopsis curta	Bardick		
Elapognathus coronatus	Western crowned snake		
Elapognathus minor	Short-nosed snake		
Furina barnardi	Yellow-naped snake		
Furina diadema	Red-naped snake		
Furina ornata	Orange-naped snake		
Glyphodon dunmalli	Dunmall's snake		
Glyphodon tristis	Brown-headed snake		
Hemiaspis damelii	Grey snake		
Hemiaspis signata	Marsh snake		
Hoplocephalus bitorquatus	Pale-headed snake		
Hoplocephalus bungaroides	Broad-headed snake		
Hoplocephalus stephensii	Stephens' banded snake		
Narophis bimaculatus	Black-naped snake		
Neelaps calonotos	Black-striped snake		
Notechis scutatus ater	Krefft's tiger snake		
Notechis scutatus humphreysi	Tasmanian tiger snake		
Notechis scutatus niger	Peninsula tiger snake		
Notechis scutatus occidentalis	Western tiger snake		
Notechis scutatus scutatus	Tiger snake		
Notechis scutatus serventyi	Chappell Island tiger snake		
Oxyuranus microlepidotus	Inland taipan		
Oxyuranus scutellatus canni	Papuan taipan		
Oxyuranus scutellatus scutellatus	Coastal taipan		
Oxyuranus temporalis	Western desert taipan		
Paroplocephalus atriceps	Lake Cronin snake		
Pseudechis australis	Mulga snake		
Pseudechis butleri	Spotted mulga snake		
Pseudechis colletti	Collett's snake		
Pseudechis guttatus	Spotted black snake		
Pseudechis pailsei	Eastern pygmy mulga snake		
Pseudechis papuanus	Papuan black snake		
Pseudechis porphyriacus	Red-bellied black snake		
Pseudechis weigeli	Western pygmy mulga snake		
Pseudonaja affinis affinis	Dugite		
Pseudonaja affinis exilis	Rottnest Island dugite		
Pseudonaja affinis tanneri	Tanner's brown snake		

CHECKLIST OF THE SNAKES OF AUSTRALIA

Scientific name	Common Name	Seen ✓	Notes
Pseudonaja aspidorhyncha	Shield-snouted brown snake		
Pseudonaja guttata	Speckled brown snake		
Pseudonaja inframacula	Peninsula brown snake		
Pseudonaja ingrami	Ingram's brown snake		
Pseudonaja mengdeni	Western brown snake		
Pseudonaja modesta	Ringed brown snake		
Pseudonaja nuchalis	Northern brown snake		
Pseudonaja textilis	Eastern brown snake		
Rhinoplocephalus bicolor	Square-nosed snake		
Simoselaps anomalus	Desert banded snake		
Simoselaps bertholdi	Jan's banded snake		
Simoselaps littoralis	West Coast banded snake		
Simoselaps minimus	Dampierland burrowing snake		
Suta dwyeri	Dwyer's snake		
Suta fasciata	Rosen's snake		
Suta flagellum	Little whip snake		
Suta gaikhorstorum	Pilbara monk snake		
Suta gouldii	Gould's hooded snake		
Suta monachus	Monk snake		
Suta nigriceps	Mitchell's short-tailed snake		
Suta ordensis	Ord curl snake		
Suta punctata	Little spotted snake		
Suta spectabilis	Mallee black-headed snake		
Suta suta	Curl snake		
Tropidechis carinatus	Rough-scaled snake		
Vermicella annulata	Bandy-bandy		
Vermicella intermedia	Intermediate bandy-bandy		
Vermicella multifasciata	Narrow-banded bandy-bandy		
Vermicella parscauda	Cape York bandy-bandy		
Vermicella snelli	Pilbara bandy-bandy		
Vermicella vermiformis	Centralian bandy-bandy		
ELAPIDAE	**SEA SNAKES**		**38 species**
Aipysurus apraefrontalis	Short-nosed sea snake		
Aipysurus duboisii	Dubois' sea snake		
Aipysurus eydouxii	Stagger-banded sea snake		
Aipysurus foliosquama	Leaf-scaled sea snake		
Aipysurus fuscus	Dusky sea snake		
Aipysurus laevis	Olive sea snake		
Aipysurus mosaicus	Mosaic sea snake		
Aipysurus pooleorum	Shark Bay sea snake		
Aipysurus tenuis	Mjoberg's sea snake		
Emydocephalus annulatus	Turtle-headed sea snake		
Emydocephalus oriarus	Western turtle-headed sea snake		
Ephalophis greyae	Mangrove sea snake		
Hydrelaps darwiniensis	Black-ringed mangrove snake		
Hydrophis atriceps	Black-headed sea snake		
Hydrophis belcheri	Belcher's sea snake		
Hydrophis caerulescens	Dwarf sea snake		
Hydrophis coggeri	Slender-necked sea snake		
Hydrophis curtus	Spine-bellied sea snake		
Hydrophis czeblukovi	Geometrical sea snake		
Hydrophis donaldi	Rough-scaled sea snake		
Hydrophis elegans	Elegant sea snake		
Hydrophis inornatus	Plain sea snake		
Hydrophis kingii	Spectacled sea snake		
Hydrophis laboutei	Laboute's sea snake		
Hydrophis macdowelli	Small-headed sea snake		
Hydrophis major	Olive-headed sea snake		
Hydrophis melanosoma	Black-banded robust sea snake		
Hydrophis ocellatus	Spotted sea snake		
Hydrophis pacificus	Large-headed sea snake		
Hydrophis peronii	Horned sea snake		
Hydrophis platurus platurus	Yellow-bellied sea snake		

CHECKLIST OF THE SNAKES OF AUSTRALIA

Scientific name	Common Name	Seen ✓	Notes
Hydrophis stokesii	Stokes' sea snake		
Hydrophis vorisi	Plain-banded sea snake		
Hydrophis zweifeli	Beaked sea snake		
Laticauda colubrina	White-lipped sea krait		
Laticauda laticaudata	Black-lipped sea krait		
Microcephalophis gracilis	Slender sea snake		
Parahydrophis mertoni	Northern mangrove snake		
HOMALOPSIDAE	**MUD SNAKES**		**5 species**
Cerberus australis	Australian bockadam		
Fordonia leucobalia	White-bellied mangrove snake		
Myron resetari	Resetar's mangrove snake		
Myron richardsonii	Richardson's mangrove snake		
Pseudoferania polylepis	Macleay's water snake		
TYPHLOPIDAE	**BLIND SNAKES**		**49 species**
Anilios affinis	Small-headed blind snake		
Anilios ammodytes	Sand-diving blind snake		
Anilios aspina	No-spined blind snake		
Anilios australis	Southern blind snake		
Anilios batillus	Shovel-snouted blind snake		
Anilios bicolor	Dark-spined blind snake		
Anilios bituberculatus	Prong-snouted blind snake		
Anilios broomi	Faint-striped blind snake		
Anilios centralis	Centralian blind snake		
Anilios chamodracaena	Cape York striped blind snake		
Anilios diversus	Northern blind snake		
Anilios endoterus	Interior blind snake		
Anilios fossor	Miner blind snake		
Anilios ganei	Gane's blindsnake		
Anilios grypus	Long-beaked blind snake		
Anilios guentheri	Top end blind snake		
Anilios hamatus	Pale-headed blind snake		
Anilios howi	Kimberley deep-soil blind snake		
Anilios insperatus	Fassifern blind snake		
Anilios kimberleyensis	Kimberley shallow-soil blind snake		
Anilios leptosoma	Murchison blind snake		
Anilios leucoproctus	Cape York blind snake		
Anilios ligatus	Robust blind snake		
Anilios longissimus	Extremely long blind snake		
Anilios margaretae	Buff-snouted blind snake		
Anilios micromma	Small-eyed blind snake		
Anilios minimus	Groote Eylandt dwarf blind snake		
Anilios nema	Thread-like blind snake		
Anilios nigrescens	Blackish blind snake		
Anilios obtusifrons	Blunt-snouted blind snake		
Anilios pilbarensis	Pilbara blind snake		
Anilios pinguis	Rotund blind snake		
Anilios proximus	Proximus blind snake		
Anilios robertsi	Roberts' blind snake		
Anilios silvia	Cooloola blind snake		
Anilios splendidus	Splendid blind snake		
Anilios systenos	Sharp-snouted blind snake		
Anilios torresianus	North-eastern blind snake		
Anilios tovelli	Darwin blind snake		
Anilios troglodytes	Sandamara blind snake		
Anilios unguirostris	Claw-snouted blind snake		
Anilios vagurima	Mornington blind snake		
Anilios waitii	Beaked blind snake		
Anilios wiedii	Brown-snouted blind snake		
Anilios yampiensis	Yampi blind snake		
Anilios yirrikalae	Yirrkala blind snake		
Anilios zonula	West Kimberley blind snake		
Indotyphlops braminus	Flowerpot blind snake*		
Ramphotyphlops exocoeti	Christmas Island blind snake		
Total species	* Denotes introduced species		**227 species + 13 subspecies**

References & Further Resources

Websites
Atlas of Living Australia www.ala.org.au
Australia's Wildlife www.australiaswildlife.com
Australian Faunal Directory www.biodiversity.org.au
Herpmapper www.herpmapper.org
Nature 4 You www.wildlifedemonstrations.com

References
Allen, G. E., Wilson, S. K., & Isbister, G. K. (2013) *Paroplocephalus* envenoming: a previously unrecognised highly venomous snake in Australia. *Medical Journal of Australia* 199 (11) 792–793.

Aplin, K. P. (1998) Three new blindsnakes (Squamata: Typhlopidae) from northwestern Australia. *Records of the Western Australian Museum* 19, 1–12.

Australian Venom Research Unit (2010) Whip Snakes (*Demansia* spp.). www.avru.org/general/general_whip.html. accessed on 11/10/2010.

Barker, D. G. & Barker, T. M. (1994) *Pythons of the World. Vol. 1, Australia.* The Herpetocultural Library, Advanced Vivarium Systems Inc., Lakeside, California.

Barker, D. G. & Barker, T. M. (2018) *Pythons of the World. Vol. 3, Pythons of Asia and the Malay Archipelago.* Eco Publishing, Arizona.

Boundy, J. (2021) *Snakes of the World, A Supplement,* CRC Press, Boca Raton, Florida.

Breenan, I., Morley, T., Hutchinson, M. & Donnellan, S. (2011) Redescription of the western desert taipan *Oxyuranus temporalis* (Serpentes: Elapidae), with notes on its distribution, diet and genetic variation. *Australian Journal of Zoology* 59 (4) 227–235.

Broad, A. J., Sutherland, S. K. & Coulter, A. R. (1978) The lethality in mice of dangerous Australian and other snake venoms. *Toxicon* (19) 661–664.

Brown, R. W. (1956) *Composition of Scientific Words.* Smithsonian Institution Press Washington DC, USA.

Brown, G. P., Shine, R. & Madsen T. R. L. (2005) Spatial ecology of Slaty-grey snakes (*Stegonotus cucullatus*, Colubridae) on a tropical Australian floodplain. *Journal of Tropical Ecology.* Vol. 21 (6) 605–612.

Brown, G. P., Madsen, T. R. L. & Shine, R. (2017) Resource availability and sexual size dimorphism: differential effects of prey abundance on the growth rates of tropical snakes. *Functional Ecology* 31 1592–1599.

Bush, B. (2017) Additions to the description of *Paroplocephalus atriceps* (Serpentes: Elapidae) with a discussion on pupil shape in it and other Australian snakes. *Zootaxa* 4344 (2): 333–34.

Bush, B. (2018) A venomous snakebite case in Australia supports the efficacy of Sutherland's original 1979 pressure immobilisation first aid. *Snakes Harmful and Harmless – ASP 110218.* 1-12.

Bush, B., Maryan, B., Browne-Cooper, R. & Robinson, D. (1995) *A Guide to the Reptiles and Frogs of the Perth region.* University of Western Australia Press, Perth.

Bush, B., Maryan, B., Browne-Cooper, R. & Robinson, D. (2007) *Reptiles and Frogs in the Bush: Southwestern Australia.* University of Western Australia Press, Perth.

Bush, B. & Maryan, B. (2011) *Field Guide to the Snakes of the Pilbara.* Western Australian Museum, Welshpool.

Canale, E., Isbister, G. K. & Currie, B. (2009) Investigating pressure bandaging for snakebite in a simulated setting: bandage type, training and the effect of transport. *Emergency Medicine Australasia* 21 (3) 184–190.

Chapple, D. G., Tingley, R., Mitchell, N. J., Macdonald, S. L., Keogh, J. S., Shea, G. M., Bowles, P., Cox, N. A. & Woinarski, J. Z. (2019) *The Action Plan for Australian Lizards and Snakes 2017.* CSIRO Publishing, Clayton South.

Cipriani, V., Debono, J., Goldenberg, J., Jackson, T. N. W., Arbuckle, K., Dobson, J., Koludarov, I., LI, B., Hay, C., Dunstan, N., Allen, L., Hendrikx, I., Kwok, H. F. & Fry, B. F. (2017) Correlation between ontogenetic shifts and venom variation in Australian brown snakes (*Pseudonaja*). *Comparative Biochemistry and Physiology*, Part C 53–60.

Cogger, H. G. (2018) *Reptiles and Amphibians of Australia* (7th edition). CSIRO Publishing, Collingwood.

Cogger, H. G., Cameron, E. E. & Cogger, H. M. (1983) *Zoological Catalogue of Australia. Vol. 1 – Amphibia and Reptilia of Australia.* Australian Government Printing Surface, Canberra.

Couper, P. J., Covacevich, J. A. & Wilson, S. K. (1998) Two new species of *Ramphotyphlops* (Squamata: Typhlopidae) from Queensland. *Memoirs of the Queensland Museum* 42 (2) 459–464.

Couper, P. J., Peck, S. R., Emery, J-P. & Keogh, J. S. (2016) Two snakes from eastern Australia (Serpentes: Elapidae); a revised concept of *Antaioserpens warro* (De Vis) and a redescription of *A. albiceps* (Boulenger). *Zootaxa* 4097 (3) 227–231.

Derez, C. M., Arbuckle, K., Ruan, Z., Xie, B., Huang, Y., Dibben, L., Shi, Q., Vonk, F. & Fry, B. G. (2018) A new species of bandy-bandy (*Vermicella*: Serpentes: Elapidae) from the Weipa region, Cape York, Australia. *Zootaxa* 4446 (1) 1–12.

Doughty, P., Maryan, B., Donnellan, S. C. & Hutchinson, M. N. (2007) A new species of taipan (Elapidae: *Oxyuranus*) from central Australia. *Zootaxa* 1422 45–58.

Ehmann, H. (1992) *Encyclopaedia of Australian Animals – Reptiles.* Collins Angus & Robertson, Sydney, NSW.

Eipper, S. C. (2012) *A Guide to Australian Snakes in Captivity: Elapids and Colubrids.* Reptile Keeper Publications: Tweed Heads, NSW.

Eipper, S. C. & Eipper, T. (2017) Observations of Black Snakes *Pseudechis* in captivity, with notes on reproduction and longevity. *Victorian Naturalist* 134(6): 199–200.

Eipper, S. C. & Eipper, T. (2023) *A Naturalist's Guide to the Snakes of Australia.* John Beaufoy Publishing, Oxford.

Eipper, S. C. & Eipper, T. (2022) *Australasian Elapids Husbandry, Ecology and Captive Care.* Edition Chimaira, Frankfurt.

Elliott, A. (2012) *A Guide to Australian Snakes in Captivity: Pythons.* Reptile Keeper Publications: Tweed Heads, NSW.

Elliott, A., Eipper, T. & Eipper, S. C. (2020) Notes of the discovery of fluorescence in Australian Scolecophidians in the genus *Anilios* Gray 1845 (Serpentes: Typhlopidae). *The Captive & Field Herpetology Journal* 4 (2) 27–30.

Ellis, R. J. (2016) A new species of blindsnake (Scolecophidia: Typhlopidae: *Anilios*) from the Kimberley region of Western Australia. *Herpetologica* 72 (3), 271–278.

Ellis, R. J. (2019) A typhlopid hotspot in the tropics: increased blindsnake diversity in the Kimberley region of Western Australia with the description of a new *Anilios* species (Serpentes: Typhlopidae). *Records of the Western Australian Museum* 34: 031–037.

Ellis, R. J., Doughty, P., Donnellan, S. C., Marin, J. & Vidal, N. (2017) Worms in the sand: systematic revision of the Australian blindsnake *Anilios leptosoma* (Robb, 1972) species complex (Squamata: Scolecophidia: Typhlopidae) from the Geraldton sandplain, with description of two new species. *Zootaxa* 4323(1) 1–24.

Ellis, R. J., Kaiser, H., Maddock, S. T., Doughty, P. & Wüster, W. (2021) An evaluation of the nomina for death adders (*Acanthophis* Daudin, 1803) proposed by Wells & Wellington (1985), and confirmation of *A. cryptamydros* Maddock et al., 2015 as the valid name for the Kimberley death adder. *Zootaxa* 4995(1) 161–172.

Esquerré, D., Donnellan, S. C., Pavón-Vázquez, C. J., Fenker, J. & Keogh, J. S. (2021). Phylogeography, historical demography and systematics of the world's smallest pythons (Pythonidae, *Antaresia*). *Molecular Physics and Evolution* 161: 107–181.

Fry, B. F., Holger, S., Van Der Weerd, L., Young, B., McNaughtan, J., Ramjan, R., Vidal, N., Poelmann, R. E. & Norman, J. A. (2008) Evolution of an Arsenal-structural and functional diversification of the venom system in the advanced snakes (Caenophidia). *Molecular & Cellular Proteomics* 7 (2) 215–246.

Gillett, A., d'Anastasi, B. & Raveneau, J. (2020) Sea Snake Stranding Response, from the Australian

Sea Snakes Facebook page accessed 8/10/2023.
Gilliam, M. W. (1979) *The genus Pseudonaja (Serpentes: Elapidae) in the Northern Territory*. Territory Parks and Wildlife Commission, Research Bulletin No. 1.
Gow, G. F. (1977) *Snakes of the Darwin Area*. Art Gallery and Museum of the Northern Territory, Darwin.
Gow, G. F. (1989) *Graeme Gow's Complete Guide to Australian Snakes*. Angus and Robertson, Sydney.
Greer, A. E. (1997) *The Biology and Evolution of Australian Snakes*. Surrey Beatty and Sons, Chipping Norton, New South Wales.
Greer, A. E. (2021) *Encyclopaedia of Australian Reptiles*. Published Online Version: 1 October 2021.
Günther, A. (1858) *Catalogue of Colubrine Snakes in the Collection of the British Museum*. London.
Günther, A. (1863) On new species of snakes in the collection of the British Museum. *Annual Magazine of Natural History* (3) 11 20–25.
Günther, A. (1864) *The Reptiles of British India*. Taylor & Francis, London.
Heatwole, H. (1999) *Sea Snakes, Australian Natural History Series*. University of NSW Press, Sydney, NSW.
Hedges, S. B., Marion, A. B., Lipp, K. M., Marin, J. & Vidal, N. (2014) A taxonomic framework for typhlopid snakes from the Caribbean and other regions (Reptilia, Squamata). *Caribbean Herpetology* 49 1–61.
Horner, P. (1998) *Simoselaps morrisi* sp. nov (Elapidae), a species of snake from the Northern Territory. *The Beagle. Records of the Museums and Art Galleries of the Northern Territory* 14, 63–70.
Horner, P. (1992) *Skinks of the Northern Territory*. Northern Territory Museum of Arts and Sciences, Darwin.
Hoser, R. T. (1998) Death adders (genus *Acanthophis*): an overview, including descriptions of five new species and one subspecies. *Monitor* 9 (2) 20–30.
Hoser, R. T. (1998a) A new snake from Queensland, Australia (Serpentes: Elapidae). *Monitor* 10 (1) 5–9.
Ineich, I. & Laboute, P. (2002) *Sea Snakes of New Caledonia*. Muséum national d'histoire naturelle, IRD Editions, Paris, France.
Ingram, G. J. & Covacevich, J. A. (1993) Two new species of striped blindsnakes. *Memoirs of the Queensland Museum* 34 (1) 181–184.
Jan, G. (1859) Plan d'une Iconographie descriptive des Ophidiens, et description sommaire de nouvelles espèces de Serpents. *Rev. Mag. Zoology* 2 11–12.
Jan, G. & Sordelli, F. (1873) *Iconographie Gènerale de Ophidiens*. Atlas 1860–1881.
Jensen, S. (2005) Symptoms and signs of snakebite in Papua New Guinea. pp. 86–101 in *Venomous Bites and Stings in Papua New Guinea – A Guide to Treatment for Health Workers and Doctors*. eds Williams, D., Jensen, S., Nimorakiotakis, B. & Winkel, K. D. Australian Venom Research Unit, Melbourne.
Jolly, C., Schembri, B. & Macdonald S. (2023) *Field Guide to the Reptiles of the Northern Territory*. CSIRO Publishing, Melbourne.
Kaiser, C. M., Kaiser, H. & O'Shea, M. (2018) The taxonomic history of Indo-Papuan groundsnakes, genus *Stegonotus* Duméril et al., 1854 (Colubridae), with some taxonomic revisions and the designation of a neotype for *S. parvus* (Meyer, 1874). *Zootaxa* 4512 1–73.
Kaiser, H., Crother, B. I., Kelly, C. M. R., Luiselli, L., O'Shea, M., Ota, H., Passos, P., Schleip, W. D. & Wüster (2013) Best practices: in the 21st century, taxonomic decisions in herpetology are acceptable only when supported by a body of evidence and published via peer-review. *Herpetological Review* 44 (1): 8–23.
Kaiser, H., Thomson, S. A., Shea, G. M. (2020) *Nawaran* Esquerré, Donnellan, Brennan, Lemmon, Lemmon, Zaher, Grazziotin & Keogh, 2020 is an invalid junior synonym of *Nyctophilopython* Wells & Wellington, 1985 (Squamata, Pythonidae): simple priority without Zoobank pre-registration. *Bionomina*, 20: 47–54.
Kinghorn, J. R. (1923) A new genus of elapine snake from North Australia. *Records of the Australian*

Museum 14: 42–45.

Kinghorn, J. R. (1964) *Snakes of Australia* (Revised edition). Angus & Robertson. Sydney.

Krefft, G. (1869) *The Snakes of Australia; an Illustrated and Descriptive Catalogue of All the Known Species*. Thomas Richards, Government Printer, Sydney.

Krell, F. T. (2021) Suppressing works of contemporary authors using the Code's publication requirements is neither easy nor advisable. *The Bulletin of Zoological Nomenclature* 78 (1), 61–67.

Kuch, U., Keogh, J. S., Weigel, J., Smith, L. A. & Mebs, D. (2005) Phylogeography of Australia's king brown snake (*Pseudechis australis*) reveals Pliocene divergence and Pleistocene dispersal of a top predator. *Naturwissenschaften* 92: 121–127.

Kuch, U. & Yuwono, F. B. (2002) First record of brown snakes *Pseudonaja* cf. *textilis* (Duméril, Bibron & Duméril, 1854) from Papua, Indonesia. *Herpetozoa* 15(1/2): 75–78.

Kuruppu, S., Fry, B. G. & Hodgson, W. C. (2006) In vitro neurotoxic and myotoxic effects of the venom from the black whip snake (*Demansia papuensis*). *Clinical and Experimental Pharmacology & Physiology* 33 (4) 364–368.

Kuruppu, S., Robinson, S., Hodgson, W. C. & Fry, B. G. (2007) The in vitro neurotoxic and myotoxic effects of the venom from the *Suta* genus (curl snakes) of elapid snakes. *Basic & Clinical Pharmacology & Toxicology*, 101 407–410.

Lang, R. D. (2013) *The Snakes of the Moluccas (Maluku), Indonesia – A Field Guide to the Land and Non-marine Aquatic Snakes of the Moluccas with Identification Key*. Edition Chimaira. Frankfurt.

Lillywhite, H. B. (2008) *Dictionary of Herpetology*. Krieger Publishing, Malabar, USA.

Leenders, T. (2019) *Reptiles of Costa Rica – A Field Guide*, Cornell University Press, Ithaca, USA.

Limpus, C. (1975) Coastal sea snakes of subtropical waters of Queensland waters (23 to 28 south latitude), pp. 173–182 in *The Biology of Sea Snakes*, ed W. A. Dunson. University Park Press, Baltimore.

Longmore, R. (ed) (1986) *Snakes – Atlas of Elapid Snakes of Australia*. Australian Government Printing Service, Canberra.

Loveridge, A. (1948) New Guinean reptiles and amphibians in the Museum of Comparative Zoology and United States National Museum. *Bulletin of the Museum of Comparative Zoology* 101 (2) 303–430.

Macleay, W. (1877) The ophidians of the Chevert Expedition. *Proceedings of the Linnaean Society of New South Wales* 2 33–41.

Macleay, W. (1887) On a new *Hoplocephalus* from the Gulf of Carpentaria. *Proceedings of the Linnaean Society of New South Wales* 2 (2) 403–404.

Maddock, S. T., Childersonte, A., Fry, B. G., Williams, D. J., Barlow, A. & Wüster, W. (2016) Multi-locus phylogeny and species delimitation of Australo-Papuan blacksnakes (*Pseudechis* Wagler, 1830: Elapidae: Serpentes). *Molecular Phylogenetics and Evolution* 107: 48–55.

Maryan, B. (1998) The Dugite or Spotted Brown Snake (*Pseudonaja affinis*) in captivity. *Monitor* 10 (1) 13–14.

Maryan, B. (2004) Notes on captive reproduction and other observations in the Pygmy Dugite *Pseudonaja affinis tanneri*. *Herpetofauna* 34 (1) 40–43.

Maryan, B., Brennan, I., Hutchinson, M. T. & Geidans, L. S. (2020) What's under the hood? Phylogeny and taxonomy of the snake genera *Parasuta* Worrell and *Suta* Worrell (Squamata: Elapidae), with a description of a new species from the Pilbara, Western Australia. *Zootaxa* 4778 (1): 1–47.

Masci, P. & Kendall, P. (1995) *The Taipan – the World's Most Dangerous Snake*. Kangaroo Press, Kenthurst.

McDowell, S. B. (1972) The species of *Stegonotus* (Serpentes: Colubridae) in Papua New Guinea. *Zoologische Mededelingen* 47(2), 6–26.

Mengden, G. A., Cogger, H. G. & Cameron, E. Identification guide to the snakes of New Guinea. Unpublished document.

Minton, S. A. (1966) A contribution to the herpetology of West Pakistan. *Bulletin of American Museum of Natural History* 134 (2) 27–184.

Mirtschin, P. J. & Davis, R. (1992) *Snakes of Australia – Dangerous & Harmless*. Hill of Content Press,

Melbourne.

Mirtschin, P. J., Rasmussen, A. R. & Weinstein, S. A. (2017) *Australia's Dangerous Snakes, Identification, Biology and Envenoming*. CSIRO Publishing, Clayton South, Victoria.

Murphy, J. C. (2007) *Homalopsid Snakes – Evolution in the Mud*. Krieger Publishing, Florida, USA.

Murphy, J. C. (2011) The nomenclature and systematics of some Australasian homalopsid snakes (Squamata: Serpentes: Homalopsidae). *The Raffles Bulletin of Zoology* 59 (2), 229–236.

Nankivell J. H., Goiran, C, Hourston, M., Shine, R., Rasmussen, A. R., Thomson, V. A. & Sanders, K. L. (2020) A new species of turtle-headed sea snake (*Emydocephalus*: Elapidae) endemic to Western Australia. *Zootaxa* 4758 (1): 141–156.

Nankivell, J. H., Maryan, B., Bush, B. G. & Hutchinson, M. (2023) Whip it into shape: revision of the *Demansia psammophis* (Schlegel, 1837) complex (Squamata: Elapidae), with a description of a new species from central Australia. *Zootaxa*. 5311 (3), 301–339.

O'Shea, M. T. (1996) *A Guide to the Snakes of Papua New Guinea*. Independent Publishing, Port Moresby.

Parker, F. (1982) *Snakes of the Western Province*. Wildlife in Papua New Guinea 82/1.

Peters, W. C. H. & Doria, G. (1878) Catalogo dei retilli e dei batraci raccolti da O. Beccari, L. M. D'Alberts e A. A. Bruijn. nella sotto-regione Austro-Malese. *Annali del Museo* Civico de Storia Naturale di Genova. ser. 1, 13: 323–450.

Phillips, B. & Shine, R. (2004) Adapting to an invasive species: toxic cane toads induce morphological change in Australian snakes. PNAS 101 (49) 17150–17155.

Pyron, R. A., Burbrink, F. T., Colli, G. R., Oca, A. N. M., Vitt, L. J., Kuczynski, C. A. & Wiens, J. J. (2011) The phylogeny of advanced snakes (Colubroidea), with discovery of a new subfamily and comparison of support methods for likelihood trees. *Molecular Phylogenetics and Evolution* 58 (2) 329–342.

Ramasamy, S., Ibister, G. K. & Hodgson, W. C. (2004) Efficacy of two antivenoms against the in vitro myotoxic effects of black snake (*Pseudechis*) venoms in the chick biventer cervicis nerve-muscle preparation. *Toxicon* 44 (8) 837–845.

Ramasamy, S., Fry, B. G. & Hodgson, W. C. (2005) Neurotoxic effects of venoms from seven species of Australasian black snakes (*Pseudechis*): efficacy of black and tiger snake antivenoms. *Clinical and Experimental Pharmacology and Physiology* 32 7–12.

Ramsay, E. P. (1877) Description of a supposed new species of *Acanthophis* from north Australia. *Proceedings of the Linnaean Society of New South Wales* 2 72–74.

Rasmussen, A. R. & Ineich, I. (2000) Sea snakes of New Caledonia and surrounding waters (Serpentes: Elapidae): first report on the occurrence of *Lapemis curtus* and a description of a new species from the genus *Hydrophis*. *Hamadryad* 25, 91–99.

Rasmussen, A. R., Sanders, K. L., Guinea, M. I. & Amey, A. P. (2014) Sea snakes in Australian waters (Serpentes: Subfamilies Hydrophiinae and Laticaudinae) – a review with an updated key. *Zootaxa* 3869 (4) 351–371.

Rawlings, L. H., Rabosky, D. L., Donnellan, S. C. & Hutchinson, M. N. (2008) Python phylogenetics: inference from morphology and mitochondrial DNA. *Biological Journal of the Linnean Society* 93 (3) 603–619.

Robertson, P. & Coventry, A. J. (2019) *Reptiles of Victoria – A Guide to Identification and Ecology*. CSIRO Publishing, Clayton South.

Rowland, P. & Eipper, S. C. (2024) *A Naturalist's Guide to the Dangerous Creatures of Australia*. John Beaufoy Publishing, Oxford.

Ruane, S., Richards, S. J., McVay, J., Tjaturadi, B., Krey, K. & Austin, C. C. (2017) Cryptic and non-cryptic diversity in New Guinea groundsnakes of the genus *Stegonotus* Duméril, Bibron and Duméril, 1854: a description of four new species (Squamata: Colubridae). *Journal of Natural History*.

Sanders, K. L., Lee, M. S. Y., Leys, R., Foster, R. & Keogh, J. S. (2008) Molecular phylogeny and divergence dates for Australasian elapids and sea snakes (Hydrophiinae): evidence from seven genes for rapid evolutionary radiations. *Journal of Evolutionary Biology* 21: 682–695.

Sanders, K. L., Rasmussen, A. R., Elmberg, J., Mumpuni, Guinea, M., Blias, P., Lee, M. S. Y & Fry, B. G.

(2012) *Aipysurus mosaicus* a new species of egg-eating sea snake (Elapidae, Hydrophiinae) with a redescription of *Aipysurus eydouxii* (Gray 1849). *Zootaxa* 3431, 1–18.

Sanders, K. L., Lee, M. S. Y, Mumpuni, Bertozi, T. & Rasmussen, A. R. (2013) Multilocus phylogeny and recent rapid radiation of the viviparous sea snakes (Elapidae, Hydrophiinae). *Molecular Genetics and Evolution* 66: 575–591.

Scanlon, J. D. (2003) The Australian elapid genus *Cacophis*: morphology and phylogeny of rainforest crowned snakes. *Herpetological Journal* 13:1–20.

Schleip, W. D. (2008) Revision of the genus *Leiopython* Hubrecht 1879 (Serpents: Pythonidae) with the redescription of taxa recently described by Hoser (2000) and the description of new species. *Journal of Herpetology* 42, 645–667.

Shea, G. M. (2015) A new species of *Anilios* (Scolecophidia: Typhlopidae) from Central Australia. *Zootaxa* 4033 (1), 103–116.

Shea, G. M. & Horner, P. (1997) A new species of *Ramphotyphlops* (Squamata: Typhlopidae) from the Darwin area, with notes on two similar species from northern Australia. *The Beagle. Records of the Museums and Art Galleries of the Northern Territory* 13, 53–60.

Shea, G. M. & Scanlon, J. D. (2007) Revision of the small tropical whipsnakes previously referred to *Demansia olivacea* (Gray, 1842) and *Demansia torquata* (Günther, 1862) (Squamata: Elapidae). *Records of the Australian Museum* (2007) 59 (2) 117–142.

Shine, R. (1991) *Australian Snakes – A Natural History.* Reed Books, Balgowlah, New South Wales.

Slater, K. R. (1962) *A Guide to the Dangerous Snakes of Papua.* Government Printer, Port Moresby.

Sleeth, M., Eipper, S. C. & Madani, G. (2022) Opportunistic observations of climbing behaviour and arboreality in Australian terrestrial elapid snakes (Elapidae: Hydrophiinae). *Herpetology Notes* 14.

Smith, H. M. & Chiszar, D. (2006) Dilemma of name recognition – why and when to use new combinations of scientific names. *Herpetological Conservation and Biology* 1 (1) 6–8.

Smith, L. A. (1982) Variation in *Pseudechis australis* (Serpentes: Elapidae) in Western Australia and description of a new species of *Pseudechis*. *Records of the West Australian Museum* 9 (2) 227–233.

Somaweera, R. (2020) *A Naturalist's Guide to the Reptiles and Amphibians of Bali.* (2nd edition) John Beaufoy Publishing, Oxford.

Sonnemann, N. & Fyfe, G. (2004) Captive breeding of Speckled Brown Snake, *Pseudonaja guttata*. *Herpetofauna* 34 (2) 92–98.

Speybroeck, J., Beukema, W., Bok, B., Van Der Voort, J. & Velikov, I. (2020) *Field Guide to the Amphibians and Reptiles of Britain and Europe.* Bloomsbury Publishing, Sydney.

Stebbins, R. C. & McGinnis, S. M. (2012) *Field Guide to Amphibians and Reptiles of California* revised edition, University of California Press, Santa Barbara, USA.

Steubing, R. (1988) Island romance: the yellow-lipped sea krait. *Malayan Naturalist* 41 (3–4) 9–11.

Storr, G. M., Smith, L. A. & Johnstone, R. E. (2002) *Snakes of Western Australia.* Western Australian Museum, Perth.

Storr, G. M. (1980) A new *Brachyaspis* (Serpentes: Elapidae) from Western Australia. *Records of the Western Australian Museum* 8 397–399.

Storr, G. M. (1981) The genus *Ramphotyphlops* (Serpentes: Typhlopidae) in Western Australia. *Records of the Western Australian Museum* 9(3), 235–271.

Storr, G. M. (1984) A new *Ramphotyphlops* (Serpentes: Typhlopidae) from central Australia. *Records of the Western Australian Museum* 11(3), 287–290.

Storr, G. M. (1989) A new *Rhinoplocephalus* (Serpentes: Elapidae) from Western Australia. *Records of the Western Australian Museum* 14 (1), 137–138.

Storr, G. M. (1989) A new *Pseudonaja* (Serpentes: Elapidae) from Western Australia. *Records of the Western Australian Museum* 14 (3), 421–423.

Strahan, R. (1992) *Encyclopedia of Australian Animals: Mammals.* Angus & Robertson, Pymble, NSW.

Sutherland, S. K. (1983) *Australian Animal Toxins.* Oxford University Press, Melbourne.

Sutherland, S. K., Coulter, A. R. & Harris, R. D. (1979) Rationalisation of first aid measures for elapid snake-bite. *Lancet* 1979 (1) 183–186.

Sutherland, S. K. & Tibballs, J. (2001) *Australian Animal Toxins* (2nd edition). Oxford University Press, Melbourne.

Swan, M. G. (2020) *Frogs and Reptiles of the Murray-Darling Basin: A Guide to their Identification, Ecology and Conservation.* CSIRO Publishing. Melbourne.

Swan, G., Shea, G. & Sadlier, R. (2022) *A Field Guide to the Reptiles of NSW.* Reed New Holland Sydney, NSW.

Swanson, S. (2017) *Field Guide to Australian Reptiles* (3rd edition). Pascal Press, Glebe, New South Wales.

The IUCN Red List of Threatened Species. *www.iucnredlist.org*. Downloaded on 20 March 2023.

Thompson, S. A. & Thompson, G. G. (2006) *Reptiles of the Western Australian Goldfields.* Goldfields Environmental Management Group, Perth.

Tiatragul, S., Brennan, I. G., Broady, E. S. & Keogh, J. S., (2023) Australian hidden radiation: Phylogenomics analysis reveals rapid Miocene radiation of blindsnakes. *Molecule Phylogenetics and Evolution* 185 107812.

Venchi, A., Wilson, S. K. & Borsboom, A. C. (2015) A new blindsnake (Serpentes: Typhlopidae) from an endangered habitat in south-eastern Queensland, Australia. *Zootaxa* 3990 (2), 272–278.

Waite, E. R. (1929) *Reptiles and Amphibians of South Australia.* Government Printer, Adelaide.

Wallach, V., Williams, K. L. & Boundy, J. (2014) *Snakes of the World: A Catalogue of Living and Extinct Species.* Taylor and Francis, CRC Press.

Wells, R. & Wellington, C. R. (1983) A synopsis of the class Reptilia in Australia. *Australian Journal of Herpetology* No. 1(3–4) 73–129.

Wells, R. & Wellington, C. R. (1985) A classification of the Amphibia and Reptilia of Australia. *Australian Journal of Herpetology* (Supplementary Series) No. 1: 1–64.

Wells, R. & Wellington, C. R. (1987) A new species of proteroglyphous snake (Serpentes: Oxyuranidae) from Australia. *Australian Herpetologist* 503, 1–8.

Welton, R. & Williams, D. (2005) Snake venom composition and activity. pp. 65–85 in *Venomous Bites and Stings in Papua New Guinea – A Guide to Treatment for Health Workers and Doctors* (eds White, J. (2013) *A Clinician's Guide to Australian Venomous Bites and Stings.* BioCSL, Parkville, Melbourne.

Williams, D., Jensen, S., Nimorakiotakis, B. & Winkel, K. D. Australian Venom Research Unit, Melbourne.

Williams, D. & Wüster, W. (2005) Snakes of Papua New Guinea. pp. 33–64 in *Venomous Bites and Stings in Papua New Guinea – A Guide to Treatment for Health Workers and Doctors*, eds Williams, D., Jensen, S., Nimorakiotakis, B. & Winkel, K. D. Australian Venom Research Unit, Melbourne.

Wilson, S. K. (2022) *A Field Guide to Reptiles of Queensland* (3rd edition). New Holland, Chatswood, Sydney.

Wilson, S. K. & Knowles, D. G. (1988) *Australia's Reptiles – A Photographic Reference to the Terrestrial Reptiles of Australia.* Cornstalk Publishing, Sydney.

Wilson, S. K. & Swan, G. (2021) *A Complete Guide to Reptiles of Australia* (6th edition). New Holland, Chatswood, Sydney.

Worrell, E. (1955) A new elapine snake from Queensland. *Proceedings of the Royal Zoological Society of New South Wales for 1953–54*: 41–43.

Worrell, E. (1961) Herpetological name changes. *Western Australian Naturalist* 8: 18–27.

Worrell, E. (1961) A new insular brown snake. *Proceedings of the Royal Zoological Society of New South Wales for 1958–59*: 56–58.

Worrell, E. (1963) *Dangerous Snakes of Australia and New Guinea.* Angus and Robertson, Sydney.

Worrell, E. (1963a) *Reptiles of Australia.* Angus and Robertson, Sydney.

Wüster, W., Dumbrell, A. J., Hay, C., Pook, C. E., Williams, D. J. & Fry, B. F. (2004) Snakes across the Strait: trans-Torresian phylogeographic relationships in three genera of Australasian snakes (Serpentes: Elapidae: *Acanthophis*, *Oxyuranus*, and *Pseudechis*). *Molecular Phylogenetics and Evolution* 34 1–14.

Acknowledgments

We, the authors, firstly thank our sons Bailey and Cody. Know that whatever we achieve in our personal and professional lives, you two boys are what we are most proud of and we love you. Tie wants to thank Cory Kereweraro for always having her back (rightly or wrongly!) at any time of the day or night and always believing she was good enough. Triple laughter cradle.

The authors wish to thank the following people who provided assistance to this project. This includes but is not limited to thought-provoking discussions, access to specimens, supply of images and help in the field. Thank you Lauren Alexander-Kay, Luke Allen, Andrew Amey, Blanche d'Anastasi, Robert Audcent, Brian Barnett, Jackson Barrett, David Berger, Shane Black, Sue Blyde, Becca Bolton, Ian 'Bushrat' Bool, Darren Boswell, Jye Brooks, Danny Brown, Clay Bryce, Eric Burke, Brian Bush, Nick Cairns, Marie Callins, Jesse Campbell, Malcolm Campbell, Nathan Clout, Hal & Heather Cogger, Allira Costa, Patrick Couper, Mark Cowan, Cami Dalrymple, Alexander Davies, Jeremy De Haan, Lauren Dibben, Emma & Zac Dixon, Steve Donnellan, Paul Doughty, Rynn Dragomirov, Vik Dunis, Graham Edgar, Bruce Edley, Harald Ehmann, Bailey Eipper, Cody Eipper, Adam Elliott, Ryan Ellis, Tessa Esparon, Lisa Farina, Jules Farquhar, Jamin Feddersen, Esmeralda Fonteijn, Ryan Francis, Bryan Fry, Glen Gaikhorst, Amber Gillett, Om Glang, Claire Goiran, Jamie Gover, Kim Green, Allen Greer, Philip Griffin, Kyle Hancock, Jason Harrison, Forrest He, Brett & Kathy Hicks, Harry Hines, Chay Hoon, Courtney Hopwood, Ash Horn, Paul Horner, Alex Hoschke, Conrad Hoskin, David Hunter, Mark Hutchinson, Cindy Jackson, Joshua Jensen, Kristy Jensen, Chris Jolly, Hirich Kaiser, Tim Karnasuta, Jai Kember, Cory Kereweraro, Scott Kickham, Paul Lambourne, Adele Leane, Mike Lyons, Michael McFadden, Ross McGibbon, Angus McNab, Keith & Theresa McPeek, Phill Mangion, Bryanna Marsden, Brad Maryan, Krystal May, Danny Melville, Jake Meney, Sergii & Svetlana Mitin, Nick Mutton, James Nankivell, Reid Newell, Hamish Noller, Mark O'Shea, Fred Parker, Aadit Patel, Michael & Joanne Payne, Rob Porter, Samuel Prakash, Jarrad Prangell, Dean Purcell, Arne Rasmussen, Wes Read, Peter Robertson, David Robinson, Peter Rowland, Mark Rosenstein, Jodi Rowley, Dan Rumsey, Adam Sapiano, Ngk Norman Sayank, The Schmidt Ocean Institute, Megan Schroder, Shawn Scott, Will Scott, Rob Seymour, Liam Shanks, Glenn Shea, Rick Shine, Mark Simpson, Ruchira Somaweera, Gary Stephenson, Judy & Terry Stollery, Jason Sulda, Gerry Swan, Mike Swan, Steve Swanson, Matt Summerville, Rhein Talbot, Ria Tan, Doris Teufel, Martin Titley, Brad Traynor, Janne Torkolla, Rob Valentic, Eric Vanderduys, Gernot Vogel, Lauren Vonahamme, Harold Voris, Mani Walsh, Richard Wells, Sanoj Wijayasekara, David Williams, Steve Wilson, Sam Wood, Justin Wright, Wolfgang Wuster, Ryan Young and Anders Zimny.

The authors would like to thank Brian Bush, Hal Cogger, Angus McNab and two other peer reviewers who wish to remain anonymous for constructive comments and suggestions to the text.

The authors would like to say a special thank you to Ngk Norman Sayank and Om Glang for showing us the stunning herpetofauna of Bali and assisting us with species specimen location. We would also like to thank the countless Balinese people who saw us out herping, stopped to chat and pointed us to where they had previously had encounters. Your friendliness and helpfulness were appreciated.

The authors would like to say thank you to Rosemary Wilkinson, Krystyna Mayer, Lucy Doncaster and the team at John Beaufoy publishing for their work on this title.

INDEX

Common Names

Australian Bockadam 283
Australian Coral Snake 96
Australian Slaty Grey Snake 71
Banded Snake, Desert 205
    De Vis' 132
    Jan's 206
Bandy Bandy 224
    Cape York 227
    Centralian 229
    Intermediate 225
    Narrow-banded 226
    Pilbara 228
Bardick 139
Black-ringed Mangrove Snake 249
Black Snake, Papuan 184
    Red-bellied 185
    Spotted 182
Black Striped Snake 113
Blind Snake, Ammodyte 297
    Barrow Island 319
    Blackish 324
    Black-tailed 311
    Blunt-snouted 325
    Brown-snouted 329
    Buff-snouted 320
    Cape York 317
    Cape York Striped 305
    Centralian 304
    Christmas Island 344
    Claw-snouted 336
    Cooloola 330
    Dark-spined 301
    Darwin 334
    Desert 306
    Faint-striped 303
    Fassifern 314
    Fat 327
    Gane's 309
    Groote Eylandt 322
    How's 313
    Kimberley 315
    Mornington 337
    Murchison 316
    North-east 333
    Northern 306
    Northern Beaked 310
    Northern Hook-snouted 312
    Pilbara Hook-snouted 326
    Prong-snouted 302
    Roberts' 329
    Robust 318
    Round-tailed 298
    Ruby Gap 308
    Sandamara 334
    Sharp-snouted 332
    Shovel-snouted 300
    Slender 323
    Small-eyed 321
    Small-headed 296
    Southern 299
    Southern Beaked 338
    Splendid 331
    West Kimberley 342
    Woodland 328
    Yampi 340
    Yirrkala 341
Broad-headed Snake 155
Brown-headed Snake 149
Brown Snake, Eastern 200
    Ingram's 195
    Northern 198
    Peninsula 194
    Ringed 198
    Speckled 193
    Strap-snouted 192
    Tanner's 191
    Western 196
Burrowing Snake, Black-naped 158
    Black-striped 150
    Dampierland 208
    North-eastern Plain-nosed 88
    Warrego 89
    West Coast 207
Carpentaria Snake 110
Carpet Python 50
    Centralian 46
    Inland, 49
    Western 48
Collett's Snake 181
Copperhead, Highland 92
    Lowland 93
    Pygmy 91
Corn Snake 69
Crowned Snake 141
    Dwarf 107
    Golden 108
    Northern 105
Curl Snake 220
    Ord River 217
Death Adder, Common 78
    Desert 82
    Floodplain 80
    Kimberley 78
    Northern 82
    Pilbara 86
    Rough-scaled 84
    Smooth-scaled 81
Dugite 188
    Rottnest Island 190
Dunmall's Snake 148
Dwyer's Snake 210
File Snake, Arafura 26
    Little 27
Flowerpot Snake 343
Grey Snake 150
Hooded Snake, Gould's 214
    Inland 215
    Pilbara 213
Keelback 73
Lake Cronin Snake 176
Little Spotted Snake 218
Macleay's Water Snake 291
Mangrove Snake, Richardson's 289
    Roebuck Bay 288
    White-bellied 285
Marsh Snake 152

INDEX 363

Masters' Snake 136
 Spotted 180
Mitchell's Short-tailed Snake 216
Mulga Snake 178
 Eastern Pygmy 183
 Western Pygmy 186
Mustard-bellied Snake 137
Orange-naped Snake 146
Ornamental Snake 133
Pale-headed Snake 154
Pink Snake 111
Port Lincoln Snake 219
Python, Black-headed 37
 Cape Yorke Spotted 33
 Children's 30
 Oenpelli 54
 Olive 42
 Papuan Spotted 34
 Pilbara Olive 41
 Pygmy 35
 Rough Scaled 47
 Southern Green 52
 Southern Spotted 32
 Southern White-lipped 44
 Stimson's 31
 Water 40
Red-naped Snake 145
Rosen's Snake 211
Rough-scaled Snake 222
Scrub Python, Australian 56
 Southern, 56
Sea Krait, Black-lipped 276
 Yellow-lipped 275
Sea Snake, Australasian Beaked 272
 Belcher's 253
 Black-banded Robust 265
 Black-headed 252
 Cogger's 255
 Dubois' 233
 Dusky 236
 Dwarf 254
 Elegant 259
 Geometrical 257
 Horned 268

Laboute's 262
Large-headed 267
Leaf-scaled 235
Mangrove 247
Marbled 234
Mjoberg's 241
Mosaic 239
Northern Mangrove 280
Olive 237
Olive-headed 264
Ornate 266
Plain 260
Plain-banded 272
Rough-scaled 258
Shark Bay 240
Short-nosed 232
Small-headed 263, 278
Spectacled 261
Spine-bellied 256
Stokes' 270
Turtle-headed 243
Western Turtle-headed 245
Yellow-bellied 269
Short-nosed Snake 142
Shovel-nosed Snake, Arnhem 101
 Eastern Narrow-banded 95
 Einasleigh 96
 North-west 95
 Northern 102
 Southern 103
 Unbanded 100
 Western Narrow-banded 99
Small-eyed Snake 112
 Northern 114
Square-nosed Snake 203
Stephens' Banded Snake 156
Taipan, Coastal 172
 Inland 170
 Papuan 171
 Western Desert 174
Tiger Snake, Chappell Island 168
 Common 166

Krefft's 162
Peninsula 164
Tasmanian 163
Western 165
Tree Snake, Brown 60
 Common 64
 Northern 63
Whip Snake, Black-necked 117
 Collared 129
 Crack-dwelling 125
 Desert 118
 Greater Black 121
 Grey 128
 Lesser Black 130
 Little 212
 Long-tailed 119
 Narrow-headed 116
 Olive 120
 Red 126
 Reticulated 124
 Shine's 127
 Sombre 123
 Yellow-faced 122
White-crowned Snake 106
White-lipped Snake 135
Wolf Snake, Common 67
Woma 38
Yellow-naped Snake 144

Scientific Names

Acanthophis antarcticus 78
   cryptamydros 79
   hawkei 80
   laevis 81
   praelongus 82
   pyrrhus 83
   rugosus 84
   wellsei 86
Acrochordus arafurae 26
   granulatus 27
Aipysurus apraefrontalis 232
   duboisii 233
   eydouxii 234
   foliosquama 235
   fuscus 236
   laevis 237
   mosaicus 239
   pooleorum 240
   tenuis 241
Anilios affinis 296
   ammodytes 297
   aspina 298
   australis 299
   batillus 300
   bicolor 301
   bituberculatus 302
   broomi 303
   centralis 304
   chamodracaena 305
   diversus 306
   endoterus 307
   fossor 308
   ganei 309
   grypus 310
   guentheri 311
   hamatus 312
   howi 313
   insperatus 314
   kimberleyensis 315
   leptosoma 316
   leucoproctus 317
   ligatus 318
   longissimus 319
   margaretae 320
   micromma 321

   minimus 322
   nema 323
   nigrescens 324
   obtusifrons 325
   pilbarensis 326
   pinguis 327
   proximus 328
   robertsi 329
   silvia 330
   splendidus 331
   systenos 332
   torresianus 333
   tovelli 334
   troglodytes 335
   unguirostris 336
   vagurima 337
   waitii 338
   wiedii 339
   yampiensis 340
   yirrikalae 341
   zonula 342
Antaioserpens albiceps 88
   warro 89
Antaresia childreni childreni 30
   childreni stimsoni 31
   maculosa maculosa 32
   maculosa peninsularis 33
   papuensis 34
   perthensis 35
Aspidites melanocephalus 37
   ramsayi 38
Austrelaps labialis 91
   ramsayi 92
   superbus 93
Boiga irregularis 60
Brachyurophis approximans 94
   australis 96
   campbelli 97
   fasciatus 98
   fasciolatus fasciolatus 98
   incinctus 100
   morris 101
   roperi 102
   semifasciatus 103
Cacophis churchilli 105
   harriettae 106

   krefftii 107
   squamulosus 108
Cerberus australis 283
Cryptophis boschmai 110
   incredibilis 111
   nigrescens 112
   nigrostriatus 113
   pallidiceps 114
Demansia angusticeps 116
   calodera 117
   cyanochasma 118
   cyanochasma 119
   olivacea 120
   papuensis 121
   psammophis 122
   quaesitor 123
   reticulata 124
   rimicola 125
   rufescens 126
   shinei 127
   simplex 128
   torquata 129
   vestigiata 120
Dendrelaphis calligastra 63
   punctulatus 64
Denisonia devisi 132
   maculata 133
Drysdalia coronoides 135
   masteri 136
   rhodogaster 137
Echiopsis curta 139
Elapognathus coronatus 141
   minor 142
Emydocephalus annulatus 243
   oriarus 245
Ephalophis greyae 247
Fordonia leucobalia 285
Furina barnardi 144
   diadema 145
   ornata 146
Glyphodon dunmalli 148
   tritis 149
Hemiaspis damelii 150
   signata 152
Hoplocephalus bitorquatus 154
   bungaroides 155

stephensii 156
Hydrelaps darwiniensis 249
Hydrophis atriceps 252
    belcheri 253
    caerulescens 254
    coggeri 255
    curtus 256
    czeblukovi 257
    donaldi 258
    elegans 259
    inornatus 260
    kingii 261
    laboutei 262
    macdowelli 263
    major 264
    melanosoma 265
    ocellatus 266
    pacificus 267
    peronii 268
    platurus platurus 269
    stokesii 270
    vorisi 272
    zweifeli 273
Indotyphlops braminus 343
Laticauda colubrina 275
    laticaudata 276
Leiopython fredparkeri 44
Liasis fuscus 40
    olivaceus barroni 41
    olivaceus olivaceus 42
Lycodon capucinus 67
Microcephalophis gracilis 278
Morelia bredli 46
    carinata 47
    imbricata 48
    spilota metcalfei 49
    spilota spilota 50
    viridis 52
Myron resetari 288
    richardsonii 289
Narophis bimaculatus 158
Neelaps calonotos 160
Notechis scutatus ater 162
    scutatus humphreysi 163
    scutatus niger 164
    scutatus occidentalis 165

    scutatus scutatus 166
    scutatus serventyi 168
Nyctophilopython oenpelliensis 54
Oxyuranus microlepidotus 170
    scutellatus canni 171
    scutellatus scutellatus 172
    temporalis 174
Pantherophis guttatus 69
Parahydrophis mertoni 280
Paroplocephalus atriceps 176
Pseudechis australis 178
    butleri 180
    colletti 181
    guttatus 182
    pailsei 183
    papuanus 184
    porphyriacus 185
    weigeli 186
Pseudoferania polylepis 291
Pseudonaja affinis affinis 188
    affinis exilis 190
    affinis tanneri 191
    aspidorhyncha 192
    guttata 193
    inframacula 194
    ingrami 195
    mengdeni 196
    modesta 198
    nuchalis 199
    textilis 200
Ramphotyphlops exocerti 344
Rhinoplocephalus bicolor 203
Simalia amethistina 56
    kinghorni 57
Simoselaps anomalus 205
    bertholdi 206
    littoralis 207
    minimus 208
Stegonotus australis 71
Suta dwyeri 210
    fasciata 211
    flagellum 212
    gaikhorstorum 213
    gouldii 214
    monachus 215

    nigriceps 216
    ordensis 217
    punctata 218
    spectabilis 219
    suta 220
Tropidechis carinatus 222
Tropidonophis mairii 73
Vermicella annulata 224
    intermedia 225
    multifasciata 226
    parscauda 227
    snelli 228
    vermiformis 229

NOTES

NOTES

NOTES

Other books about Australian Wildlife published by John Beaufoy Publishing

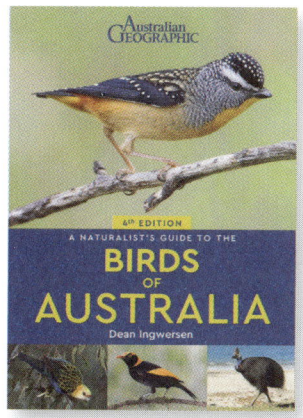

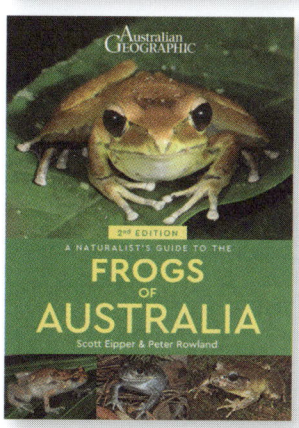
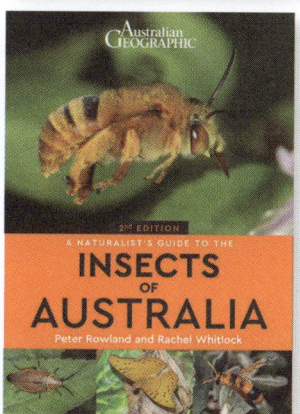
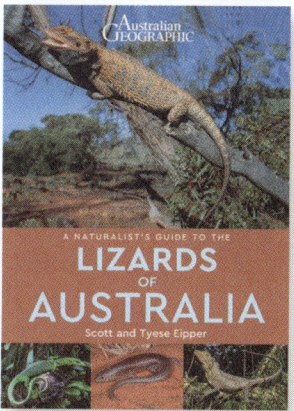
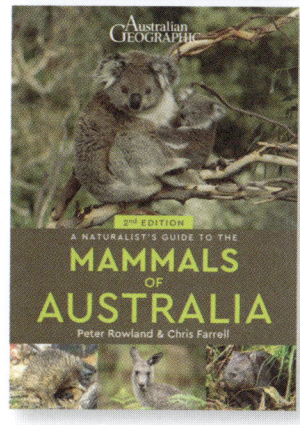
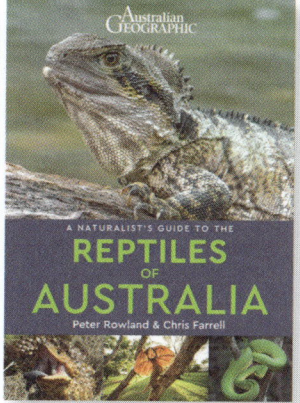
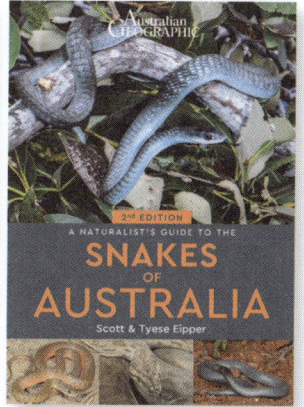

See our full range of natural history books at www.johnbeaufoy.com